Agriculture 4.0 – the Future of Farming Technology

Agriculture 4.0 – the Future of Farming Technology

Editors

Dolores Parras-Burgos
José Miguel Molina Martínez
Ginés García-Mateos

MDPI

Basel • Beijing • Wuhan • Barcelona • Belgrade • Novi Sad • Cluj • Manchester

Editors

Dolores Parras-Burgos
Department of Structures, Construction and Graphic Expression, Universidad Politécnica de Cartagena
Cartagena, Spain

José Miguel Molina Martínez
Department of Agricultural Engineering, Universidad Politécnica de Cartagena
Cartagena, Spain

Ginés García-Mateos
Department of Computer Science and Systems, University of Murcia
Murcia, Spain

Editorial Office
MDPI
St. Alban-Anlage 66
4052 Basel, Switzerland

This is a reprint of articles from the Special Issue published online in the open access journal *Applied Sciences* (ISSN 2076-3417) (available at: https://www.mdpi.com/journal/applsci/special_issues/Agriculture4).

For citation purposes, cite each article independently as indicated on the article page online and as indicated below:

Lastname, A.A.; Lastname, B.B. Article Title. *Journal Name* **Year**, *Volume Number*, Page Range.

ISBN 978-3-0365-9758-4 (Hbk)
ISBN 978-3-0365-9759-1 (PDF)
doi.org/10.3390/books978-3-0365-9759-1

Cover image courtesy of José Miguel Molina Martínez

Contents

About the Editors

Dolores Parras-Burgos

Dolores is an associate professor at the Department of Structures, Construction and Graphic Expression in the Technical University of Cartagena, Spain. Her main interest is the design and innovation of new products, systems, and technologies in the agronomic field. Design and creativity are fundamental tools in any research project, and in agriculture it is necessary to solve many challenges that require innovative solutions.

José Miguel Molina Martínez

José Miguel is a full professor at the Department of Agricultural Engineering in the Technical University of Cartagena, Spain. His interests are related to the management of water resources, the automation and control of agriculture, and the application of computer science and electronics in the agronomic field, etc. In short, the development of innovative technologies for intelligent agriculture.

Ginés García-Mateos

Ginés is a full professor at the Department of Computer Science in the University of Murcia, Spain. His current interest is the application of computer vision and deep learning technologies in agriculture. His most recent research focuses on the integration of agricultural and photovoltaic production in so-called agrovoltaic systems.

Preface

The field of agriculture is currently experiencing a significant change brought about by the integration of cutting-edge technologies and tools from the fourth industrial revolution, leading to the emergence of what is referred to as Agriculture 4.0. This Reprint is a compilation of articles that highlight several noteworthy recent developments in this domain, such as innovative methods for remote crop sensing and agricultural machinery, the creation of novel computer vision and deep learning algorithms, and progress in aquaculture technologies. It also includes a comprehensive assessment of the current state of Horticulture 4.0.

Dolores Parras-Burgos, José Miguel Molina Martínez, and Ginés García-Mateos
Editors

applied sciences

Review

Horticulture 4.0: Adoption of Industry 4.0 Technologies in Horticulture for Meeting Sustainable Farming

Rajat Singh [1], Rajesh Singh [2,3], Anita Gehlot [2,3], Shaik Vaseem Akram [2,4], Neeraj Priyadarshi [5,*] and Bhekisipho Twala [6,*]

1 School of Agriculture, Uttaranchal University, Dehradun 248007, Uttarakhand, India
2 Uttaranchal Institute of Technology, Uttaranchal University, Dehradun 248007, Uttarakhand, India
3 Department of Project Management, Universidad Internacional Iberoamericana, Campeche 24560, CP, Mexico
4 Law College Dehradun, Uttaranchal University, Dehradun 248007, Uttarakhand, India
5 Department of Electrical Engineering, JIS College of Engineering, Kolkata 741235, West Bengal, India
6 Digital Transformation Portfolio, Tshwane University of Technology, Staatsartillerie Rd, Pretoria West, Pretoria 0183, South Africa
* Correspondence: neerajrjd@gmail.com (N.P.); twalab@tut.ac.za (B.T.)

Abstract: The United Nations emphasized a significant agenda on reducing hunger and protein malnutrition as well as micronutrient (vitamins and minerals) malnutrition, which is estimated to affect the health of up to two billion people. The UN also recognized this need through Sustainable Development Goals (SDG 2 and SDG 12) to end hunger and foster sustainable agriculture by enhancing the production and consumption of fruits and vegetables. Previous studies only stressed the various issues in horticulture with regard to industries, but they did not emphasize the centrality of Industry 4.0 technologies for confronting the diverse issues in horticulture, from production to marketing in the context of sustainability. The current study addresses the significance and application of Industry 4.0 technologies such as the Internet of Things, cloud computing, artificial intelligence, blockchain, and big data for horticulture in enhancing traditional practices for disease detection, irrigation management, fertilizer management, maturity identification, marketing, and supply chain, soil fertility, and weather patterns at pre-harvest, harvest, and post-harvest. On the basis of analysis, the article identifies challenges and suggests a few vital recommendations for future work. In horticulture settings, robotics, drones with vision technology and AI for the detection of pests, weeds, plant diseases, and malnutrition, and edge-computing portable devices that can be developed with IoT and AI for predicting and estimating crop diseases are vital recommendations suggested in the study.

Keywords: horticulture; micronutrient; Industry 4.0; SDGs; IoT; blockchain; real-time prediction; AR; bigdata

Citation: Singh, R.; Singh, R.; Gehlot, A.; Akram, S.V.; Priyadarshi, N.; Twala, B. Horticulture 4.0: Adoption of Industry 4.0 Technologies in Horticulture for Meeting Sustainable Farming. *Appl. Sci.* **2022**, *12*, 12557. https://doi.org/10.3390/app122412557

Academic Editors: José Miguel Molina Martínez, Ginés García-Mateos, Dolores Parras-Burgos and Georgios Papadakis

Received: 14 September 2022
Accepted: 9 November 2022
Published: 8 December 2022

1. Introduction

According to the Food and Agriculture Organization and World Health Organization State of Food Security and Nutrition in the World (SOFI) report,45 million children die from the deadliest form of malnutrition under the age of five [1]. A chronic deficiency of essential nutrients in their nutrition has also resulted in delayed development and growth in two billion children under the age of five. This indicates that there is a need to have more emphasis on overcoming food insecurity and malnutrition due to climate extremes and economic disruption. Even the SDGs (SDG 2 and SDG 12) of the UN emphasize eradicating hunger and enhancing food security with responsible consumption and production toward sustainability [2,3]. Healthy micronutrients for overcoming malnutrition can be achieved with sustainable farming of fruits and vegetables, i.e., horticulture [4]. India is currently the world's second-largest producer of fruits and vegetables, trailing only China. Horticulture comprises fruits, root and tuber crops, mushrooms, vegetables, spices, aromatic plants, flowers, bamboo, coconut, cashew, and cocoa. Different strategies such as technology promotion, research, post-harvest management, and marketing are key for the growth of horticulture. Improving horticulture production, increasing farmer income, improving nutritional security, and improving productivity by using quality germplasm, planting material, and micro irrigation to save water are the key vision of India for the promotion of holistic horticulture growth.

Horticulture crops significantly contribute to the Indian economy by enhancing farm output, generating employment, and providing raw materials to various food-processing businesses [5]. The amount of land allotted for horticulture is minimal, but the demand for the production of horticulture is high. Therefore, meeting the demand with minimum resources is a bit challenging, as sustainable practices need to be adopted to meet a sustainable environment [6]. It has also been observed that the export rate of Asian countries such as India has increased in the past few years; however, there are a few challenges such as meeting quality standards set by international countries and payment for exports. The main challenge in fruits and vegetables is the short lifetime after plucking. Additionally, given that the majority of horticulture cultivation is processed in rural areas, it has been determined that horticulture has to be promoted in urban areas.

1.1. Problem Statement

Ref. [7] discussed the main problems that are present in horticulture, including the decline of fertile soil, global warming, rising land prices, shortage of water, and lost opportunity of low-cost labor, all of which portray serious threats to growth of horticulture. For horticultural crops, abiotic stresses like an inundation, temperature extremes, salinity, pH, and drought are major obstacles. As an outcome of global warming, there will be less precipitation all in all, and the snow cover will melt faster, resulting in extreme drought. These circumstances lead to several stress-related problems, including the deficit of fertile soil, flooding, increased evapotranspiration, low soil-oxygen levels, decreased production, leachate of nutrients, and slow planting. Alongside increasing temperatures, global warming had also contributed to record-breaking cold weather. Land salinity and irrigation-water salinity are also significant concerns for horticultural production. Along with these problems, previous studies also discussed that sustainable pest-management methods need to be adopted to protect against pathogens and pest resistance. The absence of markets to accumulate production, the substantial percentage of middlemen, the absence of marketing institutions safeguarding farmers' interests, the imperfect pricing system, and the absence of transparency in market-information systems, especially in the export market, is the main marketing problems for horticultural produce.

1.2. Motivation, Novelty, and Contribution of the Study

As discussed above, the traditional approaches lack appropriate monitoring of pest control, water management, soil management, light control, and temperature control, as these parameters are crucial for the effective production of horticulture with a sustainability aspect. The previous study found that the different issues in this group highlight the necessity for more environmentally sound processes, in addition to a rise in the utilization of automated systems and accurate processes. Industry 4.0 technology has proven its capability in various applications through the fusion of advanced technologies such as the Internet of Things, artificial intelligence, blockchain, robotics, digital twins, and big data. The amalgamation of Industry 4.0 and horticulture provides the opportunity to transform industrial agriculture into the next generation, namely Horticulture 4.0. Moreover, Industry 4.0 is capable of achieving sustainable and intelligent ecosystem with real-time management of farming, automation, intelligent decisions, real-time processing, and analysis [8]. It has proved that Industry 4.0 technologies have already delivered exceptional results in greenhouse farming for a sustainable food supply chain with the integration of hardware and software algorithms and giving rise to higher yields using fewer resources and less water. Although many studies have addressed Industry 4.0 technologies individually for fruits and vegetables, they are limited studies that merely address multiple Industry 4.0 digital technologies as single concepts for discussing their significance in horticulture. Based on the aforementioned information, this study analyzed the value and applications of Industry 4.0 technologies in horticulture for real-time monitoring of fruit-and-vegetable conditions in various environmental settings, intelligent and predictive analytics for fruit-and-vegetable growth and disease estimation, and virtual horticulture plants. The contribution of the study is as follows:

- The significance of using Industry 4.0 technologies for horticulture is discussed in the study for sustainability.
- We also discuss the significance and applications of the Internet of Things, cloud computing, artificial intelligence, blockchain, and big data for horticulture for enhancing traditional practices for disease detection, irrigation management, fertilizer management, maturity identification, marketing, and the supply chain, soil fertility, and weather patterns at pre-harvest, harvest, and post-harvest
- Finally, the article discusses the challenges and suggests recommendations for the wide adoption of any future enhancement in the vicinity of horticulture. In horticulture settings, robotics, drones with vision technology and AI for the detection of pests, weeds, plant diseases, and malnutrition, and edge-computing portable devices that can be developed with IoT and AI for predicting and estimating the disease of crops are vital recommendations suggested in the study.

1.3. Methodology of the Study

In this section, the methodology of the study is presented. Based on the limitations identified in previous studies, this study framed a research question that was employed to carry out the review. The research question was, "What is the progress and significance of Industry 4.0 technology integration in horticulture?" This research question constructed various keywords with Boolean function operators in the Scopus and Web of Science databases. The following strings were considered for obtaining publications: "horticulture AND Industry 4.0", "horticulture AND sustainability", "Industry 4.0 for fruits and vegetables", "Industry 4.0 AND digitalization", "horticulture AND digitalization", "horticulture AND IoT", "IoT AND fruits and vegetables", "disease detection AND horticulture crop", "blockchain AND horticulture", "artificial intelligence AND horticulture", "quality assessment AND artificial intelligence", "big data AND horticulture", "blockchain AND marketing of fruits and vegetables", "blockchain AND energy", "crop monitoring AND horticulture", and "soil fertility AND horticulture." During the search for articles, there was al ack of articles that merely emphasized the integration of Industry 4.0 for horticulture. Therefore, we considered a few conference articles and other supportive articles in this study. Based on the obtained articles, the study categorized four different sub-sections in which the progress and significance of the previous studies that implemented Industry 4.0 for enhancing horticulture from production to marketing are assessed. After analyzing the studies, few limitations were identified and vital recommendations for future work were suggested.

1.4. Comparative Analysis

Table 1 provides a comparative analysis of the different studies that focused on horticulture that are analyzed within the proposed study. In most previous studies, it was identified that the study only focused on some key areas in horticulture, like yield monitoring, pesticide monitoring, and quality assessment. A recent study emphasized the different issues in horticulture from the perspective of industries. Along with this, [9] reviewed the different critical parameters that are required for automation with IoT for vertical gardens. Ref. [10] carried out a review on the significance of deep-learning models in horticulture for various applications. Ref. [11] discussed the implementation of digital twins in horticulture. However, there is a possibility of emphasizing the importance of multiple Industry 4.0 technologies for addressing the different issues of horticulture from production to marketing. The current study focused on discussing the different Industry 4.0 technologies for overcoming the different issues of horticulture within the context of sustainability. This study highlights the significance and application of Industry 4.0 for horticulture. Based on the limitations identified, this study also discusses vital recommendations for future work.

Table 1. Comparative analysis of the previous studies with the current study.

Ref	Objective	Findings
[8]	The study carried out a review to analyze the critical parameters that are significant for automation in vertical gardens through IoT.	One of the study's key findings is that the majority of studies used 6–20 tiers (40%) when integrating VGs, with lettuce representing the most dominant crop (28.6%).
[9]	A brief overview of deep-learning approaches is provided in which deep-learning technologies are used in the horticultural domain.	Large datasets need to be built by collecting real-condition imaging data of plants to create robust models.
[10]	The study carried out a review to provide insight into the implementation of digital twins in horticulture.	There is a dominant focus on the cultivation process at the greenhouse level, among others, for climate control, energy management, and lighting.
[12]	Yield monitoring of annual and perennial crops of horticulture	The study contends that many commercial strategies for mapping horticultural crop yield are necessary. It was also discovered that integrating multiple technologies can provide better and more accurate results.
[13]	To determine the pesticide-residue levels in widely consumed fruits and vegetables throughout Kuwait	The results conclude that there is a high presence of pesticides in fruits and vegetables.
[14]	The Delphi Technique was utilized to recognize the most significant challenges facing the horticulture industry.	A framework was proposed for assisting researchers in identifying the key issues in horticulture within the COVID-19 context.
[15]	A real-time monitoring system was constructed to evaluate the quality losses that a perishable product can suffer during transportation and storage.	Conceptual shelf-life was minimized while stored or transported at an incorrect temperature.

The following describes how the study is structured: Section 2 discusses an overview of Horticulture 4.0 and Industry 4.0 technologies, Section 3 addresses the intervention of Industry 4.0 technologies in horticulture, and Section 4 covers the discussion and recommendations.

2. Overview of Horticulture and Industry 4.0

High temperatures have two main effects on crop production: they inhibit vegetative growth and have a detrimental effect on fruit production. Prolonged transpiration along with subjection to extreme temperatures limits vegetable crops that are susceptible to considerable transpiration deficits [16,17]. Another problem is that the maximum time for the fruit set is proportional to humidity levels. Extremely high temperatures can occur. The obvious limitation encountered by cold temperatures is the freezing of plant tissues. Rapidly pushing freezing temperatures at a phase of rapid growth might cause damage to a variety of plant tissues [18,19]. Some plants can adapt to cold temperatures provided there is sufficient time and the appropriate circumstances, whereas others cannot. Changes in regional precipitation patterns may cause an increase in drought conditions in many different parts of the world as atmospheric CO_2 levels to rise [20]. Although the current belief is that leaf photosynthetic responds to high CO_2 as well as soil-water deficiency, the connections between CO_2 enrichment and drought stress remain unexplored [21]. These are the few problems that have been addressed by previous studies in the area of horticulture. Below, we discuss the significance of Industry 4.0 technologies and their importance in enhancing practices in horticulture. Industry 4.0, often known as the Fourth Industrial Revolution, is the phenomenon that results from the impact of the IoT on information and communication technology. With the help of technologies like the integration of cyber-physical systems (CPS), IoT, and the real-time interaction between machinery, software, and people, it aims to revolutionize the industry through smart factories that will enable greater flexibility in production needs, efficient resource allocation, and process integration from device monitoring to final delivery [22].

Figure 1 depicts the technologies that can be intervened in horticulture, such as artificial intelligence (AI), blockchain, big data, the Internet of things (IoT), and cloud computing, in industry 4.0. Figure 1 illustrates the implementation and application area in which Industry 4.0 can be adopted for horticulture. In horticulture, there are different stages, including pre-harvesting, harvest, and post-harvest. Industry 4.0 technologies cover the following parameters for horticulture: automation of actuators, disease detection, irrigation management, fertilizer management, maturity identification, marketing and supply chain, soil fertility, and weather patterns. Along with these applications, Industry 4.0 integration in horticulture enables quality assessment and grading of horticulture products. The quality assessment of horticulture products is different in the three stages. The following are aspects through which the quality and grading of a horticulture product are decided, and they are typical cultivar ripeness, absence of defects and blemishes, a non-harmful number of pesticide residues, freshness, and other chemicals. Therefore, the quality assessment and grading of horticulture products mainly rely on the parameters discussed above. The monitoring of each parameter and utilizing the data generated from different wireless sources with Industry 4.0 technologies boost the performance of the horticulture industry for the future.

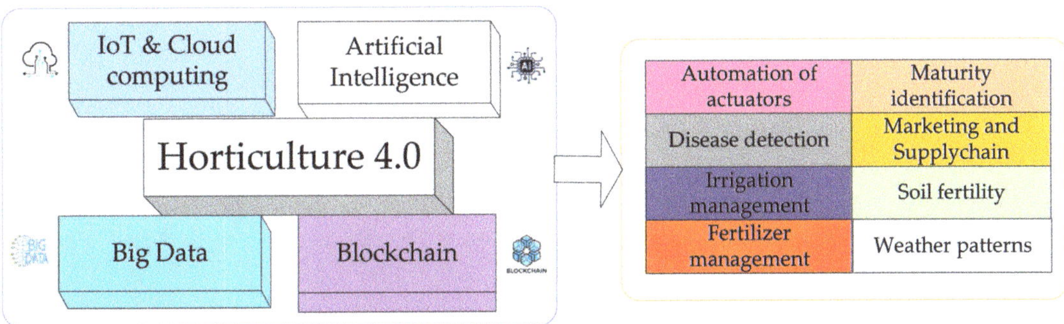

Figure 1. Horticulture 4.0.

Overview of Technologies and TheirFunctions

IoT is the primary technology that enables the provision of real-time information that is required for other technologies to implement their functions. IoT is an open network of intelligent devices that may self-organize; exchange information, data, and resources; and respond and act according to circumstances and environmental changes [22]. Fundamentally, IoT comprises the following layers in its architecture: perception layer, access layer, network layer, middleware layer, and application layer [23]. The perception layer is a key layer in which sensors, actuators, and identification technologies empower the transmission of real-time information and also enable control of things from remote locations through internet connectivity. The network layer empowers the transmission of packets of data with the assistance of communication protocols that are embedded in the architecture. Wireless fidelity (Wi-Fi), Bluetooth, and Zigbee are widely used communication protocols to establish a wireless sensor network and connect it with the internet to form the IoT [24].

The limitations of these communication protocols are the short transmission range and high-power consumption. In view of the IoT, most IoT devices are resource-constrained (work on battery power), so it demands communication protocols that consume less energy and transmit to a long range. A low-power–wide-area network (LPWAN) enables these challenges to be overcome, as it meets the requirements of the IoT [25]. Long-range (LoRa),

Sigfox, and narrowband (NB-IoT) are LPWAN communication technologies. As part of the application layer, a cloud server is suitable for visualization, real-time monitoring, and applying further analytics on the data available in it.AI is the science of enabling computers to accomplish intelligent tasks that could only be completed by humans.

AI is a multidisciplinary technology capable of combining machine learning, cognition, human-computer interaction, emotion recognition, data storage, and decision-making [26]. The bottleneck of AI was overcome with the advancement of processing power, and the development of deep learning and enhanced learning based on huge data progressed. With the constant advancement of GPUs has come the successful development of customized processors and increased computer capacity; this has established the groundwork for AI's rapid progress. Artificial neural networks (ANNs), decision-support systems (DSSs), genetic algorithms (Gas), support-vector machines, and computer vision are some of the AI techniques that are widely applied in the field of agriculture for the management of soil, crops, disease, and weeds [27]. Big data is defined as data sets that are too massive or complicated to be acquired, maintained, and processed in a low-latency manner by conventional database systems [28]. Big data has characteristics such as high volume, high velocity, and high variety. Big-data analytics can ultimately enable better and faster decision-making, model, and forecast future outcomes, and improve business intelligence [29]. A distributed, unchangeable database called blockchain makes it simpler to track assets and record transactions in a corporate network [30]. Distributed ledger technology, immutable records, and smart contracts are the key elements of blockchain. Each transaction is logged as a "block" of data as it occurs. Each block is linked to the ones that come before and after it. Transactions are linked in an irreversible chain

Horticulture is utilizing the IoT to collect data from field planting and horticultural facilities for production, management, and service. Robots, drones, remote sensors, and computer imagery are used in horticulture as part of the IoT to monitor crops, survey, and map fields, as well as to provide data to farmers for logical farm-management strategies that will reduce both time and cost [31]. AI is facilitating various sectors in agriculture to enhance productiveness and performance and to overcome traditionally demanding situations in each field. The intervention of AI in horticulture is helping farmers to improve their farming efficiency and reduce hostile environmental impacts [32]. Blockchain horticulture empowers information to be traced throughout the food supply chain to improve food safety. The ability of blockchain to store and manage data creates traceability, which is utilized to enhance the creation and deployment of innovations for intelligent farming and index-based horticultural insurance. Improved quality control and food safety are advantages of applying blockchain to gardening. Increased supply-chain traceability of farmers' productivity will lead to more equitable payments for farmers [33].

Farmers can learn in-depth information about topics like rainfall patterns, water cycles, fertilizer requirements, and more thanks to large datasets. Companies can utilize this information to choose which crops to produce and when to harvest them in order to maximize their profits [34]. Data on horticulture are gathered, examined, and stored using cloud computing. Farmers can better understand crop conditions by using cloud-connected wireless sensors to collect data from the field and machine-learning algorithms to analyze that real-time information [35]. Precision horticultural farming heavily relies on augmented reality. Farmers can use augmented reality in horticulture to boost production, reduce crop waste, and teach other farmers [36].

3. Intervention of Industry 4.0 in Horticulture

In this section, we discuss the various Industry 4.0 technologies in horticulture intending to transform into Horticulture 4.0.

3.1. IoT Intervention in Horticulture

The potential of pests and plant diseases is inseparably associated in particular with-specific weather characteristics. Previous research concluded that humidity and rainfall have a significant impact on pathogen spread and propagation [37]. Pests and diseases are more inclined to spread as a result of the wind. According to the information presented above, pests and diseases are prevalent in crops, and monitoring of the related parameters can be achieved with the IoT. An intelligent monitoring system was proposed on the basis of the IoT with a global packet for radio service (GPRS) and Zigbee communication protocol for pest warning, planting works, and production-quality checks of apples with the assistance of soil sensors, meteorological sensors, and cameras [38]. The feasibility, yield, and irrigation water-use efficiency of an IoT-based precision-irrigation system with LoRaWAN technology on fresh-market tomato production was analyzed [39]. ET, Watermark 200SS-5 soil-metric potential sensors, MP40, and a decision-support system were used to design and test four irrigation-scheduling treatments. A study was conducted with the integration of a camera module, and the images collected were used to analyze the water-management system as well as detect plant disease within a green environment [40].

Researchers established a technology platform using the IoT for environmental data acquisition, disaster warning, transmission, remote control, and an information push in vegetable greenhouses in real-time and lessen the influence of climate catastrophes on vegetable development [41]. The data collected by an IoT board are expected to be utilized to train machine-learning models for the development of intelligent automated indoor microclimate horticulture crops [42]. A database contains the results of the sensor analysis and it is also linked with data from the Indonesian Weather Agency, which contains daily meteorological data from the cultivation location [43]. To ensure the proper operation of the greenhouse automation system, multiple measuring stations are needed in a modern greenhouse to identify the local climate parameters in various areas of a large-scale greenhouse [44]. The IoT paradigm is allowing the scientific community to establish integrated settings where data could be automatically transferred among many distinctive networks to provide consumers with specific relevant information [45]. The security of the foods supplied can be ensured by utilizing the provenance data that are kept throughout the supply chain of vegetables, including during planting, harvesting, government oversight, testing, transportation, customs clearance, warehousing, repackaging, and internal testing [46,47].

Every kind of pest and disease is believed to be harmful to plants and to have a serious unfavorable effect on horticulture. To decrease the frequent use of pesticides and fungicides and to anticipate when pests will arise, the IoT system was created [48]. Soft-computing technologies are used to identify fruits by combining the three essential characteristics of an object—color, shape, and texture. This method reduces the dimensions of the feature vector. As a result, the combined and normalized image features produce better classification accuracy with fewer training data [49]. Real-time supply-chain monitoring can give stakeholders insight into perishable food to better manage to price and take appropriate action to uphold quality requirements [50]. Farmers confront a variety of challenges when growing vegetables, including issues with seeds, managing pests and diseases, commodity costs, and product marketing. Farmers can use the internet and the concept of mobile cloud computing to access information and engage in an interactive dialogue about vegetable production through mobile learning [51]. A framework for papaya grading based on the Artificial Bee Colony algorithm was proposed to classify papaya fruits from digital photographs [52,53].

Figure 2 illustrates an architecture that was implemented by previous studies for monitoring different parameters in horticulture. As discussed in Section 2, one of the key components of the IoT is sensors. Sensors such as temperature, soil moisture, humidity, light intensity, pH, NPK, and water level are crucial for obtaining important information about the soil and environment. Along with these sensors, the camera module is used to obtain data from the horticulture field. The sensors remain the same for the indoor and outdoor environments of horticulture cultivation. Data on the soil and environment are mapped to obtain better productivity by smart controlling of irrigation and fertilizer. Based on the soil data, the crop-yield analysis is analyzed, and fertilizers can be used as needed. All of these sensors send the soil and environment data to the computing unit, at which point the data-sensor processor communicates with the cloud server through a wireless-communication protocol and gateway. In the cloud server, the data are visualized on the graphical user interface.

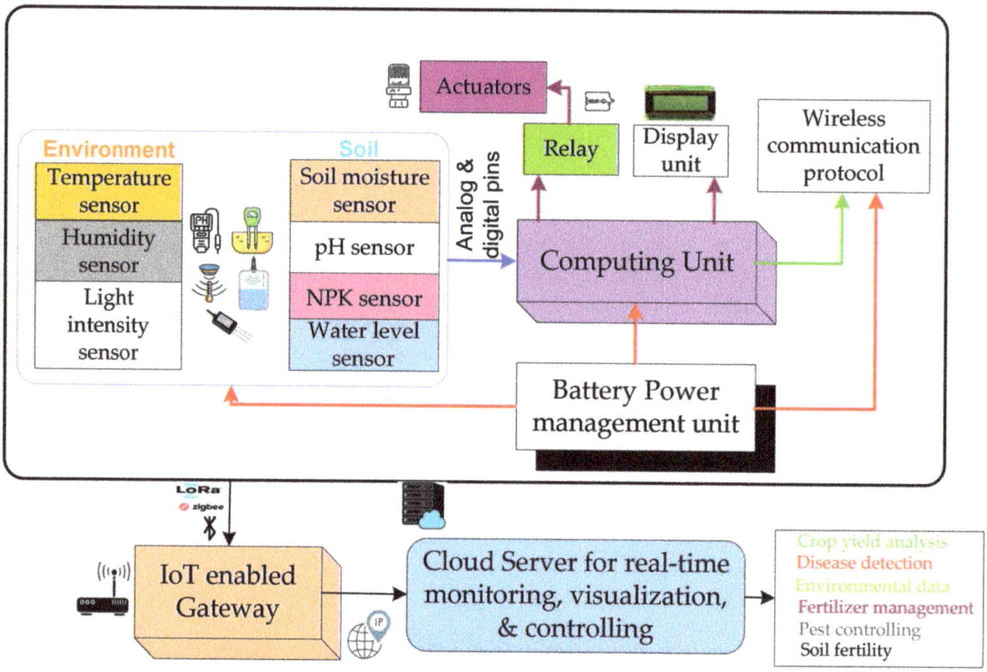

Figure 2. IoT for horticulture.

3.2. AI in Horticulture

Fresh and wholesome food is essential for feeding the expanding world population, and greenhouses and indoor agricultural techniques play a crucial role in this. In the past two decades, hyperspectral imaging research has developed, and in the decades to come, its use in horticulture is expected to grow. There are still challenges to the applicability. The automated detection of pests and diseases in plants empowers the effective monitoring of scalable fruit-and-vegetable crops. The detection of pests and diseases at the right time enhances pest- and disease-management systems [54]. The previous study concluded that AI algorithms can be implemented in horticulture for distinct applications, including fruit detection, pest, disease detection, weed detection, plant-stress detection, and yield prediction through the spectroscopy-and-camera system [55]. A study was implemented to identify common pests and diseases in apple fruit using sparse coding. Computer-vision techniques can identify pest- and disease-damaged fruits and provide data to assist in the detection and treatment of diseases and pests in the early stages [56]. Soil-organic-carbon (SOC) monitoring is a crucial characteristic of soil quality because it directly determines soil fertility and enables sustainable soil-nutrient management [57]. To improve SOC prediction, artificial neural networks (ANN), cubist regression, support-vector machines (SVM), multiple linear regression (MLR), and random forests (RF) were applied to the data of soil-nutrient indicators, total catchment area, and the topographic-wetness index.

Automatic detection of plant pests and diseases can aid in the monitoring of large farms and gardens. The application of AI in the drying process of fruits and vegetables has received a lot of attention because it can generate better-dried fruit-and-vegetable products when combined with an efficient physical field [58,59].An IoT-based tool can determine whether a climacteric fruit has been artificially ripened or not. To determine whether the fruit is ripened artificially or naturally, machine-learning algorithms are applied [60]. We discuss these techniques and discuss the significance of combining computer-vision techniques with an autonomous robotic system that uses the deep-learning concept of artificial intelligence [61]. In order to reduce food waste, one study used sensors and analyzed gases produced by certain food products to detect rotten food at an early stage and boost accuracy. To forecast how frequently food will degrade, the researchers used machine learning, the IoT, and sensors. To assess the freshness of food, this study used ML and IoT. The implementation of vision-based hardware in robots enables the realization of intelligent spraying, crop-yield prediction and price forecasting, predictive insights, and disease diagnosis (Figure 3). In addition, during the supply chain, the AI-based IoT system enables the indoor environment conditions to be adjusted on the basis of external climatic conditions and travel time to avoid the spoilage of fruit and vegetables.

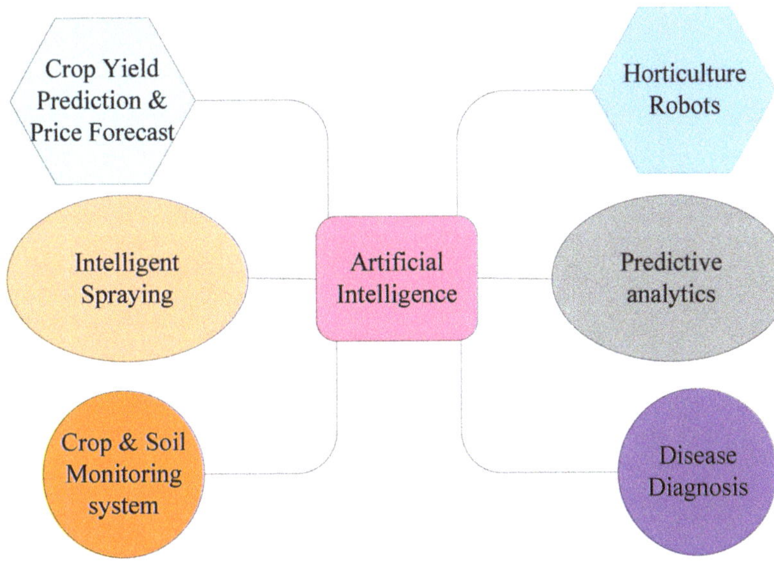

Figure 3. AI in horticulture and farming.

Table 2 illustrates the different studies that implement AI for horticulture crops for disease detection, quality assessment, and grading. The table provides the different algorithms that have been implemented for feature selection, feature extraction, classification, and regression for identifying defects, bruising quality assessment, and grading of the fruit. Different studies have used different feature-selection and feature-extraction techniques such as random frog, random forest, linear discriminant analysis-based fully convolutional networks, competitive adaptive reweighted sampling–successive projection algorithms, uninformative variable elimination, successive projection algorithm, gray histograms, and gray-level co-occurrence matrices. Classification and regression techniques include least-squares support-vector machines, support vector machines, convolutional neural networks, logistic regression, random forest, multilayer perceptron, linear discriminant analysis, k-nearest neighbors, and backpropagation in the neural network.

Table 2. AI for horticulture crops.

Ref	Function	Feature Selection/Extraction	Classification/Regression
[62]	A classification model is created to separate hard and soft blueberries.	Random frog	Least-squares support-vector machine, active learning algorithms
[63]	To accurately identify internal bruising in blueberries	Random forest, linear discriminant analysis-based fully convolutional networks	Support-vector machine
[64]	Detection of internal mechanical damage of blueberry	Customized convolutional neural networks	Convolutional neural networks, logistic regression, random forest, multilayer perceptron
[65]	Accurate assessment of bruising damage	Relief	Linear discriminant analysis, random forest, support vector machine, k-nearest neighbors
[66]	Assessment of strawberry ripeness using hyperspectral imaging	Competitive adaptive reweighted sampling–successive projection algorithm, Uninformative variable elimination, the successive projection algorithm	Random forest, support vector machine, partial least-squares regression, support-vector regression, random forest regression
[67]	Detection and analysis of decaying blueberries	Gray histogram, gray-level co-occurrence matrix	Backpropagation in neural network

Table 3 illustrates AI implementation in horticulture with accuracy and data-acquisition parameters. The previous studies illustrate that AI has been implemented for multiple applications such as quality assessment of fruit by evaluating moisture levels in the fruits. Following these studies, AI was used to forecast the yield of fruits such as bananas and blueberries. A few studies implemented AI for the detection and segmentation of apple fruits and branches in orchards through fused convolutional features, ResNet-101, clustering-RCNN, and CNN. The accuracy of different classifiers is illustrated in the table, and SVM, KNN, and DT achieved 100%. During the classification of tomato diseases, CNN achieved 99.18% accuracy on the dataset formed from 14,828 images. In the detection of citrus fruits, the CNN applied to the UAV images achieved 96.24% accuracy.

Table 3. AI implementation for horticulture with accuracy and data-acquisition sources.

Ref	Objective	Data Acquisition Source	Algorithm	Accuracy
[68]	Quality assessment of fruits through moisture content	Non-invasive sensing	SVM, KNN, and DT	100% for days 1 and 4 by three classifiers with 80% moisture content
[69]	Forecasted banana-harvest yields	Agrarian-reform beneficiary (ARB) cooperative of Dapco in Davao del Norte, Philippines	Long–short-term memory	Single and multiple LSTM layers with 43.5% and 44.95% error rates.
[70]	A data-processing pipeline is built to count blueberries, assess maturity, and assess compactness.	Four southern highbush blueberry cultivars	Mask R-CNN	Detection: 90.6% and segmentation: 90.4%
[71]	Classification of the different tomato diseases	The dataset contains 14,828 images of tomatoes leave infected with nine diseases.	CNN	99.18%
[72]	An apple monitoring system was developed with an edge-detection network.	Spherical video camera and two personal computers	Fused convolutional features	F1 score: 53.1% (FCF)
[73]	Detection and segmentation of apples in an orchard	Vision sensor	ResNet-101	F1 score: 0.832 for the detection of apples
[74]	Fruit detection in apple orchards	Not mentioned	Clustering-RCNN	0.853
[75]	Detection of citrus fruits	Unmanned aerial vehicle images	CNN	96.24

3.3. *Blockchain*

Blockchain is a distributed-ledger technology with the advantage of being tamper-resistant to information. It is anticipated to be able to address the issue of resource allocation for transactions among numerous unreliable parties in the supply chain for fresh fruit [76]. One potential method for supply-chain traceability in the pineapple industry is blockchain technology. The fruit-chain protocol that was introduced has identical consistency and liveliness qualities as assuming an honest majority of computer power [77] and is roughly fair with an overwhelming probability. Although blockchain might be viewed as a viable option for food-chain traceability, it was determined that [78] the goal of the investigation was to learn more about blockchain technology and its potential applications in the retail industry. Additionally, potential blockchain uses that might help the retail sector and the wider industry are foreseen. The study also underlined the crucial role that blockchain technology plays in the retail sector for fruit as well as the connections between these aspects [79].

Figure 4 shows blockchain technology in horticulture. The blockchain enables a distributed network to be established among the different entities in the supply chain for real-time monitoring and tracking of the activities from any location that is immutable and transparent. Blockchain empowers digital and secure trading to be created by incorporating smart contracts between entities. In addition, blockchain enables the realization of secured transactions of the export of fruits and vegetables in the international market. The quality and standards of the fruits and vegetables that are set by international bodies can be protected with secured hash cryptography.

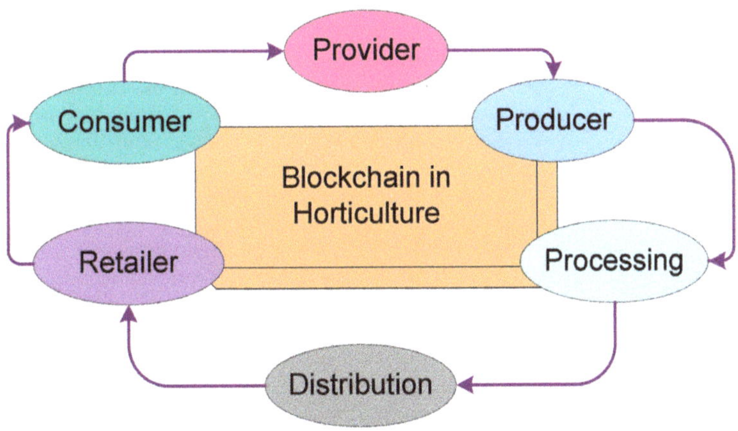

Figure 4. Blockchain technology in horticulture.

Consumers are driving an entirely different transformation in food procurement as a consequence of the growth in global food catastrophes that are triggering health insecurity. Consumers have called for transparency, traceability, and attentiveness along the entire fruit supply chain. The importance of the supply chain in this industry is increased by the fact that the products are perishable and have a limited shelf life. Yields are impacted by inconsistent delivery and a lack of fertilizers and insecticides as a result of dependence on middlemen, market instability, and other factors. Increased costs for input and transportation, post-harvest losses, and problems with safety and quality dominating supply-chain losses are key challenges involved in the fruit supply chain. Blockchain integration in the fruit supply chain (Figure 5) allows for post-harvest and inventory management streamlining, increasing operational effectiveness and lowering losses. End-to-end traceability with QR codes on the fruits presents the final customer with an honest and reliable narrative. A fair price for the producers is ensured by the grouping and collaboration of all stakeholders on a single platform, which fosters confidence and transparency. Real-time data collection allows for simple tracking and tracing, which helps with recall management. In addition to this, blockchain is used for monitoring the pre-harvest process for yield and quality. Post-harvest management for monitoring the crucial phases to prevent losses and boost output, monitoring a set of procedures for confirming sustainable practices, and digital records that cannot be altered and display accurate information are necessary to meet legal requirements.

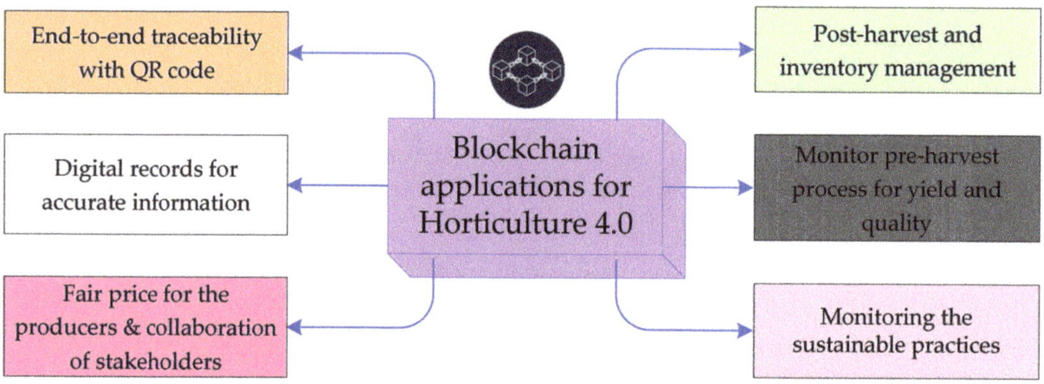

Figure 5. Applications of blockchain for Horticulture 4.0.

3.4. *Big Data in Horticulture*

A multi-sensor network system was established to accumulate smaller ecological statistical data on vegetable-growing regions. Researchers were able to identify the critical components pushing pest spread using multidimensional information such as environment, soil, meteorology of vegetable fields, and ultimately the vegetable-pest warning system premised on multidimensional big data [80]. The distinctive nutrition-based vegetable-production and -distribution system utilizes the inventive big-data framework and its multiple benefits to provide a healthful food recommendation to the end customer as well as various predictive analyses to boost system efficacy [45]. The new ICT horticulture project will heavily rely on big-data methodologies; therefore, it is important to understand how to manage them and how they could affect everyday business [81] Big-data intervention in horticulture is presented in Figure 6. Because fruits and vegetables are produced in such large quantities, the sensor data that are available and can be used in horticulture are now considered big data. The big data can be transmitted to the cloud server and made available in a distribution box and control box through wireless-communication protocols.

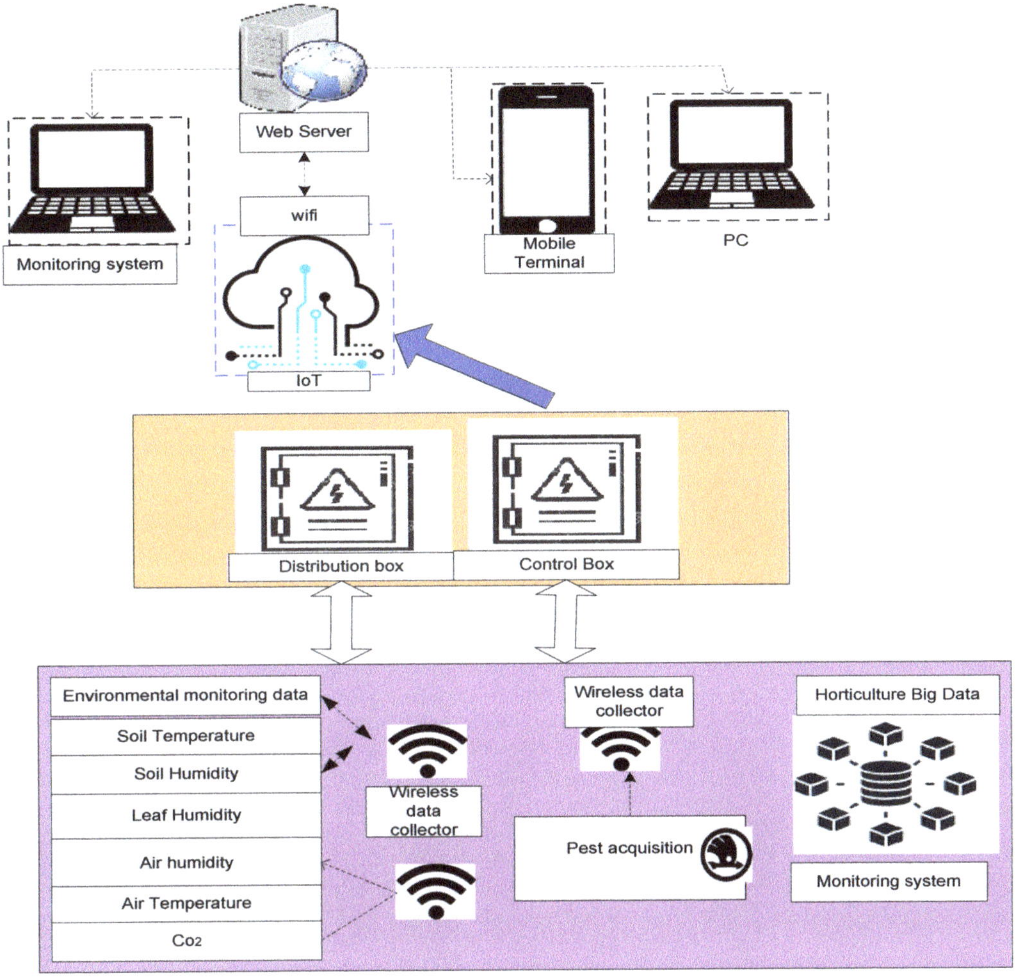

Figure 6. Big-data overview in horticulture.

4. Results

In this section, the results identified from the analysis of previous studies based on Industry 4.0 integration for horticulture monitoring are discussed.

- The IoT is predominantly used in horticulture to monitor various pests and diseases that are harmful to plants and have a serious negative impact on horticulture. The Internet of Things system was configured to minimize the frequency with which pesticides and fungicides are used and to predict when pests will appear. Along with this, IoT and cloud computing are employed to identify fruits by combining an object's three essential characteristics—color, shape, and texture. Farmers face a variety of challenges when growing vegetables, including seed issues, pest-and-disease management, commodity costs, and product marketing. Farmers can access information and engage in an interactive process about vegetable production through the internet and the concept of the mobile cloud.
- AI in horticulture enables the detection of diseases, quality assessment, and grading of horticulture crops. Linear discriminant analysis-based fully convolutional networks, random forest, competitive adaptive reweighted sampling–successive projection al-

gorithm, uninformative variable elimination, successive projection algorithm, and gray-level co-occurrence matrix have been used for feature extraction and selection. Least-squares support-vector machine, support-vector machine, convolutional neural networks, logistic regression, random forest, multilayer perceptron, and k-nearest neighbors are the key techniques used for classification and regression.

- Yields are impacted by inconsistent delivery and a lack of fertilizers and insecticides as a result of dependence on middlemen, market instability, and other factors. Increased input and transportation costs, post-harvest losses, and safety and quality issues dominate supply-chain losses. However, the incorporation of blockchain improves pre-harvest and post-harvest management through real-time tracing and secure transactions in the distributed network.

5. Discussion and Recommendation

In this section, based on the discussion of the analysis above, we discuss the recommendations for applications in horticulture as part of future work. A few vital recommendations are as follows:

- When considering long-term benefits like improved production and automation, the use of the IoT in horticulture is strongly advised and very successful. The Internet of Drones enables horticulture operations to be improved. The Internet of Drones is an architecture intended to focus on providing UAVs, also commonly known as drones, with coordinated connectivity to controlled airspace. Drones for horticulture are used to monitor fields, plant crops, irrigate fields, spray pesticides, and also to assess the health of plants [82,83]. Moreover, drones with vision-based technology and AI empower the detection of weeds and identify the stages of the growth of plants in pre-harvest, harvest, and post-harvest. Drones can be used to monitor the quality assessment and grading of horticulture crops to minimize the damage and loss of the product during transportation and storage.
- Robotics leverage AI technology to improve harvest quality and precision. In horticulture settings, robotics with vision technology and AI for the detection of pests, plant diseases, and malnutrition. Before choosing which herbicide to apply in a region, AI can detect and target weeds through the visuals obtained from the camera module inserted into robots. A wide range of issues in the farming sector can be resolved by using an amalgamation of robots and AI approaches [84]. In addition to this, currently, different robots with AI technology are used for weed removal and plucking, sorting, and packaging of fruits [85].
- Globally, horticulture supply chains have an exciting opportunity to improve transactional efficiency, reduce resistance, and promote traceability thanks to blockchain technology [80]. Blockchain technology can help the horticulture and food industry deal with and control known risks while maintaining affordability across the ecosystem. Blockchain in horticulture enables the linking of various horticultural entities for the visualization of data from production to supply on distributed-ledger networks. Granular information on rainfall patterns, water cycles, fertilizer requirements, and more are made available to farmers through big data. This enables them to make wise decisions, such as which plants to sow for greater profitability and when to harvest. Making the proper choices ultimately increase crop output. The fundamental goal of precision horticulture is to guarantee profitability, efficacy, and sustainability [86].
- In horticulture, edge computing is employed for different activities. Edge computing can be utilized to analyze the data at the edge of a network for better monitoring and productivity. Mobile edge computing (MEC) is an architecture that provides cloud-computing services at the edge of networks leveraging mobile base stations [87,88]. Currently, different study areas have implemented edge-computing-based portable devices in order to predict what is useful for the user in enhancing their practices [89,90]. In the scenario of horticulture, an edge-computing-based portable device can be developed with IoT and AI for predicting and estimating disease in a crop. Based on the

analysis and prediction, it can also communicate to the user on the cloud server and the respective horticulture authorities to suggest a solution in real-time.

- Figure 7 illustrates a hybrid architecture that is implemented in horticulture with the amalgamation of the IoT, cloud computing, AI/ML, blockchain, and big data. This architecture enables a smart and intelligent ecosystem to be achieved in horticulture with multiple features such as blockchain-assisted marketing, prediction of international markets, and the quality of fruits and vegetables based on real-time environmental data. The generated output and sensor data can then be distributed in the peer-to-peer network of any location.

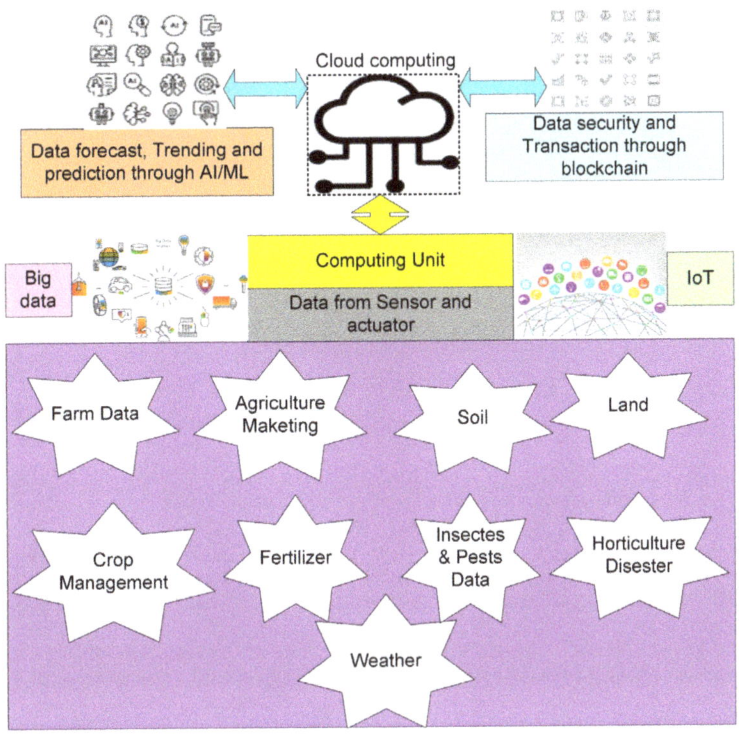

Figure 7. Hybrid architecture for Horticulture 4.0.

6. Conclusions

Horticulture is the field of cultivation of fruits and vegetables. It ensures production and consumption by minimizing malnutrition in the current scenario addressed by United Nations. Recently, Industry 4.0 technologies have delivered the ability of digitalization and realize the SDGs set by the United Nations. The previous studies did not highlight the significance and application of Industry 4.0 for distinct issues of horticulture. Based on this limitation, the current study addressed the significance and application of Industry 4.0 technologies such as the Internet of Things, cloud computing, artificial intelligence, blockchain, and big data for horticulture to enhance traditional practices of disease detection, irrigation management, fertilizer management, maturity identification, marketing, and supply chain, soil fertility, and weather patterns at pre-harvest, harvest, and post-harvest.

The findings of the study are as follows: in horticulture, the IoT is primarily used to monitor various pests and diseases that are harmful to plants and have a serious negative impact on horticulture. In addition, IoT and cloud computing is used to identify fruits by combining three essential characteristics of an object: color, shape, and texture.AI in horticulture has enabled the detection of diseases, quality assessment, and crop grading.

For feature extraction and selection, we used fully convolutional networks, a random forest, and a competitive adaptive reweighted sampling–successive projection algorithm based on linear discriminant analysis. The key techniques used for classification and regression are the least-squares support-vector machine, support-vector machine, convolutional neural networks, logistic regression, multilayer perceptron, and k-nearest neighbors. Inconsistent delivery and a lack of fertilizers and insecticides as an outcome of dependence on middlemen, market instability, and other factors have a negative impact on yields. Increased input and transportation costs, post-harvest losses, and safety and quality issues that dominate supply-chain losses can be overcome with blockchain during-harvest and post-harvest management. Finally, the study suggested vital recommendations for future works, such as robotics; drones with vision technology and AI for the detection of pests, weeds, plant diseases, and malnutrition; and edge computing portable devices developed with the IoT and AI for predicting and estimating disease in crops.

Author Contributions: Conceptualization, R.S. (Rajesh Singh) and R.S. (Rajat Singh); methodology, A.G.; formal analysis, S.V.A.; data curation, S.V.A.; writing—original draft preparation, R.S. (Rajat Singh); writing—review and editing, N.P. and B.T.; visualization, R.S. (Rajesh Singh) and A.G.; funding acquisition, N.P. and B.T. All authors have read and agreed to the published version of the manuscript.

Funding: The APC was funded by Tshwane University of Technology, South Africa.

Institutional Review Board Statement: Not applicable.

Informed Consent Statement: Not applicable.

Data Availability Statement: Data sharing not applicable.

Conflicts of Interest: The authors declare no conflict of interest.

References

1. UN Report: Global Hunger Numbers Rose to as Many as 828 Million in 2021. Available online: https://www.fao.org/newsroom/detail/un-report-global-hunger-SOFI-2022-FAO/en (accessed on 6 September 2022).
2. Goal 2: Zero Hunger-United Nations Sustainable Development. Available online: https://www.un.org/sustainabledevelopment/hunger/ (accessed on 6 September 2022).
3. Goal 12. Department of Economic and Social Affairs. Available online: https://sdgs.un.org/goals/goal12 (accessed on 6 September 2022).
4. Keatinge, J.D.H.; Virchow, D.; Schreinemachers, P. Horticulture for sustainable development: Evidence for impact of international vegetable research and development. In Proceedings of the International Symposia on Tropical and Temperate Horticulture-ISTTH2016 1205, Cairns, Australia, 20 November 2016; pp. 179–190.
5. Qingxue, L.; Wu, H. Research on vegetable growth monitoring platform based on facility agricultural IoT. In *International Conference on Geo-Informatics in Resource Management and Sustainable Ecosystem*; Springer: Singapore, 2016.
6. Cappelli, I.; Fort, A.; Pozzebon, A.; Tani, M.; Trivellin, N.; Vignoli, V.; Bruzzi, M. Autonomous IoT Monitoring Matching Spectral Artificial Light Manipulation for Horticulture. *Sensors* **2022**, *22*, 4046. [CrossRef]
7. Lamm, K.; Powell, A.; Lombardini, L. Identifying Critical Issues in the Horticulture Industry: A Delphi Analysis during the COVID-19 Pandemic. *Horticulturae* **2021**, *7*, 416. [CrossRef]
8. Bhavana, H.; Bhagwan, A. Review on: Role of robotics in horticulture. *J. Pharm. Phytochem.* **2021**, *10*, 306–309.
9. Yang, B.; Xu, Y. Applications of deep-learning approaches in horticultural research: A review. *Hortic. Res.* **2021**, *8*, 123. [CrossRef]
10. Ariesen-Verschuur, N.; Verdouw, C.; Tekinerdogan, B. Digital Twins in greenhouse horticulture: A review. *Comput. Electron. Agric.* **2022**, *199*, 107183. [CrossRef]
11. Halgamuge, M.N.; Bojovschi, A.; Fisher, P.M.; Le, T.C.; Adeloju, S.; Murphy, S. Internet of Things and autonomous control for vertical cultivation walls towards smart food growing: A review. *Urban For. Urban Green* **2021**, *61*, 127094. [CrossRef]
12. Longchamps, L.; Tisseyre, B.; Taylor, J.; Sagoo, L.; Momin, A.; Fountas, S.; Manfrini, L.; Ampatzidis, Y.; Schueller, J.K.; Khosla, R. Yield sensing technologies for perennial and annual horticultural crops: A review. *Precis. Agric.* **2022**, 1–42. [CrossRef]
13. Jallow, M.F.A.; Awadh, D.G.; Albaho, M.S.; Devi, V.Y.; Ahmad, N. Monitoring of Pesticide Residues in Commonly Used Fruits and Vegetables in Kuwait. *Int. J. Environ. Res. Public Health* **2017**, *14*, 833. [CrossRef]
14. Bulgari, R.; Petrini, A.; Cocetta, G.; Nicoletto, C.; Ertani, A.; Sambo, P.; Ferrante, A.; Nicola, S. The Impact of COVID-19 on Horticulture: Critical Issues and Opportunities Derived from an Unexpected Occurrence. *Horticulturae* **2021**, *7*, 124. [CrossRef]
15. Rehman, M.U.; Rather, G.H.; Gull, Y.; Mir, M.R.; Mir, M.M.; Waida, U.I.; Hakeem, K.R. Effect of climate change on horticultural crops. In *Crop Production and Global Environmental Issues*; Springer: Cham, Switzerland; pp. 211–239.

16. Eftekhari, M.S. Impacts of Climate Change on Agriculture and Horticulture. In *Climate Change*; Bandh, S.A., Ed.; Springer: Cham, Switzerland, 2022. [CrossRef]
17. Saran, P.L.; Singh, S.; Solanki, V.; Choudhary, R.; Manivel, P. Evaluation of Asparagus adscendens accessions for root yield and shatavarin IV content in India. *Turk. J. Agric. For.* **2021**, *45*, 8. [CrossRef]
18. Zulfiqar, F. Effect of seed priming on horticultural crops. *Sci. Hortic.* **2021**, *286*, 110197. [CrossRef]
19. Grygorieva, O.; Klymenko, S.; Kuklina, A.; Vinogradova, Y.; Vergun, O.; Sedlackova, V.H.; Brindza, J. Evaluation of Lonicera caerulea L. genotypes based on morphological characteristics offruits germplasm collection. *Turk. J. Agric. For.* **2021**, *45*, 14. [CrossRef]
20. Mall, M.; Kumar, R.; Akhtar, M.Q. Horticultural crops and abiotic stress challenges. In *Stress Tolerance in Horticultural Crops*; Woodhead Publishing: Sawston, UK, 2021; pp. 1–19.
21. Alireza, A.; Dennis, R.; Ludena, A. Big data application to the vegetable production and distribution system. In Proceedings of the 2014 IEEE 10th International Colloquium on Signal Processing and Its Applications, Kuala Lumpur, Malaysia, 7–9 March 2014.
22. Kaburuan, E.R.; Jayadi, R. A design of IoT-based monitoring system for intelligence indoor micro-climate horticulture farming in Indonesia. *Procedia Comput. Sci.* **2019**, *157*, 459–464. [CrossRef]
23. Alem, Č.; Hadžialić, M. Internet of Things (IoT): A review of enabling technologies, challenges, and open research issues. *Comput. Netw.* **2018**, *144*, 17–39.
24. Mohammad Nuruzzaman, B.; Rahman, M.; Billah, M.; Saha, D. Internet of things (IoT): A review of its enabling technologies in healthcare applications, standards protocols, security, and market opportunities. *IEEE Internet Things J.* **2021**, *8*, 10474–10498.
25. Kais, M.; Bajic, E.; Chaxel, F.; Meyer, F. A comparative study of LPWAN technologies for large-scale IoT deployment. *ICT Express* **2019**, *5*, 1–7.
26. Arya, P.S.; Gangwar, M. A Proposed Architecture: Detecting Freshness of Vegetables using Internet of Things (IoT) & Deep Learning Prediction Algorithm. In Proceedings of the 2021 3rd International Conference on Advances in Computing, Communication Control and Networking (ICAC3N), Greater Noida, India, 17–18 December 2021; pp. 718–723. [CrossRef]
27. Caiming, Z.; Lu, Y. Study on artificial intelligence: The state of the art and future prospects. *J. Ind. Inf. Integr.* **2021**, *23*, 100224.
28. Sagiroglu, S.; Sinanc, D. Big data: A review. In Proceedings of the 2013 International Conference on Collaboration Technologies and Systems (CTS), San Diego, CA, USA, 20–24 May 2013; pp. 42–47. [CrossRef]
29. Yosra, H.; Boulila, W.; Farah, I.R.; Romdhani, I.; Hussain, A. Big data and IoT-based applications in smart environments: A systematic review. *Comput. Sci. Rev.* **2021**, *39*, 100318.
30. Mohd, J.; Haleem, A.; Singh, R.P.; Khan, S.; Suman, R. Blockchain technology applications for Industry 4.0: A literature-based review. *Blockchain Res. Appl.* **2021**, *2*, 100027.
31. Alireza, A.; Ludena, R.D.A. Big Data approach to a novel nutrition-based vegetable production and distribution system. In Proceedings of the 2013 IEEE International Conference on Computational Intelligence and Cybernetics (CYBERNETICSCOM), Yogyakarta, Indonesia, 3–4 December 2013.
32. Kumar, P.C.; Bhatnagar, R. Social internet of things in agriculture: An overview and future scope. In *Toward Social Internet of Things (SIoT): Enabling Technologies, Architectures and Applications*; Springer: Berlin/Heidelberg, Germany, 2020; pp. 317–334.
33. Sun, Z.; Hui, X.; Wensheng, W. An architecture for the agricultural machinery intelligent scheduling in cross-regional work based on cloud computing and internet of things. In *International Conference on Computer and Computing Technologies in Agriculture*; Springer: Berlin/Heidelberg, Germany, 2010.
34. Coble, K.H.; Mishra, A.K.; Ferrell, S.; Griffin, T. Big data in agriculture: A challenge for the future. *Appl. Econ. Perspect. Policy* **2018**, *40*, 79–96. [CrossRef]
35. Ruthie, M. Big data in agriculture and nutrition. *Agric. Improv. Nutr. Seizing Momentum* **2019**, *142*, 1–15.
36. Kamilaris, A.; Fonts, A.; Prenafeta-Boldύ, F.X. The rise of blockchain technology in agriculture and food supply chains. *Trends Food Sci. Technol.* **2019**, *91*, 640–652. [CrossRef]
37. Codeluppi, G.; Cilfone, A.; Davoli, L.; Ferrari, G. VegIoT garden: A modular IoT management platform for urban vegetable gardens. In Proceedings of the 2019 IEEE International Workshop on Metrology for Agriculture and Forestry (MetroAgriFor), Portici, Italy, 24–26 October 2019; IEEE: Piscataway, NJ, USA, 2019; pp. 121–126.
38. Salgado-Salazar, C.; Shiskoff, N.; Daughtrey, M.; Palmer, C.L.; Crouch, J.A. Downy mildew: A serious disease threat to rose health worldwide. *Plant Dis.* **2018**, *102*, 1873–1882. [CrossRef] [PubMed]
39. Feng, C.; Wu, H.R.; Zhu, H.J.; Sun, X. The design and realization of apple orchard intelligent monitoring system based on internet of things technology. In *Advanced Materials Research*; Trans Tech Publications Ltd.: Bäch, Switzerland, 2012; Volume 546, pp. 898–902.
40. Zhang, H.; He, L.; Di Gioia, F.; Choi, D.D.; Heinemann, P. Internet of things (IoT)-based precision irrigation with LoRaWAN technology applied to vegetable production. In *2021 ASABE Annual International Virtual Meeting*; American Society of Agricultural and Biological Engineers: St. Joseph, MI, USA, 2021; p. 1.
41. Khan, F.A.; Ibrahim, A.A.; Zeki, A.M. Environmental monitoring and disease detection of plants in smart greenhouse using internet of things. *J. Phys. Commun.* **2020**, *4*, 055008. [CrossRef]
42. Bhujel, A.; Basak, J.K.; Khan, F.; Arulmozhi, E.; Jaihuni, M.; Sihalath, T.; Lee, D.; Park, J.; Kim, H.T. Sensor Systems for Greenhouse Microclimate Monitoring and Control: A Review. *J. Biosyst. Eng.* **2020**, *45*, 341–361. [CrossRef]

43. Wayangkau, I.H.; Mekiuw, Y.; Rachmat, R.; Suwarjono, S.; Hariyanto, H. Utilization of IoT for soil moisture and temperature monitoring system for onion growth. *Emerg. Sci. J.* **2020**, *4*, 102–115. [CrossRef]
44. Alper, M.A.; Sokullu, R. An IoT-based greenhouse monitoring system with Micaz motes. *Procedia Comput. Sci.* **2017**, *113*, 603–608.
45. Dennis, A.L.R.; Ahrary, A.; Horibe, N.; Yang, W.S. IoT-security approach analysis for the novel nutrition-based vegetable production and distribution system. In Proceedings of the 3rd IIAI International Conference on Advanced Applied Informatics, IIAI-AAI 2014, Kokura Kita-ku, Japan, 31 August–4 September 2014.
46. Jie, Y.; Zhang, X.; Lu, Q.; Xin, C.; Liu, C.; Chen, Z. IOT based provenance platform for vegetables supplied to Hong Kong. In *Recent Advances in Computer Science and Information Engineering*; Springer: Berlin/Heidelberg, Germany, 2012; pp. 591–596.
47. Harish, M.P.C.S.P.; Sanghavi, K.B.P.A. Smart refrigerator and vegetable identification system using image processing and IOT. *Open Access Int. J. Sci. Eng.* **2021**, *7*, 4.
48. Kumar, M.; Pal, Y.; Gangadharan, S.M.P.; Chakraborty, K.; Yadav, C.S.; Kumar, H.; Tiwari, B. Apple Sweetness Measurement and Fruit Disease Prediction Using Image Processing Techniques Based on Human-Computer Interaction for Industry 4.0. *Wirel. Commun. Mob. Comput.* **2022**, *2022*, 1–12.
49. Rajasekar, L.; Sharmila, D. Performance analysis of soft computing techniques for the automatic classification of fruits dataset. *Soft Comput.* **2019**, *23*, 2773–2788. [CrossRef]
50. John, K.; Haw, S.; Su, N.; Mulaafer, S. Kiwi Fruit IoT Shelf Life Estimation During Transportation with Cloud Computing. In Proceedings of the 2021 IEEE International Conference on Artificial Intelligence in Engineering and Technology (IICAIET), Kota Kinabalu, Malaysia, 13–15 September 2021.
51. Shwetapadma, P.; Sethy, P.K. Post-Harvest Grading Of Carica Papaya Fruit Using Image Segmentation And Soft Computing. *Int. J. Adv. Res. Comput. Sci.* **2017**, *8*, 7.
52. Yaya, W.; Nugraha, E. Development Mobile Learning For Vegetable Farming In Indonesia Based On Mobile Cloud Computing. In Proceedings of the 2020 6th International Conference on Science in Information Technology (ICSITech), Palu, Indonesia, 21–22 October 2020.
53. Narendra, V.G.; Ancilla, J.P. Defects Detection in Fruits and Vegetables Using Image Processing and Soft Computing Techniques. In *International Conference on Harmony Search Algorithm*; Springer: Singapore, 2020.
54. Silke, H.; de Zwart, F.; Elings, A.; Petropoulou, A.; Righini, I. Cherry Tomato Production in Intelligent Greenhouses—Sensors and AI for Control of Climate, Irrigation, Crop Yield, and Quality. *Sensors* **2020**, *20*, 6430. [CrossRef]
55. Kingsley, J.; Isong, I.A.; Kebonye, N.M.; Ayito, E.O.; Agyeman, P.C.; Afu, S.M. Using Machine Learning Algorithms to Estimate Soil Organic Carbon Variability with Environmental Variables and Soil Nutrient Indicators in an Alluvial Soil. *Land* **2020**, *9*, 487. [CrossRef]
56. Yousef, A.-G.; Aghabara, A.; Davari, M.; Maja, J.M. Feasibility of Using Computer Vision and Artificial Intelligence Techniques in Detection of Some Apple Pests and Diseases. *Appl. Sci.* **2022**, *12*, 906. [CrossRef]
57. Diaz-Gonzalez, F.A.; Vuelvas, J.; Correa, C.A.; Vallejo, V.E.; Patino, D. Machine learning and remote sensing techniques applied to estimate soil indicators–Review. *Ecol. Indicators* **2022**, *135*, 108517. [CrossRef]
58. Maione, C.; Araujo, E.M.; Santos-Araujo, S.N.D.; Boim, A.G.F.; Barbosa, R.M.; Alleoni, L.R.F. Determining the geographical origin of lettuce with data mining applied to micronutrients and soil properties. *Sci. Agric.* **2021**, *79*, e20200011. [CrossRef]
59. Chen, J.; Zhang, M.; Xu, B.; Sun, J.; Mujumdar, A.S. Artificial intelligence assisted technologies for controlling the drying of fruits and vegetables using physical fields: A review. *Trends Food Sci. Technol.* **2020**, *105*, 251–260. [CrossRef]
60. Sreeraj, M.; Joy, J.; Kuriakose, A.; Sujith, M.R.; Unni, H. CLadron*: AI assisted device for identifying artificially ripened climacteric fruits. *Procedia Comput. Sci.* **2020**, *171*, 635–643. [CrossRef]
61. Bittu, S.; Aziz, N.A.; Siddiqi, M.A.M. AI Based Robotic Systems for the quality control of Date Palm Fruits-A Review. In Proceedings of the 2019 International Conference on Digitization (ICD), Sharjah, United Arab Emirates, 18–19 November 2019.
62. Hu, M.H.; Zhai, G.T.; Zhao, Y.; Wang, Z.D. Uses of selection strategies in both spectral and sample spaces for classifying hard and soft blueberry using near infrared data. *Sci. Rep.* **2018**, *8*, 6671. [CrossRef]
63. Zhang, M.; Jiang, Y.; Li, C.; Yang, F. Fully convolutional networks for blueberry bruising and calyx segmentation using hyperspectral transmittance imaging. *Biosyst. Eng.* **2020**, *192*, 159–175. [CrossRef]
64. Wang, Z.; Hu, M.; Zhai, G. Application of Deep Learning Architectures for Accurate and Rapid Detection of Internal Mechanical Damage of Blueberry Using Hyperspectral Transmittance Data. *Sensors* **2018**, *18*, 1126. [CrossRef]
65. Kuzy, J.; Jiang, Y.; Li, C.Y. Blueberry bruise detection by pulsed thermographic imaging. *Postharvest. Biol. Technol.* **2018**, *136*, 166–177. [CrossRef]
66. Shao, Y.Y.; Wang, Y.X.; Xuan, G.T.; Gao, Z.M.; Hu, Z.C.; Gao, C.; Wang, K.L. Assessment of strawberry ripeness using hyperspectral imaging. *Anal. Lett.* **2020**, *54*, 1547–1560. [CrossRef]
67. Qiao, S.C.; Tian, Y.W.; Song, P.; He, K.; Song, S.Y. Analysis and detection of decayed blueberry by low field nuclear magnetic resonance and imaging. *Postharvest. Biol. Technol.* **2019**, *156*, 110951. [CrossRef]
68. Ren, A.; Zahid, A.; Zoha, A.; Shah, S.A.; Imran, M.A.; Alomainy, A.; Abbasi, Q.H. Machine Learning Driven Approach Towards the Quality Assessment of Fresh Fruits Using Non-Invasive Sensing. *IEEE Sens. J.* **2020**, *20*, 2075–2083. [CrossRef]
69. Rebortera, M.A.; Fajardo, A.C. An enhanced deep learning approach in forecasting banana harvest yields. *Int. J. Adv. Comput. Sci. Appl.* **2019**, *10*, 275–280. [CrossRef]

70. Ni, X.; Li, C.; Jiang, H.; Takeda, F. Deep learning image segmentation and extraction of blueberry fruit traits associated with harvestability and yield. *Hortic. Res.* **2020**, *7*, 110. [CrossRef] [PubMed]
71. Brahimi, M.; Boukhalfa, K.; Moussaoui, A. Deep learning for tomato diseases: Classification and symptoms visualization. *Appl. Artif. Intell.* **2017**, *31*, 299–315. [CrossRef]
72. Wang, D.; Li, C.; Song, H.; Xiong, H.; Liu, C.; He, D. Deep Learning Approach for Apple Edge Detection to Remotely Monitor Apple Growth in Orchards. *IEEE Access* **2020**, *8*, 26911–26925. [CrossRef]
73. Kang, H.; Chen, C. Fruit detection and segmentation for apple harvesting using visual sensor in orchards. *Sensors* **2019**, *19*, 4599. [CrossRef]
74. Kang, H.; Chen, C. Fast implementation of real-time fruit detection in apple orchards using deep learning. *Comput. Electron. Agric.* **2020**, *168*, 105108. [CrossRef]
75. Csillik, O.; Cherbini, J.; Johnson, R.; Lyons, A.; Kelly, M. Identification of citrus trees from unmanned aerial vehicle imagery using convolutional neural networks. *Drones* **2018**, *2*, 39. [CrossRef]
76. Rafael, P.; Shi, E. Fruitchains: A fair blockchain. In Proceedings of the ACM Symposium on Principles of Distributed Computing, Washington, DC, USA, 25–27 July 2017.
77. Zhang, Y.; Chen, L.; Battino, M.; Farag, M.A.; Xiao, J.; Simal-Gandara, J.; Jiang, W. Blockchain: An emerging novel technology to upgrade the current fresh fruit supply chain. *Trends Food Sci. Technol.* **2022**, *124*, 1–22. [CrossRef]
78. Arsenii, S. Traceability technology: Fruits and vegetables trader case study. In Proceedings of the International Conference on Digital Technologies in Logistics and Infrastructure (ICDTLI 2019), St. Petersburg, Russia, 4–5 April 2019; Atlantis Press: Hong Kong, China, 2019.
79. Sharif, K.I.M.; Hassan, M.G.; Miraz, M.H.; Zulkifly, E.; Udin, Z.M.; Omar, M. Factors Affecting Blockchain in Fruit Retail Market: An Unveiling Myth of Blockchain. In *Intelligent Manufacturing and Mechatronics*; Springer: Singapore, 2021; pp. 1095–1106.
80. Zhang, C.; Cai, J.; Xiao, D.; Ye, Y.; Chehelamirani, M. Research on vegetable pest warning system based on multidimensional big data. *Insects* **2018**, *9*, 66. [CrossRef] [PubMed]
81. Puri, V.; Nayyar, A.; Raja, L. Agriculture drones: A modern breakthrough in precision agriculture. *J. Stat. Manag. Syst.* **2017**, *20*, 507–518. [CrossRef]
82. Tripicchio, P.; Satler, M.; Dabisias, G.; Ruffaldi, E.; Avizzano, C.A. Towards smart farming and sustainable agriculture with drones. In Proceedings of the 2015 International Conference on Intelligent Environments, Prague, Czech Republic, 15–17 July 2015; pp. 140–143.
83. Gharibi, M.; Boutaba, R.; Waslander, S.L. Internet of Drones. *IEEE Access* **2016**, *4*, 1148–1162. [CrossRef]
84. Rajan, K.; Saffiotti, A. Towards a science of integrated AI and Robotics. *Artif. Intell.* **2017**, *247*, 1–9. [CrossRef]
85. Vivekanand, P.; Pant, M.; Snasel, V. Blockchain technology in food supply chains: Review and bibliometric analysis. *Technol. Soc.* **2022**, *69*, 101954.
86. Abbas, N.; Zhang, Y.; Taherkordi, A.; Skeie, T. Mobile Edge Computing: A Survey. *IEEE Internet Things J.* **2018**, *5*, 450–465. [CrossRef]
87. Shi, W.; Cao, J.; Zhang, Q.; Li, Y.; Xu, L. Edge computing: Vision and challenges. *IEEE Internet Things J.* **2016**, *3*, 637–646. [CrossRef]
88. Satyanarayanan, M. The Emergence of Edge Computing. *Computer* **2017**, *50*, 30–39. [CrossRef]
89. Chang, Z.; Liu, S.; Xiong, X.; Cai, Z.; Tu, G. A Survey of Recent Advances in Edge-Computing-Powered Artificial Intelligence of Things. *IEEE Internet Things J.* **2021**, *8*, 13849–13875. [CrossRef]
90. Guillén, M.A.; Llanes, A.; Imbernón, B.; Martínez-España, R.; Bueno-Crespo, A.; Cano, J.-C.; Cecilia, J.M. Performance evaluation of edge-computing platforms for the prediction of low temperatures in agriculture using deep learning. *J. Supercomput.* **2021**, *77*, 818–840. [CrossRef]

applied sciences

MDPI

Perspective

Agriculture 4.0: Is Sub-Saharan Africa Ready?

Nugun P. Jellason [1,*], Elizabeth J. Z. Robinson [1] and Chukwuma C. Ogbaga [2,3]

1 School of Agriculture, Policy and Development, University of Reading, Reading RG6 6EU, UK; e.j.robinson@reading.ac.uk
2 Department of Biological Sciences, Nile University of Nigeria, Airport Road, Abuja 900001, Nigeria; chukwumaogbaga@gmail.com
3 Department of Microbiology and Biotechnology, Nile University of Nigeria, Airport Road, Abuja 900001, Nigeria
* Correspondence: n.p.jellason@reading.ac.uk

Abstract: A fourth agricultural revolution, termed agriculture 4.0, is gradually gaining ground around the globe. It encompasses the application of smart technologies such as artificial intelligence, biotechnology, the internet of things (IoT), big data, and robotics to improve agriculture and the sustainability of food production. To date, narratives around agriculture 4.0 associated technologies have generally focused on their application in the context of higher-income countries (HICs). In contrast, in this perspective, we critically assess the place of sub-Saharan Africa (SSA) in this new technology trajectory, a region that has received less attention with respect to the application of such technologies. We examine the continent's readiness based on a number of dimensions such as scale, finance, technology leapfrogging, institutions and governance, education and skills. We critically reviewed the challenges, opportunities, and prospects of adopting agriculture 4.0 technologies in SSA, particularly with regards to how smallholder farmers in the region can be involved through a robust strategy. We find that whilst potential exist for agriculture 4.0 adoption in SSA, there are gaps in knowledge, skills, finance, and infrastructure to ensure successful adoption.

Keywords: agriculture 4.0; internet of things (IoT); precision agriculture; robotics; smallholders; sub-Saharan Africa (SSA)

Citation: Jellason, N.P.; Robinson, E.J.Z.; Ogbaga, C.C. Agriculture 4.0: Is Sub-Saharan Africa Ready?. *Appl. Sci.* **2021**, *11*, 5750. https://doi.org/10.3390/app11125750

Academic Editors: Ginés García-Mateos, Dolores Parras-Burgos and José Miguel Molina Martínez

Received: 28 April 2021
Accepted: 18 June 2021
Published: 21 June 2021

1. Introduction

Agricultural systems across the globe have evolved over the years. The literature identifies a "first agricultural revolution" or "Agriculture 1.0" [1–4], that involved hunting, gathering and settled farming; and a "second agricultural revolution", or the British agricultural revolution [5], in the 18th century, which saw an increase in agricultural production due to improved land productivity from mechanised agriculture [3]. Agricultural systems then evolved as a "third" Asian Green Revolution was introduced with a technology package of hybrid seeds, irrigation, modern pest control and synthetic fertilisers [6] in the 1960s; and more recently to what is termed agriculture 4.0 [7,8], the "fourth agricultural revolution". Agriculture 4.0, a recent and potentially game-changing transition lacks a universally accepted definition. However, it encompasses the adoption of high technology (High-Tech) solutions such as the internet of things (IoT), biotechnology innovations, cloud computing, precision agriculture, smart farming, drones, sensors, and robotics [9–13]. It is also underpinned by the idea of sustainable intensification, covering concepts that are in line with sustainable food production and better agricultural systems [14,15].

The Asian Green Revolution, which ended in 1990, has been credited with resolving food crises, reducing poverty and offering potentially important lessons for sub-Saharan African (SSA) countries [16]. Yet, the Asian Green Revolution has also proven to be controversial. Criticisms include the inability of smallholder farmers to compete with larger farms, which led to increased inequality amongst farmers; and an increase in fertiliser use that led to eutrophication of streams and lakes [17–19].

Taking a cue from the Asian Green Revolution, African agriculture policymakers, with the support of donors like the Bill and Melinda Gates Foundation, the Rockefeller Foundation, UK Aid, the United States Agency for International Development (USAID) and the Mastercard Foundation, proposed an African Green Revolution where improved seeds and fertilisers were to drive this process [20,21]. This led to the creation of the Alliance for a Green Revolution in Africa (AGRA) in 2006 [22]. The broad goals of AGRA were to provide African smallholder farmers with high-yielding agricultural practices that would allow them to double their yields [18]. Yet to date the evidence suggests that this, albeit ambitious, target has not been achieved. Critics variously suggest that AGRA did not achieve its goals due to the lack of consultations with farmers; the imposition of Western-type technologies not appropriate for SSA's farming systems; high input costs that were not offset by sufficiently high yields; and a focus on chemical-intensive monoculture cropping that leads to loss of crop and diet diversity [22].

This article assesses the challenges and prospects for agriculture 4.0 adoption in SSA. Most of the literature addressing agriculture 4.0 takes a higher-income country (HIC) viewpoint. To the best of our knowledge, few works exist on this topic from an SSA perspective, and only some of the technologies associated with agriculture 4.0 that have been introduced in higher-income countries appear to be being implemented in SSA [12]. In Section 2 we consider the potential of agriculture 4.0 technologies in SSA, by focusing on characteristics of the region, and of the technologies. We explore SSA's readiness for, and ability to embrace, agriculture 4.0, focusing on the challenges and opportunities for SSA's farmers to tap into the agriculture 4.0 revolution. In Section 3, we explore how SSA can leverage the opportunities for adopting agriculture 4.0. Finally, Section 4 concludes, by reflecting on the implication of SSA losing out in the new wave of an agricultural revolution.

2. Prospects and Challenges for Agriculture 4.0 Adoption in Sub-Saharan Africa

Agriculture in SSA is mostly rainfed, and in many countries is dominated by smallholder farmers who tend to have low levels of irrigation [23], and face biophysical and institutional challenges on top of the recent neglect of the sector [24]. These challenges include lack of insurance [25] and inefficient credit markets [26,27]; degraded soils, biodiversity loss [28]; and climate change [29–31]. Together these challenges exacerbate efforts to reduce food insecurity [7,32].

Different strategies have been employed by smallholder farmers across SSA to tackle these various challenges. For example, sustainable intensification of agriculture using new technologies has been highlighted in the literature as essential for producing additional food without reducing biodiversity and other ecosystem services [11,33]. In some African countries, this has included the promotion and adoption of integrated pest management, conservation agriculture, agroforestry, and system improvements [34,35]. Conservation agriculture (CA) which involves minimum tillage, diversified crop rotations and soil surface cover has been widely practiced in southern parts of Africa [36–38]. This broad suite of technologies and practices has economic (such as reduced labour requirements), agronomic (such as improved water conservation), and environmental (such as reduced soil erosion) benefits. These practices also reduce soil degradation and improve climate change management through carbon sequestration.

To address a lack of irrigation in arid and semi-arid areas, in the western Sudano-Sahelian zones, rainwater harvesting techniques such as *Zai* (planting pits) are utilised in Burkina Faso [39]; in northern Nigeria, integrated crop-livestock systems in a circular economy act as insurance against failure of one system and at the same time, improve the water holding capacity of soils for efficient food production [40,41]. In eastern Africa, genetically modified (GM) water-efficient maize seed has been promoted among smallholder farmers by the International Maize and Wheat Improvement Centre (CIMMYT) in partnership with Monsanto PLC to adapt to rainfall variability [42]. Finally, farmers in SSA do increasingly have access to rural micro, small and medium-sized enterprises (MSMEs) that seek to provide financing [26], better credit, and weather-indexed-based insurance against climate variability [25].

In the rest of this section, we explore the extent to which agriculture 4.0 technologies can complement, replace or improve upon these and other strategies that focus on sustainable increases in agricultural production; the scarcity of natural resources; adapting to climate change; and avoiding food waste [4,14]. We first consider scale, in particular, the reality that many farmers in SSA are smallholders, many intercrop, and within-field crop diversity is common, which may hinder the broad adoption of agriculture 4.0 technologies [43]. Second, we focus on finance, the reality that agriculture 4.0 technologies tend to be financially intensive [19], and the historically poor access to capital that smallholder farmers in SSA have faced. Our third area of exploration is the extent to which African countries have the digital infrastructure in place. Fourth, we address the institutions, governance, and ethics surrounding agricultural technology adoption; and finally, we explore the extent to which Africa's farmers have the education and skills needed to embrace agriculture 4.0.

2.1. Scale

Smallholders continue to dominate agriculture in SSA, with 80% of agricultural output produced by farmers with landholdings less than 2 hectares on average, [27]. Some scholars have suggested that improvements in agricultural productivity in SSA, just like in the case of the Asian Green Revolution, are likely to be driven by smallholders who currently control most of the landholdings [44]. Others have argued that relying on the conventional smallholder model of agriculture is inefficient because African smallholder productivity is very low, e.g., [20,45]. Either way, understanding the scale at which agriculture 4.0 technologies are likely to operate is particularly important for SSA; and the reality is that, at least in the short to medium term, smallholder farmers will remain important in the sector.

Agriculture 4.0 has often been associated with large-scale farming [13,46]. For example, unmanned aerial vehicles (UAVs) are increasingly being used for fertilisers and chemical spraying. UAVs, in combination with smartphone platforms to provide remote sensing data, also use global positioning systems (GPS) for digital soil mapping [10,11,14,15] for various environmental and agricultural development purposes [47]. However, uptake of this technology was reported in a South African study to be low [47].

Modalities already exist in SSA where smallholder farmers could rent these or similar services or jointly source such "lumpy" technologies, thereby overcoming any "scale" constraint. For example, in recent years, SSA countries have seen elements of the 'fourth agricultural revolution' technology such as the mobile application for tractor hiring being promoted by international organisations including the International Crops Research Institute for the Semi-Arid Tropics (ICRISAT) and the private sector [12]. One specific example is a blockchain-enabled application that links tractor providers and smallholder farmers to promote agricultural mechanisation [48]. This application, called Hello Tractor, is a software that provides a platform for renting tractors for land cultivation from owners by smallholders. More broadly, ICRISAT's innovation Hub (iHub) has been developed to support activities along the crop production value chains from seed certification, product manufacturing and retail [49], many of which are scale neutral and as such could be of direct benefit to smallholder farmers in SSA.

Agriculture 4.0 technologies could have an important role in African countries where fertiliser use has long been low [50], despite many initiatives to increase its use. For example, in June 2006, the African Union (AU) member states' Ministers of agriculture, at a summit in Abuja, Nigeria, resolved to increase the level of fertiliser use from 8 to 50 kg per hectare on average by the year 2015. This was despite a lack of scientific evidence to support the value of such an arbitrary increase [51], prompting concerns that the move would likely be counterproductive, leading to an increased cost of production. Agriculture 4.0 technologies such as soil mapping for soil analysis could offer more sustainable alternatives by identifying areas of nutrient deficiencies to inform precision fertiliser application that is cost-efficient and environmentally sound [14].

Remote sensing data has long been used in agriculture, whether to simply map cropland, or to determine biomass, yields, and crop stress [52]. Further, it is becoming more affordable, in part due to the availability of free high-resolution satellite imagery. Digital soil mapping has the potential to identify the location of high-value agricultural land by providing detailed information on soils and is already being used in SSA [47]. However, the scale and heterogeneity of Africa's agricultural landscapes are likely to continue to pose challenges, particularly for smallholder farmers [43]. For example, Lowder, Skoet [27] highlight, in particular, the difficulty of using satellite data to determine crop yields in smaller farms.

The livestock sector has experienced its own "revolution", which has paralleled what has been referred to as "Industry 2.0", the mass production of goods to increase productivity [3]. Particularly in higher-income countries, livestock farming has shifted away from home-based animal husbandry towards intensive, often large-scale, farming [3]. Agriculture 4.0 technologies can be found in, for example, livestock monitoring and biosensing, which can be used to detect and identify infectious diseases, drug residues, and ovulation prediction [53]. Many of these technologies are linked to intelligent animal health monitoring systems, in which animals may wear the technology [3]. Agriculture 4.0 approaches have also been linked to ecologically friendly efficient farming approaches, that could include monitoring emissions from livestock [54].

In many SSA countries, the scale and intensity of the livestock sector differ considerably from the large-scale farming found in HICs. However, although in some higher-income countries per capita consumption of meat is plateauing or even falling, in SSA and other low- and middle-income countries (LMIC), consumption continues to increase, with pork, beef, and chicken being the most consumed meats [54], and with this increase in consumption comes increases in GHG emissions. This suggests that in the future, agriculture 4.0 technologies relevant to livestock will be increasingly important for SSA.

2.2. Finance and Capital for Investment, Research and Development

Financial constraints and poor literacy are likely to affect the uptake of agriculture 4.0 technologies. Much of SSA's low-income rural population works in highly uncertain environments, with little access to capital, input and output markets, or crop insurance. These realities limit the ability of smallholder farmers to invest in "modern" technologies, especially those requiring upfront high-cost investments [55]. In addition, investment in agricultural research and development (R&D) as a percent of agricultural gross domestic product (GDP) in SSA is relatively low at 0.38% (2001–2013 average) and falling, and considerably lower than other regions in the world when measured per hectare of cropland or per agricultural labourer [7]. Importantly for agriculture 4.0, a digital infrastructure funding gap of about one billion euros exists in SSA [25]. This suggests a high degree of underfunding of agricultural research in Africa [24] with implications for technology adoption.

Efforts by African governments to increase investments in agriculture are yet to reach targeted goals. For instance, the African governments launched the Comprehensive African Agricultural Development Program (CAADP) in 2006 that commits them to spend at least 10% of their total budget on agriculture by 2025. As of 2014, only Malawi, Zimbabwe and Mozambique had met or surpassed this target, with Zambia, Rwanda and Niger close to reaching the target [24]. And in 2017, only 20 out of the 47 reporting African Union (AU) countries remained on track to meet the 2025 commitments [25].

Agriculture 4.0 technologies are generally perceived to be financially intensive [11], thereby making their deployment likely to be difficult for Africa's smallholder farmers. For example, "next-generation" nanosensor technology is being used to increase sustainable food production and better agricultural system in the United States of America and Australia [56,57]. A fast speed method has recently been shown to increase chloroplast transformation in vitro with the potential of improving crop yields for crops such as arugula, watercress, spinach and tobacco [56]. These same technologies are being used for plant phenotyping and disease monitoring in the field [58]. Yet their cost of deployment makes them currently almost certainly inaccessible to African smallholder farmers.

A study on the perception of Brazilian farmers regarding the adoption of agriculture 4.0 technologies such as precision agriculture integrated with data from remote sensors on smartphones found that the cost of the machines, software, equipment and connectivity was a constraint to adoption [59]. More generally, capital availability has also been found to be a pre-requisite to the adoption of conservation agriculture technologies, especially where new equipment is needed [60].

Financial and agricultural advisory services technology start-ups such as Esoko in Ghana, Farmcrowdy in Nigeria, and EcoFarmer in Zimbabwe, which offer farmer advice on input use, credit and weather-indexed insurance, have received considerable attention in SSA [25]. For example, tech start-ups in these areas received investments of about 335 million Euros in 2018 [25]. These investments will create employment opportunities and increase uptake of the digital technologies [25]. This also suggests that financing agriculture 4.0 technologies, whilst still a constraint, may be less so in the future.

2.3. Leapfrog Technology Opportunities and Digital Infrastructure Availability

The reality for many African countries has been poorly developed physical infrastructure, including a low density of fixed telephone landlines and physical bank branches. This might historically have constrained economic development. However, increasingly African countries have proven able to leapfrog these technologies. That is, rather than follow what might be considered traditional technology adoption pathways, taken earlier by HICs, African countries may be able to skip an intermediate technology stage.

Examples of this technology leapfrogging that have already occurred that have particularly benefitted smallholder farmers relate to telecoms and banking [61]. For example, many African countries have rapidly transitioned from having a very low penetration of fixed landlines and traditional bricks and mortar banking infrastructure to a relatively high proportion of households with access to mobile telephony and mobile banking. New and modern information and communication technologies (ICTs) have played critical roles in bridging key agricultural extension and development infrastructure gaps. The mobile money transfer service application, M-Pesa, was pioneered in Kenya by Vodafone and expanded into other SSA countries such as Tanzania and South Africa [62]. M-Pesa is transforming agriculture in SSA, in part by providing mobile payment systems to farmers that have enabled many to boost crop production and move out of subsistence agriculture [63]. Other agriculture 4.0 mobile-technology-supported financial and agricultural technologies such as M-Farm and Esoko have been widely adopted in SSA countries such as Kenya and Ghana respectively [62]. Esoko provides farmers with daily market prices for many key agricultural crops, using Short Message Services (SMS) sent to farmers' mobile phones, in addition to information on weather, market situation and farming tips [64,65].

Many agriculture 4.0 technologies rely on the fourth generation (4G) of broadband cellular network technology, cloud computing, and big data analytics. Around 60% of SSA smallholder farmers have access to mobile connection with an increasing network of 4G connections and Data Centres being developed in SSA [25], suggesting that Africa's farmers are, in this respect, relatively well placed to take advantage of mobile phone-driven agriculture 4.0 technologies, that could also leverage the breakthroughs already recorded in ICTs' adoption in SSA and leapfrog the conventional technologies in agricultural advisory and financial services provision for the underserved [25].

2.4. Institutions, Governance, and Ethics

Agriculture 4.0 will almost certainly change the way of work for farmers and the culture around traditional ways of farming [15], whether in higher or lower-income countries. Some scholars are already asking whether agriculture 4.0 is the way forward for society in general, expressing concern that the social and ethical implications of adopting such technologies have not received adequate consideration in the design and implementation processes [6,13].

These concerns are likely to be just as relevant for African smallholder farmers as for those in higher-income countries, and indeed concerns over institutions, governance and the ethical implications of new technologies are nothing new. There have long been concerns over the ethics of hybrid seeds, and plant breeders' rights, e.g., [66]. Efforts to increase mechanisation in African farming have in the past been hampered by rent-seeking through elite capture and the lack of access to spare parts [67]. And sustainable intensification of agricultural systems across African countries has been hampered by weak institutions, whether due to poorly functioning markets or property rights over land [68].

2.5. Education and Skills

Younger and more educated farmers have been found to be more likely to adopt agriculture 4.0 technologies [69], and more broadly farmers are likely to need ICT skills and capabilities to fully embrace agriculture 4.0 [70]. Yet farmers in SSA are less literate than those in HICs; farm workers less literate than non-farm workers [7,71], and many farmers lack the skills to operate sophisticated technologies and to collect and manage the high volume of data used to improve decision-making [72]. This suggests that lack of education and skills in data management as a barrier to agriculture 4.0 adoption is likely to be an issue in some African countries. The gap in the quality of education and the need for more training in Africa have previously been highlighted in several African Union Commission Policy frameworks, see [72–74].

3. Going Forward: Opportunities for SSA to Tap into the Agriculture 4.0 Revolution

3.1. Reasons for Optimism

The surge in food demand in SSA due to the rise in population and the demand for more nutritious food by an increasing number of middle-income families is leading to greater opportunities for smallholder farmers in SSA to improve their production and incomes through the diversification and enhancement of their production systems using agriculture 4.0 technologies [12]. More broadly, agriculture 4.0 technologies have the potential to increase job creation, improve the revenues of farmers, and increase self-sufficiency and food exports [75]. However, if this fourth revolution bypasses Africa's farmers, African countries are likely to increasingly deplete their available natural habitats or rely on food imports, to the detriment of valuable biodiversity, long-term food security status, rural livelihoods, and poverty reduction.

Smallholder farmers in SSA tend to operate on small patches of land, making use of family labour or 'community work parties' to cultivate their lands, with little access to credit or insurance, and with high uncertainty of weather around agro-climatic conditions, prices, and access to markets. Despite these challenges, there seems to be considerable scope for the adoption of agriculture 4.0 technologies in SSA. Already the continent has demonstrated leadership with regards to mobile banking and the use of mobile telephony for increasing market efficiency; and access to 4G networks is relatively good and growing. Also, funding for tech start-ups has received boosts recently from foreign investors and multinational organisations to bridge the funding gaps. Extension information is currently accessible to an increasing number of farmers through mobile applications [25]. However, more training to harness the benefits of ICT education is needed to consolidate these gains [72].

One overarching constraint for Africa's farmers is that many must deal with the multiple constraints of scale, finance, and poorly functioning institutions. Innovative approaches that deal with multiple constraints include the Hello Tractor application that addresses scale and lumpy investments associated with mechanisation, access to credit and rental markets for machinery, and takes advantage of mobile supported technologies. Similarly, African farmers are benefiting from access to soil mapping, aerial chemical and fertiliser applications, that combine large and small-scale technology, and that increase the productivity and income of smallholder farmers. Yet overall technology adoption is still very limited across the continent, for example, the use of tractors remains low, leading to poor productivity and low yields [7]. As such, it is imperative to consider the potential of site-specific technology adoption to fill the low yield gap experienced in SSA [44].

Agriculture 4.0 suites of technologies may find application in an SSA context when smallholder farmers organise themselves into clusters. Clustering can promote interaction with stakeholders along agricultural value chains and has been shown to enhance the adoption of technology [76]. Clustering can also ensure that technology companies can be profitable when they deploy the agriculture 4.0 technologies for the benefit of smallholder farmers due to the advantage of scale, compared to dealing with individual farmers working on less than a hectare on average [25]. Such clusters are also likely to enhance access to more profitable markets by farmers, thereby increasing farmers' bargaining power to secure the best deals for their clusters.

There is evidence that SSA as a whole is already embracing agriculture 4.0-associated technology systems as part of the new wave of the fourth industrial revolution [61] and this offers the continent an opportunity to advance in diverse areas such as climate insurance services provision, agritech-financing, agricultural advisory services provision and farmer-supply chain linkages using blockchain technologies [25,77]. The positive impacts anticipated of this agriculture 4.0 technology adoption include yield increases and a reduction of carbon footprint in line with the Sustainable Development Goals (SDGs), low cost of attracting and maintaining farmers in the supply chain, and price transparency [25].

Adoption of technologies promoted by agriculture 4.0 that lead to improvements in the environment may be particularly relevant for smallholders. For example, adopting agriculture 4.0 technologies for precision agriculture is likely to lead to fertilisers and pesticides being applied in the required doses and concentration, informed by data on the soil condition and nutrient requirements for each crop linked to the weather forecast for rainfall [12]. It is estimated that about 60% of conventional fertilisers are lost to the environment on application, leading to pollution [14], particularly of water bodies. These agriculture 4.0 technologies, therefore, enable farmers to both reduce their input costs and improve their environmental sustainability. The use of nanofertilisers, an example of nanotechnology precision agriculture, in which nutrients are released slowly resulting in exact dosages [14], also has the potential to improve farm economic and environmental sustainability.

In urban areas, many agriculture 4.0 technologies are associated with the automation of agricultural operations using high-tech solutions [12] supported by well-developed ICT infrastructure, for instance, the rise in wireless communication technologies that use low power wide area network (LPWAN) such as Sigfox, LoRa and NB-IoT [78]. This is not the case in rural areas where smallholders tend to be located [79]. Nevertheless, platform-based ICT and mobile technologies are most suited to the smallholder context and are likely to support SSA with regard to food security and agricultural sustainability [25].

3.2. Reasons for Caution

Despite the potential for optimism with regards to agriculture 4.0 in Africa, there are reasons to remain cautious. First, the future of agriculture, as it has been in the past, will be driven by technology and innovation, though increasingly in the context of climate change. Yet despite the challenges affecting SSA's food production and security, the continent's readiness to take advantage of this technology revolution is still in doubt. In many African countries, current levels of investment in critical infrastructure are far below the threshold required for the continent to benefit from the high tech revolution [61]. For example, in the spread of mobile phones for agricultural advisory and banking services, much of SSA could miss out again on another agricultural revolution if the required conditions for adoption such as technology skills are not in place. Moreover, as pressure on the natural environment increases there will similarly be an increasing need for agricultural advances and innovations to be underpinned by good agricultural practices. Agriculture

4.0 technologies uptake, when done well, should be capable of promoting data-driven agriculture for economically and environmentally sustainable farming. However, these technologies do not exist in a vacuum, and the extent to which agriculture 4.0 does indeed align with sustainable development depends on the broader institutional environment within which relevant technologies are introduced, how they are introduced and taken up.

Second, though the expectation is that agriculture 4.0 technologies will improve agriculture and food security in an efficient and sustainable way, it is possible that these new technologies and ways of working could alter agricultural systems for the worse [80]. For example, increased technology adoption could lead to a disregard of experiential knowledge and disconnect the farmer from the landscape [15], something of potentially particular relevance for SSA agriculture.

Third, several studies focusing on technology adoption have found that despite the significance of technology in improving productivity, unwillingness on the part of farmers to adopt could be a constraint to adoption [7,31]. This could be due to poor awareness, lack of insurance to manage the risks involved in up taking a new idea or lack of skills and knowledge about the appropriateness of technology, as reported by a study conducted in Brazil [59]. Lack of acceptance of technology has also been reported as a barrier to the adoption of agriculture 4.0 technologies in a HIC context [11]. In Nigeria, despite the benefits of agricultural biotechnology advancement in food and non-food production from genetically modified organisms (GMOs) such as pest control and early maturation, the technology has not been accepted due to health, legal and ethical concerns [81].

Fourth, the benefits of this revolution in technology are not likely to be evenly spread across the globe, and it remains uncertain as to with whom and where the benefits will reside [11]. Further, without a clear understanding of how agriculture 4.0 will affect societies, particularly in lower-income countries, these new technologies have the potential to create more problems than they solve due to ethical concerns linked to the deployment of such technology. As such, there is a need for policymakers and technology companies to work together with farmers and communities more broadly in SSA to ensure that the benefits of this suite of technologies are not only optimised for productivity and efficiency but that both environmental and social impacts are addressed explicitly.

4. Conclusions

This perspective provides one of the first assessments of agriculture 4.0 that focuses specifically on sub-Saharan Africa. It highlights the key challenges that the continent faces; the extent to which the region is, or is not, ready to embrace this technology revolution and the risks of missing out. It also emphasises the importance of understanding the benefits and potential costs of agriculture 4.0 technologies in the African context; the relationship with the existing diverse agricultural development pathways that focus on sustainable food and agricultural systems [13].

There is an alternative drive to promote agroecology principles that encompass both social and natural sciences that underscore systems philosophy and ecological thinking [13,82]. This is based on the evidence that agroecology practices increase yields in a sustainable and affordable way [22,51]. Whilst agroecology principles have shown great promise for sustainable yield improvement, this approach is likely to be insufficient to meet the food need of the growing SSA population. Therefore it is imperative to identify and adopt suitable technologies that are context-specific and in line with current realities [83].

Author Contributions: Conceptualization, N.P.J., E.J.Z.R. and C.C.O.; resources, N.P.J. and E.J.Z.R.; writing—original draft preparation, N.P.J. and C.C.O.; writing—review and editing, N.P.J., E.J.Z.R. and C.C.O.; supervision, E.J.Z.R.; project administration, N.P.J.; funding acquisition, E.J.Z.R. All authors have read and agreed to the published version of the manuscript.

Funding: This research was funded by UK Research and Innovation through the Global Challenges Research Fund programme, "Growing research capability to meet the challenges faced by developing countries" ("Grow"), grant number ES/P011306/1 and The APC was funded by the University of Reading through the Social and Environmental Trade-Offs in African Agriculture (Sentinel) project.

Institutional Review Board Statement: Not applicable.

Informed Consent Statement: Not applicable.

Data Availability Statement: Not applicable.

Acknowledgments: Special thanks to the anonymous reviewers for the useful comments on improving this perspective paper; thanks to Beth Downe, International Institute for Environment and Development (IIED) for executive project support.

Conflicts of Interest: The authors declare no conflict of interest. The funders had no role in the design of the study; in the collection, analyses, or interpretation of data; in the writing of the manuscript, or in the decision to publish the results.

References

1. Meliala, J.; Hubeis, M.; Jahroh, S.; Maulana, A. Position of farmers in agriculture 4.0: Finding from farmers partner of aggregator online vegetables commodity in Indonesia. *Arch. Agric. Environ. Sci.* **2019**, *4*, 300–306. [CrossRef]
2. Rapela, M.A. Post-Malthusian Dilemmas in Agriculture 4.0. In *Fostering Innovation for Agriculture 4.0: A Comprehensive Plant Germplasm System*; Rapela, M.A., Ed.; Springer International Publishing: Cham, Germany, 2019; pp. 1–16.
3. Liu, Y.; Ma, X.; Shu, L.; Hancke, G.P.; Abu-Mahfouz, A.M. From Industry 4.0 to Agriculture 4.0: Current Status, Enabling Technologies, and Research Challenges. *IEEE Trans. Ind. Inform.* **2021**, *17*, 4322–4334. [CrossRef]
4. Zhai, Z.; Martínez, J.F.; Beltran, V.; Martínez, N.L. Decision support systems for agriculture 4.0: Survey and challenges. *Comput. Electron. Agric.* **2020**, *170*, 105256. [CrossRef]
5. Simpson, J. European Farmers and the British "Agricultural Revolution". In *Exceptionalism and Industrialisation: Britain and its European Rivals, 1688–1815*; Cambridge University Press: Cambridege, UK, 2004; pp. 69–85. [CrossRef]
6. Rose, D.C.; Chilvers, J. Agriculture 4.0: Broadening Responsible Innovation in an Era of Smart Farming. *Front. Sustain. Food Syst.* **2018**, *2*, 87. [CrossRef]
7. Fuglie, K.; Gautam, M.; Goyal, A.; Maloney, W.F. *Harvesting Prosperity: Technology and Productivity Growth in Agriculture*; Bank, T.W., Ed.; The World Bank: Washington, DC, USA, 2020.
8. Araújo, S.O.; Peres, R.S.; Barata, J.; Lidon, F.; Ramalho, J.C. Characterising the Agriculture 4.0 Landscape—Emerging Trends, Challenges and Opportunities. *Agronomy* **2021**, *11*, 667. [CrossRef]
9. Frankelius, P.; Norrman, C.; Johansen, K. Agricultural Innovation and the Role of Institutions: Lessons from the Game of Drones. *J. Agric. Environ. Ethics* **2019**, *32*, 681–707. [CrossRef]
10. Klerkx, L.; Jakku, E.; Labarthe, P. A review of social science on digital agriculture, smart farming and agriculture 4.0: New contributions and a future research agenda. *NJAS Wagening. J. Life Sci.* **2019**, *90–91*, 100315. [CrossRef]
11. Barrett, H.; Rose, D.C. Perceptions of the Fourth Agricultural Revolution: What's In, What's Out, and What Consequences are Anticipated? *Sociol. Rural.* **2020**. [CrossRef]
12. Cotter, M.; Asch, F. Editorial: Smallholder targeted Agriculture 4.0 in temperature limited cropping systems. *J. Agron. Crop Sci.* **2020**, *206*, 421–422. [CrossRef]
13. Klerkx, L.; Rose, D. Dealing with the game-changing technologies of Agriculture 4.0: How do we manage diversity and responsibility in food system transition pathways? *Glob. Food Secur.* **2020**, *24*, 100347. [CrossRef]
14. De Clercq, M.; Vats, A.; Biel, A. Agriculture 4.0-The Future of Farming Technology. World Government Summit. Available online: https://www.oliverwyman.com/content/dam/oliver-wyman/v2/publications/2018/February/Oliver-Wyman-Agriculture-4.0.pdf (accessed on 15 November 2020).
15. Rose, D.C.; Wheeler, R.; Winter, M.; Lobley, M.; Chivers, C.-A. Agriculture 4.0: Making it work for people, production, and the planet. *Land Use Policy* **2021**, *100*, 104933. [CrossRef]
16. Hazell, P.B.R. *The Asian Green Revolution*; International Food Policy Research Institute (IFPRI): Washington, DC, USA, 2009.
17. Das, R.J. The green revolution and poverty: A theoretical and empirical examination of the relation between technology and society. *Geoforum* **2002**, *33*, 55–72. [CrossRef]
18. Holmén, H. *A Green Revolution for Africa-Does It Need to Be so Controversial?* International Centre for Economic Research (ICER): Lund, Sweden, 2003; p. 36.
19. Glaeser, B. *The Green Revolution Revisited: Critique and Alternatives*; Allen & Unwin: London, UK, 1987; pp. 1–170.
20. Kariuki, J.G. *The Future of Agriculture in Africa*, 15th ed.; Najam, A., Barakatt, C., Eds.; Boston University: Boston, MA, USA, 2011.
21. AGRA. Our Story-Alliance for a Green Revolution in Africa (AGRA). Available online: https://agra.org/our-story/ (accessed on 5 January 2021).
22. Wise, T.A. *Failing Africa's Farmers: An Impact Assessment of the Alliance for a Green Revolution in Africa*; Global Development and Environment Institute, Tufts University: Medford, OR, USA, 2020; p. 38.
23. Jayne, T.S.; Mather, D.; Mghenyi, E. Principal Challenges Confronting Smallholder Agriculture in Sub-Saharan Africa. *World Dev.* **2010**, *38*, 1384–1398. [CrossRef]
24. Goyal, A.; Nash, J. *Reaping Richer Returns: Public Spending Priorities for African Agriculture Productivity Growth*; The World Bank: Washington, DC, USA, 2017.

25. CTA. *The Digitalisation of African Agriculture Report 2018–2019*; The Technical Centre for Agricultural and Rural Cooperation (CTA): Wageningen, The Netherlands, 2019; p. 241.
26. Abdallah, A.-H. Does credit market inefficiency affect technology adoption? Evidence from Sub-Saharan Africa. *Agric. Financ. Rev.* **2016**, *76*, 494–511. [CrossRef]
27. Lowder, S.K.; Skoet, J.; Raney, T. The Number, Size, and Distribution of Farms, Smallholder Farms, and Family Farms Worldwide. *World Dev.* **2016**, *87*, 16–29. [CrossRef]
28. Van Ittersum, M.K.; van Bussel, L.G.J.; Wolf, J.; Grassini, P.; van Wart, J.; Guilpart, N.; Claessens, L.; de Groot, H.; Wiebe, K.; Mason-D'Croz, D.; et al. Can sub-Saharan Africa feed itself? *Proc. Natl. Acad. Sci. USA* **2016**, *113*, 14964. [CrossRef]
29. Boko, M.; Niang, I.; Nyong, A.; Vogel, C.; Githeko, A.; Medany, M.; Osman-Elasha, B.; Tabo, R.; Yanda, P. *Africa Climate Change 2007: Impacts, Adaptation and Vulnerability: Contribution of Working Group II to the Fourth Assessment Report of the Intergovernmental Panel on Climate Change*; IPCC: Cambridge, UK, 2007; pp. 433–467.
30. Jellason, P.N.; Baines, N.R.; Conway, S.J.; Ogbaga, C.C. Climate Change Perceptions and Attitudes to Smallholder Adaptation in Northwestern Nigerian Drylands. *Soc. Sci.* **2019**, *8*, 31. [CrossRef]
31. Jellason, N.P.; Conway, J.S.; Baines, R.N. Understanding impacts and barriers to adoption of climate-smart agriculture (CSA) practices in North-Western Nigerian drylands. *J. Agric. Educ. Ext.* **2020**, *27*, 55–72. [CrossRef]
32. World Bank. *World Development Report 2008: Agriculture for Development*; The World Bank: Washington, DC, USA, 2007; p. 390.
33. Vanlauwe, B.; Coyne, D.; Gockowski, J.; Hauser, S.; Huising, J.; Masso, C.; Nziguheba, G.; Schut, M.; Van Asten, P. Sustainable intensification and the African smallholder farmer. *Curr. Opin. Environ. Sustain.* **2014**, *8*, 15–22. [CrossRef]
34. Pretty, J.; Toulmin, C.; Williams, S. Sustainable intensification in African agriculture. *Int. J. Agric. Sustain.* **2011**, *9*, 5–24. [CrossRef]
35. Pretty, J.; Bharucha, Z.P. Integrated Pest Management for Sustainable Intensification of Agriculture in Asia and Africa. *Insects* **2015**, *6*, 152–182. [CrossRef] [PubMed]
36. Giller, K.E.; Witter, E.; Corbeels, M.; Tittonell, P. Conservation agriculture and smallholder farming in Africa: The heretics' view. *Field Crop. Res.* **2009**, *114*, 23–34. [CrossRef]
37. Baudron, F.; Tittonell, P.; Corbeels, M.; Letourmy, P.; Giller, K.E. Comparative performance of conservation agriculture and current smallholder farming practices in semi-arid Zimbabwe. *Field Crop. Res.* **2012**, *132*, 117–128. [CrossRef]
38. Vanlauwe, B.; Wendt, J.; Giller, K.E.; Corbeels, M.; Gerard, B.; Nolte, C. A fourth principle is required to define Conservation Agriculture in sub-Saharan Africa: The appropriate use of fertilizer to enhance crop productivity. *Field Crop. Res.* **2014**, *155*, 10–13. [CrossRef]
39. Vohland, K.; Barry, B. A review of in situ rainwater harvesting (RWH) practices modifying landscape functions in African drylands. *Agric. Ecosyst. Environ.* **2009**, *131*, 119–127. [CrossRef]
40. Mortimore, M.J. Dryland Development: Success Stories from West Africa. *Environment* **2005**, *45*, 10–21. [CrossRef]
41. Jellason, N.P.; Conway, J.S.; Baines, R.N.; Ogbaga, C.C. A review of farming challenges and resilience management in the Sudano-Sahelian drylands of Nigeria in an era of climate change. *J. Arid. Environ.* **2021**, *186*, 104398. [CrossRef]
42. Whitfield, S.; Dixon, J.L.; Mulenga, B.P.; Ngoma, H. Conceptualising farming systems for agricultural development research: Cases from Eastern and Southern Africa. *Agric. Syst.* **2015**, *133*, 54–62. [CrossRef]
43. Chivasa, W.; Mutanga, O.; Biradar, C. Application of remote sensing in estimating maize grain yield in heterogeneous African agricultural landscapes: A review. *Int. J. Remote. Sens.* **2017**, *38*, 6816–6845. [CrossRef]
44. Larson, D.F.; Otsuka, K. Introduction: Why an African Green Revolution Is Needed and Why It Must Include Small Farms. In *Pursuit of an African Green Revolution*; Springer: Tokyo, Japan, 2016; pp. 1–11.
45. Collier, P.; Dercon, S. African Agriculture in 50 years: Smallholders in a Rapidly Changing World? In Proceedings of the Expert Paper for the FAO Conference on "How to Feed the World in 2050?", Rome, Italy, 12–13 October 2009.
46. Bronson, K. Looking through a responsible innovation lens at uneven engagements with digital farming. *NJAS Wagening. J. Life Sci.* **2019**, *90–91*, 100294. [CrossRef]
47. Van Zijl, G. Digital soil mapping approaches to address real world problems in southern Africa. *Geoderma* **2019**, *337*, 1301–1308. [CrossRef]
48. Hello Tractor. Use Hello Tractor to Make Extra Cash with Your Farm Equipment. Available online: https://hellotractor.com/ (accessed on 1 November 2020).
49. ICRISAT Innovation Hub. Innovation Hub opens for Agri-Tech Entrepreneurs. Available online: https://www.icrisat.org/innovation-hub-opens-for-agri-tech-entrepreneurs/ (accessed on 1 November 2020).
50. FAO. Digital Innovation for promoting Agriculture 4.0 in the Near East and North Africa. In Proceedings of the FAO Regional Conference for the Near East, Muscat, Oman, 2–4 March 2020.
51. Tittonell, P. Ecological intensification of agriculture—Sustainable by nature. *Curr. Opin. Environ. Sustain.* **2014**, *8*, 53–61. [CrossRef]
52. Paliwal, A.; Jain, M. The Accuracy of Self-Reported Crop Yield Estimates and Their Ability to Train Remote Sensing Algorithms. *Front. Sustain. Food Syst.* **2020**, *4*, 1–10. [CrossRef]
53. Velasco-Garcia, M.N.; Mottram, T. Biosensors in the livestock industry: An automated ovulation prediction system for dairy cows. *Trends Biotechnol.* **2001**, *19*, 433. [CrossRef]
54. Huh, J.-H.; Kim, K.-Y. Time-Based Trend of Carbon Emissions in the Composting Process of Swine Manure in the Context of Agriculture 4.0. *Processes* **2018**, *6*, 168. [CrossRef]

55. Alliance for a Green Revolution in Africa (AGRA). *Progress Towards Agricultural Transformation*; Alliance for a Green Revolution in Africa (AGRA): Nairobi, Kenya, 2016; pp. 105–124.
56. Kwak, S.-Y.; Wong, M.H.; Lew, T.T.S.; Bisker, G.; Lee, M.A.; Kaplan, A.; Dong, J.; Liu, A.T.; Koman, V.B.; Sinclair, R.; et al. Nanosensor Technology Applied to Living Plant Systems. *Annu. Rev. Anal. Chem.* **2017**, *10*, 113–140. [CrossRef]
57. Srivastava, A.K.; Dev, A.; Karmakar, S. Nanosensors and nanobiosensors in food and agriculture. *Environ. Chem. Lett.* **2018**, *16*, 161–182. [CrossRef]
58. Lew, T.T.S.; Sarojam, R.; Jang, I.-C.; Park, B.S.; Naqvi, N.I.; Wong, M.H.; Singh, G.P.; Ram, R.J.; Shoseyov, O.; Saito, K.; et al. Species-independent analytical tools for next-generation agriculture. *Nat. Plants* **2020**, *6*, 1408–1417. [CrossRef]
59. Bolfe, É.; Jorge, L.; Sanches, I.; Junior, A.; Costa, C.; Victoria, D.; Inamasu, R.; Grego, C.; Ferreira, V.; Ramirez, A. Precision and Digital Agriculture: Adoption of Technologies and Perception of Brazilian Farmers. *Agriculture* **2020**, *10*, 653. [CrossRef]
60. Knowler, D.; Bradshaw, B. Farmers' adoption of conservation agriculture: A review and synthesis of recent research. *Food Policy* **2007**, *32*, 25–48. [CrossRef]
61. Dosso, M.; Nwankwo, C.I.; Travaly, Y. The Readiness of Innovation Systems for the Fourth Industrial Revolution (4IR) in Sub-Saharan Africa. In *Entrepreneurship, Technology Commercialisation, and Innovation Policy in Africa*; Daniels, C., Dosso, M., Amadi-Echendu, J., Eds.; Springer International Publishing: Cham, Germany, 2021; pp. 13–32.
62. Wamboye, E.; Tochkov, K.; Sergi, B. Technology Adoption and Growth in sub-Saharan African Countries. *Comp. Econ. Stud.* **2015**, *57*, 136–167. [CrossRef]
63. Maree, J.; Piontak, R.; Omwansa, T.; Shinyekwa, I.; Njenga, K. *Developmental Uses of Mobile Phones in Kenya and Uganda*; Capturing the Gains The University of Manchester: Manchester, UK, 2013; p. 37.
64. Odunze, D.; Mthitwa, H. An Analysis of the Impact of the Use of Mobile Communication Technologies by Farmers in Zimbabwe. A Case Study of Esoko and EcoFarmer Platforms. In Proceedings of the Conversations on Development: Can ICTs Make a Difference in Climate, Political and Health Disturbances? Munster, Germany, 26 May 2015.
65. Van Schalkwyk, F.; Young, A.; Verhulst, S. Esoko–Leveling the Information Playing Field for Smallholder Farmers in Ghana. Available online: https://odimpact.org/files/case-esoko.pdf (accessed on 24 January 2021).
66. Kinchy, A. *Seeds, Science, and Struggle: The Global Politics of Transgenic Crops*; MIT Press: Cambridge, MA, USA, 2012.
67. Daum, T.; Birner, R. The neglected governance challenges of agricultural mechanisation in Africa-insights from Ghana. *Food Secur.* **2017**, *9*, 959–979. [CrossRef]
68. Schut, M.; van Asten, P.; Okafor, C.; Hicintuka, C.; Mapatano, S.; Nabahungu, N.L.; Kagabo, D.; Muchunguzi, P.; Njukwe, E.; Dontsop-Nguezet, P.M.; et al. Sustainable intensification of agricultural systems in the Central African Highlands: The need for institutional innovation. *Agric. Syst.* **2016**, *145*, 165–176. [CrossRef]
69. Monteleone, S.; Moraes, E.A.D.; Maia, R.F. Analysis of the variables that affect the intention to adopt Precision Agriculture for smart water management in Agriculture 4.0 context. In Proceedings of the 2019 Global IoT Summit (GIoTS), Aarhus, Denmark, 17–21 June 2019; pp. 1–6.
70. Pogorelskaia, I.; Várallyai, L. Agriculture 4.0 and the role of education. *J. Agric. Inform.* **2020**, *11*, 45–51. [CrossRef]
71. Gollin, D.; Lagakos, D.; Waugh, M. Agricultural Productivity Differences across Countries. *Am. Econ. Rev.* **2014**, *104*, 165–170. [CrossRef]
72. AUC. *Continental Educational Strategy for Africa (CESA) (2016–2025)*; African Union Commission (AUC): Addis Ababa, Ethiopia, 2016; p. 44.
73. AUC. *Science, Technology and Innovation Strategy for Africa (STISA) 2024*; African Union Commission (AUC): Addis Ababa, Ethiopia, 2014; p. 56.
74. AUC. *Continental Strategy for Technical and Vocational Education and Training (TVET)*; African Union Commission (AUC): Addis Ababa, Ethiopia, 2018; p. 40.
75. Tompkins, S. Getting Ready for Agriculture 4.0. Available online: https://www.thestar.com.my/opinion/letters/2020/07/16/getting-ready-for-agriculture-40 (accessed on 19 November 2020).
76. Joffre, O.M.; Poortvliet, P.M.; Klerkx, L. To cluster or not to cluster farmers? Influences on network interactions, risk perceptions, and adoption of aquaculture practices. *Agric. Syst.* **2019**, *173*, 151–160. [CrossRef]
77. AUC. The Digital Transformation Strategy For Africa (2020–2030). Available online: https://au.int/sites/default/files/documents/38507-doc-dts-english.pdf (accessed on 12 January 2021).
78. Mekki, K.; Bajic, E.; Chaxel, F.; Meyer, F. A comparative study of LPWAN technologies for large-scale IoT deployment. *ICT Express* **2019**, *5*, 1–7. [CrossRef]
79. Khaled, S. Are you ready for the Artificial Intelligence in Ag? *Western* **2020**, *2021*. Available online: https://www.western-irrigation.com/blog/artificial-intelligence/ (accessed on 28 April 2021).
80. Fielke, S.J.; Garrard, R.; Jakku, E.; Fleming, A.; Wiseman, L.; Taylor, B.M. Conceptualising the DAIS: Implications of the 'Digitalisation of Agricultural Innovation Systems' on technology and policy at multiple levels. *NJAS Wagening. J. Life Sci.* **2019**, *90–91*, 100296. [CrossRef]
81. Agozie, E. *Nigerian Scientists, Experts Disagree on GM Foods*; Premium Times: Abuja, Nigeria, 2019.
82. Francis, C.; Lieblein, G.; Gliessman, S.; Breland, T.A.; Creamer, N.; Harwood, R.; Salomonsson, L.; Helenius, J.; Rickerl, D.; Salvador, R.; et al. Agroecology: The Ecology of Food Systems. *J. Sustain. Agric.* **2003**, *22*, 99–118. [CrossRef]
83. Juma, C. *The New Harvest: Agricultural Innovation in Africa*, 2nd ed.; Oxford University Press: Oxford, UK, 2015.

applied sciences

Article

Development of a Precision Feeding System with Hierarchical Control for Gestation Units Using Stalls

Jingjing Xia [1,2], Jichen Xu [1], Zhixiong Zeng [1], Enli Lv [1,*], Feiren Wang [2], Xinyuan He [1] and Ziwei Li [1]

1 College of Engineering, South China Agricultural University, Guangzhou 510642, China; xiajingjing@gdmec.edu.cn (J.X.); zhixzeng@scau.edu.cn (Z.Z.)
2 Schools of Automobile, Guangdong Mechanical and Electrical Polytechnic, Guangzhou 510550, China; 2020010018@gdmec.edu.cn
* Correspondence: enlilv@scau.edu.cn; Tel.: +86-20-8528-2860

Abstract: To obtain good productive performance, sows have different nutrition requirements at different gestation periods. However, in gestation stalls, conventional feeders have large relative errors, management is difficult because of the large numbers of sows, and there are shortcomings in feeding precision and data management. In order to achieve precision feeding and enhance the control of multiple feeders for gestating sows housed in stalls, this study was carried out to investigate a precision feeding system that could be controlled at multiple levels. This system consisted of an electronic sow feeder (ESF), controller area network (CAN), personal digital assistant (PDA), central controller, and Internet of Things platform (IoTP). The results of the experiment showed that relative errors of 60 ESFs delivering feed were within ±2.94%, and the coefficient of variation was less than 1.84%. When the received signal strength indicator (RSSI) ranged from −80 dbm to −70 dbm, the packet loss rate of the PDA was 3.425%. When the RSSI was greater than −70 dbm, no packet loss was observed, and the average response time was 556.05 ms. The IoTP was at the performance bottleneck when the number of concurrent threads was greater than 1700. These experimental results indicated that the system was not only highly accurate in delivering feed, but was also highly reliable in the transmission of information, and therefore met the production requirements of an intensive gestation house.

Keywords: gestating sow; precision feeding; CAN; PDA; IoT platform

Citation: Xia, J.; Xu, J.; Zeng, Z.; Lv, E.; Wang, F.; He, X.; Li, Z. Development of a Precision Feeding System with Hierarchical Control for Gestation Units Using Stalls. *Appl. Sci.* **2023**, *13*, 12031. https://doi.org/10.3390/app132112031

Academic Editors: Dolores Parras-Burgos, José Miguel Molina Martínez, Ginés García-Mateos and Yutaka Ishibashi

Received: 5 September 2023
Revised: 28 October 2023
Accepted: 30 October 2023
Published: 4 November 2023

1. Introduction

Globally, there is an increasing demand for animal products (e.g., meat, dairy products, eggs, wool, leather, etc.) [1–4]. African swine fever [5] and excessive population growth [6,7] have brought new challenges, requiring pig husbandry to further develop in the direction of intensification, specialization, and refinement [8,9].

Feed costs account for a major proportion of the total cost of raising pigs [10]. Reducing feed waste helps to mitigate the world food crisis, as well as reduce the costs of pork production [11,12]. The appropriate feeding allowance should be determined at various gestation stages [13,14], as inadequate or excessive feeding can compromise the reproductive performance of sows [15]. The gestation period of sows is typically categorized into three stages. During early gestation, there is no clear benefit observed from increased feeding [16], and excessive nutrient intake may lead to higher embryo mortality and lower litter sizes [17]. In mid-gestation, gilts are still growing and developing, and sows need to recover body reserves lost from the previous breeding cycle [18]. Late gestation is a period for improving the birth weight of piglets [19,20]. Increasing the energy intake of sows has a positive effect on individual piglets, but may also elevate the stillborn rate [21].

The implementation of precision feeding has a positive impact on the sustainability of pig husbandry [22,23]. Compared with conventional feeding, precision feeding decreases the intake of lysine and protein and reduces feed costs without adversely affecting growth and reproductive performance [24–26]. Additionally, in terms of environmental protection, precision feeding reduces nitrogen and phosphorus excretion [27,28]. Buis et al. [29] observed that gilts subject to precision feeding ate more and lost less weight during subsequent lactation. Quiniou [30] found that precision feeding decreased the risk of sows being too fat or too thin, thereby decreasing the risk of impaired farrowing or milk production, and also that there was less variability in backfat thickness when sows farrowed in the same batch. Iida et al. [31] showed that the risk of fetal displacement in sows could be identified early by measuring feed intake and time spent at an ESF.

The integrated application of automation and Internet of Things (IoT) technologies in pig husbandry can contribute to more accurate data records [32,33] which in turn reduce the burden on practitioners, lower the cost of farming, and raise productivity. The IoT connects physical objects and spaces into a local area network or internet and enables efficient monitoring and management by integrating data objects [34–36]. Zeng et al. devised a three-layer wireless sensor network system based on ZigBee to monitor four environmental parameters (temperature, relative humidity, and concentrations of carbon dioxide and ammonia) in commercial gestation units; real-time monitoring of the microclimate and timely intervention were achieved by analyzing the temporal and spatial characteristics of the pigsty [37]. Lee et al. developed a monitoring system using wireless broadband leaky coaxial cable to collect data from Bluetooth Low-energy (BLE) tags attached to individual pigs. These data were used to compute and estimate the position and movement of each pig [38].

Group housing and conventional stall housing have their own advantages and disadvantages [39,40]; however, because of different degrees of development and economic factors, stalls are still heavily used outside Europe. As part of the transition of intensive pig farms towards automation and informatization, the development of a precision feeding system suitable for stall housing represents a significant step forward. Liu and Xiong et al. designed an electronic feeder for the precision feeding of sows housed in stalls [41,42]. However, there is a scarcity of research examining the shortcomings and defects of precision feeding systems within a real intensive pig farm. We identified these following challenges: (1) it is inconvenient for breeders to adjust the parameters of the ESF, because they have to raise their heads or stand on tiptoe to click buttons attached to ESFs; (2) close contact between breeders, sows, or feed should be avoided to prevent African swine fever; (3) pig farms are usually built in the suburbs with poor wireless network signals, resulting in the frequent disconnection of communication between the ESF and the internet; and (4) the unified management of the ESF is completely reliant on the internet, which is prone to data loss, leading to production errors.

In order to overcome the abovementioned difficulties, we devised a precision feeding system suitable for gestation units using stalls which combined automation, CAN, PDA, central controller, and IoT technologies. In this system, the average performance of 60 lines, and potential correlation between the feeding performance and the location of the feeding line were focused on. The three-level hierarchical control method was introduced: (1) for a Center Controller and all ESFs in a gestation unit, and CAN-based LAN was utilized to unite them; (2) for an individual ESF and a PDA, and WIFI-based LAN was used to establish a connection; (3) for a Center Controller and the IoTP, and the Internet took responsibility to share production data. Live trials were conducted to evaluate the three-level hierarchical control method's performance and reliability.

2. Materials and Methods

2.1. Gestation Sow Housing

The use of stalls is common for intensive pig farms in China. Our experiments were conducted at a pig farm in Sichuan Province, China. The sows were housed in a gestation unit from weaning to 110 days of breeding. The gestation unit consisted of 30 columns, each containing 60 stalls 2.2 m long and 0.65 m wide, and an automatic feed line 2.0 m above the floor. The feeding system was applied to one of the columns, and each stall was equipped with an ESF. A total of 60 ESFs were used, together with a central controller and a PDA. Figure 1 illustrates part of the environment in which the precision feeding system was implemented.

Figure 1. Part of the environment in which the precision feeding system was implemented.

2.2. System Design

2.2.1. Overall Architecture

The precision feeding system for gestating sows was constructed using ESFs, a CAN, a central controller, a PDA, and an IoTP. The ESFs ensured accurate, timed, and ration feeding; the CAN connected all the ESFs and provided a unified management method; the central controller managed the CAN through a graphical interface and uploaded feeding data to the IoTP; the PDA operated individual ESF via an application and WLAN communication; and the IoTP received and displayed device information and feeding data. Figure 2 shows the overall architecture of the system.

2.2.2. Electronic Sow Feeder

The ESFs comprised a delivery mechanism and a control circuit. The delivery mechanism consisted of a plastic bucket, a motor (37GA370SH283011, 24V, 20RPM, Zhejiang Youtuo Motor Co., Ltd., Jinhua, China), and a feeding auger. The electronic circuit included an MCU (STM32F103, ARM), a CAN transceiver module (ISO1050DUBR, Texas Instruments Semiconductor Technology Co., Ltd., Shanghai, China), a WLAN transceiver module (ESP-075, Ai-Thinker Technology Co., Ltd., Shenzhen, China), memory, a buzzer, etc.

Each ESF implemented the following functions: (1) it fed sows at regular intervals according to the programmed scheme; (2) it maintained a connection with the CAN, reported operation status, and received commands from the central controller; and (3) it operated the socket server at the appointed port to communicate with the PDA on a one-to-one basis.

Figure 3 shows the delivery mechanism and the electronic circuit.

2.2.3. Controller Area Network

Because of the large number of ESFs in the gestation unit, there was a need for a control strategy which could cope with multiple devices and manage the devices uniformly and efficiently. CAN is an efficient serial bus communication protocol with characteristics including high communication efficiency, strong anti-interference ability, a node arbitration mechanism, and self-diagnosis of errors. Because gestation units are complex environments with a lot of interference noise, the development of a CAN is highly beneficial for the management of ESF clusters.

2.2.4. Central Controller

The CAN controlled the ESFs via data streams and communication protocols but these could not be intuitively understood by the breeders. The central controller implemented a graphical interactive interface to manage the CAN using a touchscreen, thereby improving the user experience for the breeders. The central controller was equipped with an Intel Celeron J1900 CPU, 4GB RAM and a touchscreen and ran the Windows 7 operating system. The human–computer interaction was developed using the C++ programming language and the Qt5.9 framework. Qt is a cross-platform GUI framework for desktop, embedded, or mobile platforms.

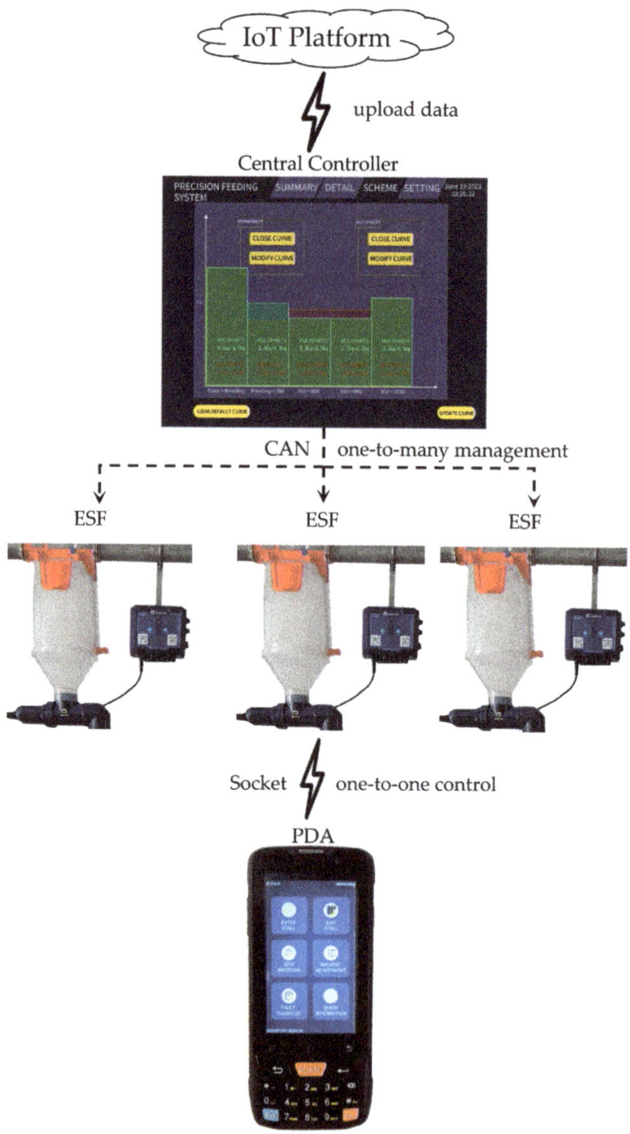

Figure 2. Overall architecture of the precision feeding system with hierarchical control for gestation units using stalls.

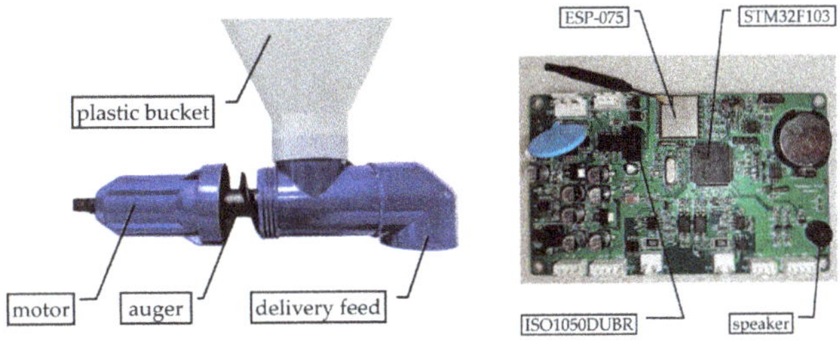

(**a**) (**b**)

Figure 3. (**a**) Delivery mechanism and (**b**) electronic circuit.

The central controller fulfilled the following functions: (1) it connected the CAN to manage all the ESFs as a master node; (2) it collected the operation status of the ESF as well as the feeding data which were uploaded to the IoTP in HTTP; and (3) it displayed the status of the ESF, presented information, adjusted the feeding scheme, and calibrated the delivery speed. Figure 4 shows parts of the screens.

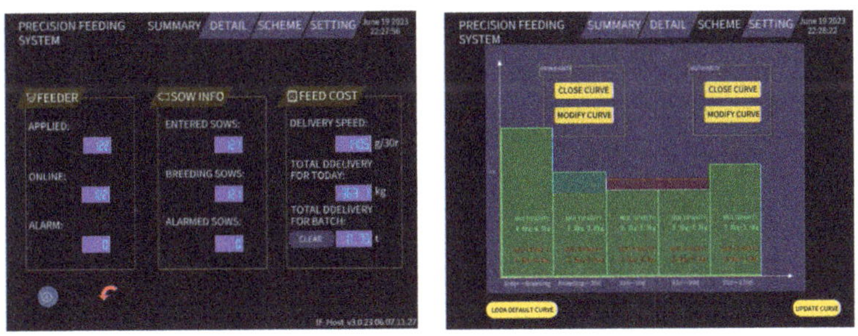

(**a**) (**b**)

Figure 4. (**a**) Operating status page and (**b**) feeding scheme page.

2.2.5. PDA

The feeding system was regulated as a whole by the CAN, but the ESFs also required a convenient, one-to-one management method. As a type of GUI software with excellent performance and high visibility, Android application technology was used to implement the key functions related to activity, service, broadcast, content provider, and other components. The application running on the PDA (equipped with a laser scanner and Android 8.0 operating system) was developed to manage an individual ESF.

The application includes "ScanReceiver", "ConnectService", communication protocol, function page, and so on. "ScanReceiver" extended "BroadcastReceiver", received the scanning broadcast from PDA, and parsed out the valid content. "ConnectService" extended service, connected to WLAN according to SSID, and established a one-to-one socket pipeline. The communication protocol was employed to safeguard against malicious attacks through message parsing and validation. The function pages served as the gateway to ESF management. Figure 5 illustrates the workflow of using the PDA to manage the ESF. The steps involved for breeders to manage the ESF using the PDA were as follows:

(1) The PDA scanned the QR code attached to the ESF and transferred the byte content to the APP via broadcast.
(2) The content was parsed to extract the Service Set Identifier (SSID), representing the unique WLAN generated by an ESF.
(3) Based on the SSID, the app established a socket pipeline with the ESF.
(4) Breeders could manage the ESF through function pages such as Entry, Exit, Breeding, Backfat Adjustment, Diagnosis and Test, and Information Query pages.

(5) Connectivity between the app and ESF had to be maintained. If the ESF did not receive a heartbeat from the app every 3 s, it would interrupt the connection to prevent data inconsistency.

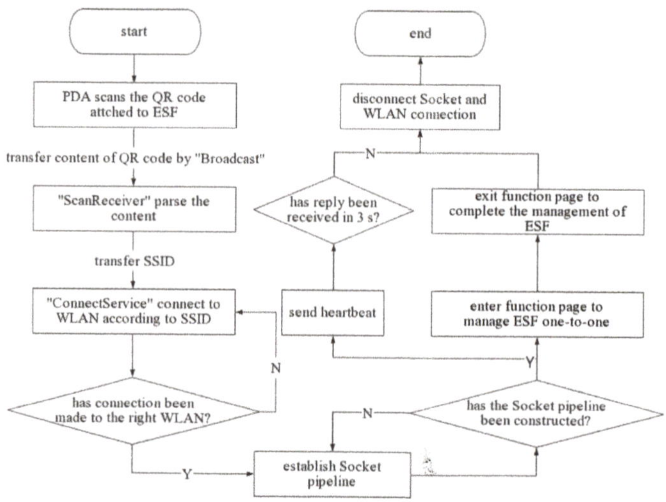

Figure 5. Workflow of PDA management of ESF.

Figure 6 displays some of the PDA screens, each with specific responsibilities:

(1) The "ENTER STALL" page facilitated the association between a sow and an ESF.
(2) The "EXIT STALL" page allowed for the disassociation between a sow and an ESF.
(3) The "SOW BREEDING" page enabled inputting mating date to adjust the feeding scheme accordingly.
(4) The "BACKFAT ADJUSTMENT" page allowed input of the backfat thickness to notify the ESF to update the feeding scheme.
(5) The "FAULT DIAGNOSIS" page provided a means to query fault information related to the ESF.
(6) The "QUERY INFORMATION" page provided functionality to inquire about the sow's status.

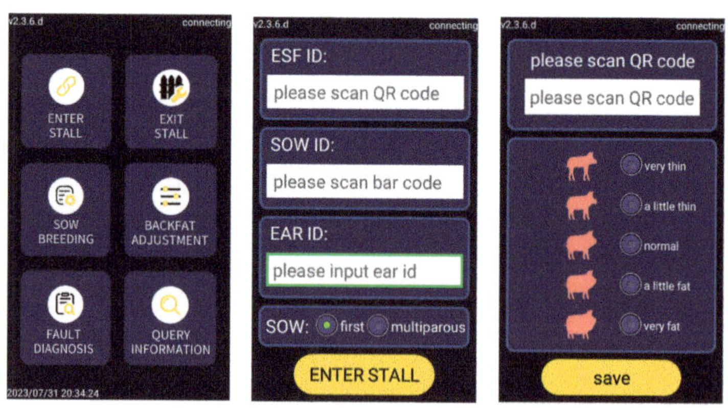

Figure 6. (**a**) Home page, (**b**) "ENTER STALL" page, and (**c**) "BACKFAT ADJUSTMENT" page.

2.2.6. IoT Platform

The IoTP allowed the operating status and feeding data from multiple production lines, units, and farms to be gathered from the ESFs and enabled terminal devices (smart phone and computers) to browse these data remotely. The IoTP ran on the Alibaba ecs.n4.small cloud server, which had a single-core VCPU, 2 GB RAM, 100 GB hard disk, and 0.5 Gb/s network bandwidth.

The IoTP could be divided into three parts: receiving, data, and display layers. The receiving layer was responsible for collecting and parsing the messages uploaded by the central controller, as well as verifying the legitimacy of the data in order to prevent malicious attacks. The data layer stored the ESF operating status and feeding data provided an interface for the display layer to query the data, and also made scientific and appropriate suggestions according to the feeding situation of the sows. The display layer used visualization to enable users to view real-time and historical data, including information relating to the feeding situation, feeding scheme, and the ESF operating status.

The receiving layer was constructed using the Spring Boot backend framework and Java programming language, and data were stored in a MySQL database. The data-browser software in the display layer was built using the Vue frontend framework and JavaScript programming language. Coupling was decreased and scalability advanced because of the separation of the frontend and backend. Figure 7 shows the layered architecture of the IoTP as well as the data-browser software viewed on a smart phone.

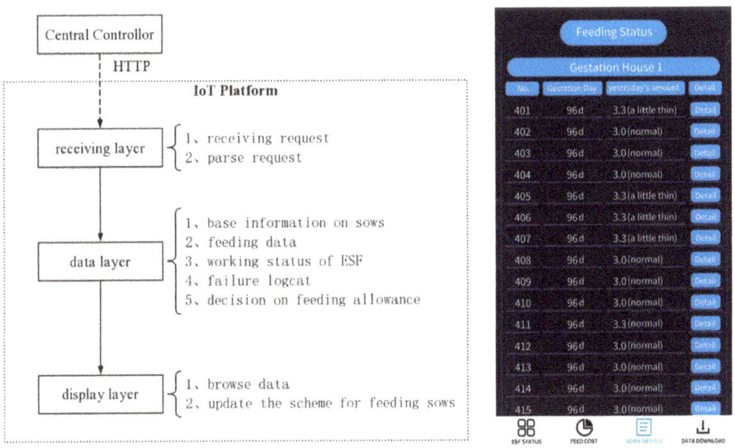

(a) (b)

Figure 7. (**a**) IoTP layered architecture and (**b**) the data-browser software viewed on a smart phone.

2.3. Statistical Analysis

2.3.1. Feed Delivery Accuracy of the Electronic Sow Feeders

The accuracy of ESFs in delivering feed is crucial for ration feeding and preventing feed waste. The relative error is calculated as shown in Equation (1), and the smaller the relative error, the higher the feeding accuracy.

$$\delta = \frac{M - M_0}{M_0} \times 100\% \tag{1}$$

where δ is the relative error; M is the weight of actual feed in g; and M_0 is the weight of expected feed in g.

The coefficient of variation is calculated as shown in Equation (2), and the smaller this is, the better the feeding stability.

$$CV = \frac{S}{X} \times 100\% \quad (2)$$

where CV is the coefficient of variation; S is the sample standard deviation; and X is the mean feed weight.

2.3.2. PDA Communication Reliability

The communication reliability of the PDA had a significant impact on its dependability for controlling ESFs and the motivation of breeders to use a PDA. The reliability of wireless communication between PDA and ESF can be effectively evaluated using the packet loss rate (PLR) and response time (PDA_RT). A smaller PLR indicates more reliable wireless communication, and this is calculated as shown in Equation (3).

$$PLR = \frac{S_a - R_b}{S_a} \times 100\% \quad (3)$$

where Sa is the number of times packet-a was sent and Rb is the number of times packet-b was received.

The shorter the PDA_RT, the more rapid the wireless communication, and this is calculated as shown in Equation (4).

$$PDA\_RT = \begin{cases} T_r - T_s, \text{ successfullyreveivepacket} - b \\ 5000 \quad , \text{ losepacket} - b \end{cases} \quad (4)$$

where T_r is the timestamp of packet-a sent by PDA and T_s is the timestamp of packet-b received by PDA.

2.3.3. Data Insertion Performance of the IoT Platform

The IoTP was responsible for maintaining communication with multiple central controllers as well as ensuring data security. In the performance test of the IoTP, throughput rate (TR) and response time (IoTP_RT) were important indicators. TR is calculated as shown in Equation (5) and refers to the number of requests processed per second. A larger TR indicates that the IoTP can connect with a greater number of central controllers.

$$TR = \frac{NoR}{T_c} \quad (5)$$

where NoR is the number of requests processed by the IoTP and T_c is the total time taken to process all requests in ms.

IoTP_RT is calculated as shown in Equation (6) and refers to the time taken to process a request. The smaller the response time, the faster the IoTP processes the request.

$$IoT\_RT = T_f - T_s \quad (6)$$

where T_f is the time in ms at which IoTP finishes processing the request and T_s is the time in ms at which the IoTP starts processing the request.

3. Results

3.1. *ESF Feed Delivery Experiment*

The ESF ration feeding depended on both the speed of feed delivery and the rotation angle of the auger. The speed of feed delivery refers to the mass of feed delivered by a revolution of the motor. The rotation angle of the auger was electronically calculated to feed rations of different sizes. In the gestation unit, the feed was transported via a feed line and chain, with longer paths resulting in increased abrasion. In order to investigate the effect of abrasion on feeding performance and the stability of the system, the speed of feed delivery was measured using rations of different sizes. The trials were conducted on a sample feed line with 60 consecutive ESFs. The tools included electronic scales (EHA28, Guangdong Xiangshan Weighing Instrument Group Co., Ltd., Zhongshan, China) and containers.

For the speed measurement, a PDA was used to control the 60 ESFs to rotate augers sequentially, and the mass of feed delivered was measured. The error from 30 revolutions was smaller than that from one revolution, so the former was chosen for the speed measurement. The results are shown in Figure 8. The data were essentially centered between 1350 g and 1450 g, with a standard deviation of 21.60, a mean value of 1395 g, and a coefficient of variation of 1.55%. Fifty-seven of the feeders had similar delivery speeds, whereas feeder numbers 5, 31, and 59 deviated further, which could be due to variations in the auger, motor, plastic box, or even the installation environment. The data indicated that there was no obvious relationship between feed abrasion and delivery speed on a feed line, and 60 ESFs had good feeding uniformity.

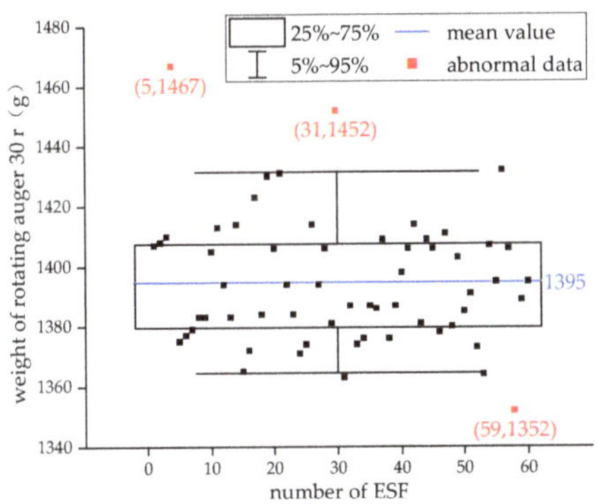

Figure 8. Weight of feed delivered by 60 consecutive ESFs with 30 revolutions of the augers.

For the ration feeding, six experimental groups were established: 1.6 kg, 2.0 kg, 2.4 kg, 2.8 kg, 3.2 kg, and 3.6 kg. In each experimental group, 60 ESFs were controlled to deliver feed quantitatively, and the weight of feed actually delivered was measured separately. The results in Figure 9 show that the relative error of the feeding system consistently remained within ±2.94% and the coefficient of variation was less than 1.84% at different expected feed weights. In terms of overall feeding performance, the feeding system was shown to have a low relative error, a low coefficient of variation, high accuracy, and strong stability, which met the requirements for the accurate control of feed allowances for sows.

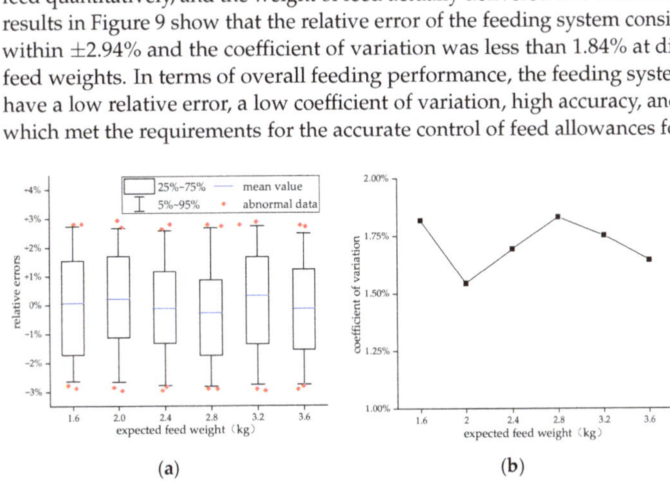

Figure 9. Relationship between (**a**) the relative error and the expected feed weight and (**b**) the coefficient of variation and the expected feed weight.

3.2. PDA Communication Experiment

The RSSI of the PDA affects its communication ability and can lead to socket pipeline rupture and information loss, in addition to interfering with the normal operation of the feeding system. In order to investigate the relationship between RSSI and PDA_RT, five ESF were randomly selected as samples and divided into eight groups with RSSI as the variable. The test process was as follows: (1) the PDA sent packet-a to the ESF, and recorded the time of sending; (2) the ESF received packet-a and replied with packet-b; (3) the PDA received packet-b and recorded the time of receipt. The number of times packet-a was sent, the number of times packet-b was received, the times at which packet-a was sent, and the times of receipt of packet-b were then counted, and the PLR and PDA_RT calculated.

During the communication request process, to ensure the uniqueness of the packet-b response, the PDA could only run one request at a time. The PDA released the socket channel for other requests only after a complete process was finished. If the PDA sent packet-a but did not receive packet-b, the socket channel could be continuously occupied, causing subsequent communication requests to be blocked. Therefore, the upper limit of the communication duration was set at 5000 ms. If packet-b was not received within 5000 ms, the communication request would be regarded as timed out and terminated.

Figure 10 shows the results. When the RSSI was from −80 dbm to −70 dbm, the PDA sent 146 packet-as and received 141 packet-bs, with a PLR of 3.425%. When RSSI was from −90 dbm to −80 dbm, the PDA sent 47 packet-as and received 32 packet-bs, with a PLR of 31.91%, and the communication pipeline was easy to disconnect, which explained the small sample size at this interval. When the RSSI was from −70 dbm to −10 dbm, there was no packet loss, the average PDA_RT was 556.05 ms, and the communication between the PDA and ESF was stable. Therefore, the power of the ESF WLAN transceiver in the ESF should be adjusted to an RSSI between −70 dbm and −10 dbm to meet the requirements of the application used by the gestation unit.

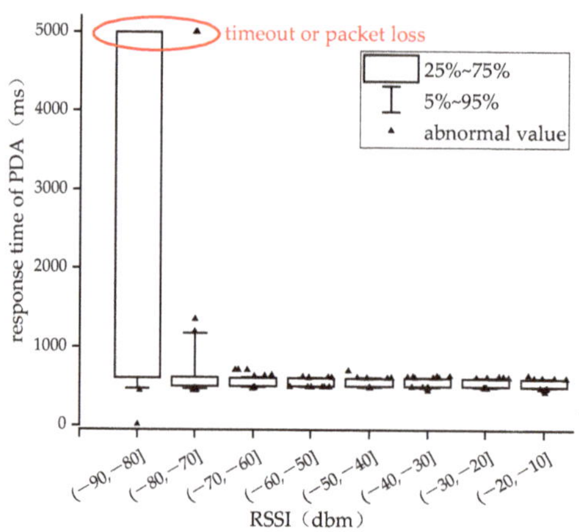

Figure 10. The relationship between PDA_RT and RSSI.

3.3. Data Insertion Experiment for the IoT Platform

The IoTP had to handle multiple concurrent data insertion requests simultaneously, and its performance in data insertion was more critical to the system's reliability than its data reading performance. As the number of central controllers deployed increased, IoTP concurrency also increased, raising the risk of the server encountering issues such as excessive memory usage, inadequate CPU processing power, and blocked IO resources. These problems could eventually result in the service program crashing.

In order to explore the optimum number of concurrent data insertions that the IoTP could load under the existing server configuration, we developed a Web application using

Java programming language and executed it on a Tomcat server. The performance test of the web application running on Tomcat was carried out using the software JMeter 5.6.2. The detailed process was as follows: (1) We established 22 thread groups, each representing a different number of concurrent threads (ranging from 0 to 2200 threads in increments of 100). These thread groups were set up to simulate the task where the Center Controller sent HTTP requests to the IoTP. Each thread simulated a Center Controller. (2) The packet size of each Http request was set as fixed 364 bytes. When IoTP received an HTTP request, it was parsed into a database row structure, and then inserted into the database. The IoTP would respond with a message indicating the successful data insertion. All threads initiated within a span of 5 s. JMeter was used to generate an aggregated report, allowing the analysis of the system's response time and throughput rate.

The experimental results are shown in Figure 11. For concurrent threads below 1700, there was a gradual increase in both the TR and the average IoTP_RT with the escalation of concurrent threads. However, once the concurrent threads exceed 1700, both the TR and average IoTP_RT stabilized at approximately 254 requests per second (rqs) and 4548 ms, respectively. The cause was the server being configured with one virtual cpu core and 2 GB memory. When the number of concurrent threads reached 1700, the resource usage reached the maximum and the request was blocked. This behavior could be attributed to the server configuration, which consisted of a single virtual CPU core and 2 GB of memory. As the concurrent threads reached 1700, resource usage peaked, resulting in blocked requests. Consequently, it was evident that the IoTP experienced a performance bottleneck when the number of concurrent threads exceeded 1700, highlighting the need to avoid overloading it. On the other hand, when the concurrent threads remained below 1700, the IoTP exhibited improved performance and ensured the availability of additional computational resources to handle unforeseen circumstances.

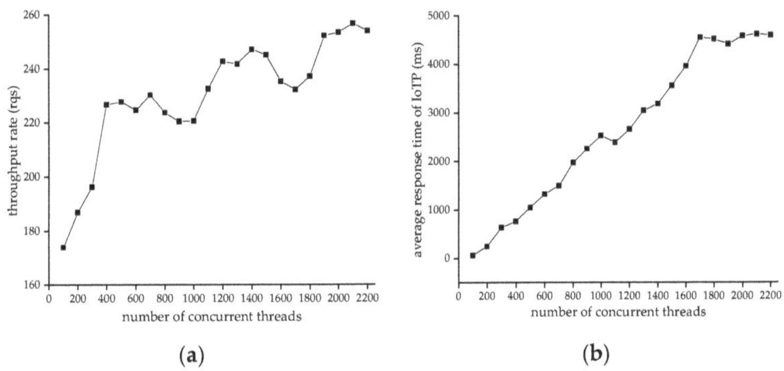

Figure 11. (**a**) Relationship between TR and concurrent threads and (**b**) relationship between average IoT_RT and concurrent threads.

4. Discussion

In Europe, there is increased protection of animal welfare. However, in many developing countries and regions, stalls are still used to raise sows in gestation units. Most studies focus on the design of precision feeding equipment in group housing, but few have considered the problems that need to be overcome for precision feeding in stall housing systems: (1) conventional feeders are prone to over-adjustment leading to large errors in the delivery of feed; (2) there are many feeders installed in the stalls but a lack of efficient hierarchical control; and (3) with high levels of bio-security, the production area and information department are isolated from each other, leading to the slow transfer of production data.

In this research, we conducted a feed delivery experiment using 60 ESFs installed at various positions along the feed transmission line. The speed at which each ESF delivered feed was measured. The data revealed a central tendency between 1350 g and 1450 g, with an average value of 1395 g and a coefficient of variation at 1.55%. Notably, our analysis did not reveal discernible relationship between feed abrasion and delivery speed. To evaluate the accuracy of feed delivery, we based our calculations on an average speed of 1395 g per 30 revolutions, and we investigated the relative error between the actual delivered feed mass and the expected delivery mass. Our findings consistently demonstrated that the relative error for all 60 ESFs remained within the range of ±2.94%, with a coefficient of variation of less than 1.84% across different expected feed weights. It is worth noting that Chen et al. [43] conducted a similar experiment, but their focus was on a single ESF, resulting in a maximum relative error of +2%. Given our emphasis on the collective performance of multiple ESFs, the ±2.94% range is deemed an acceptable indicator. The structure of the feeding auger stands out as the optimal choice for the widespread implementation of ESFs, as it ensures uniform and consistent feeding performance.

In the scientific literature, there is a conspicuous absence of data communication of ESFs in large-scale gestation units using stalls. In a study involving the use of the feeding station Nedap Velos, conducted by Thomas et al. [44], a significant challenge arose—feed intake data had to be manually extracted on a daily basis due to the lack of long-term data storage capabilities. Additionally, the study reported instances of lost RFID tags in group housing, resulting in potential inaccuracies in data recording. In our system, each ESF is dedicated to a specific sow, which addressed the issues mentioned above. Furthermore, through the CAN, ESF data are seamlessly transmitted to the Center Controller, facilitating efficient data collection and distribution. This not only enhances data management but also leads to significant cost savings in terms of labor.

Sow production management can be categorized into two approaches: normal management and batch flow management [45]. In batch flow management, timed artificial insemination and the administration of various hormones are employed to regulate the timing of estrus, ovulation, and farrowing [46]. However, normal management necessitates individualized care for each sow. To facilitate this, we designed the PDA for the convenient adjustment of individual ESFs. The PDA features an intuitive user interface that provides essential functionalities for pig farming, effectively reducing the cognitive burden on breeders by minimizing comprehension costs. Through a wired connection with ESFs, close contact between breeders and ESFs is prohibited to ensure bio-safety. In practical applications, maintaining the power of the ESF's WIFI module at levels greater than −70 dbm within an appropriate operating distance is advisable to ensure a stable communication effect.

IoTP was implemented to integrate data from various regions and dismantle information barriers among departments. A server (equipped with one core VCPU and 2 GB of RAM) can manage up to 1700 concurrent threads. Exceeding this limit may result in slow data links, blockages, and, in severe cases, cloud service crashes. To address this, we recommend scaling the number of servers and implementing load balancing algorithms when surpassing 1700 concurrent threads. This approach will enhance system concurrency and fault tolerance. Additionally, considering varying GRPS signal strengths in different pig farms, utilizing Ethernet cables is a viable option to improve the central controller's network signal.

While the system has been successfully implemented on a pig farm in Sichuan Province, China, it proved to be challenging to collect additional data and analyze the feeding status of the sows due to strict access policies aimed at preventing African swine fever. In practice, individuals not directly associated with the pig farm are restricted from prolonged stays or even entry.

In the next stage of our research, we will investigate the effects of this feeding system on the performance of gestating sows and explore the potential associations between gestation days, feed intake, backfat thickness, fetal abortion, litter size, and weaning performance. Furthermore, our team's endeavors may revolve around leveraging IoT platforms along with Big Data analytics and Foundation Models technology, for conducting more comprehensive analyses of production data.

5. Conclusions

In order to achieve precision feeding in gestation units using stalls, and to enhance the management ability of feeding systems, this paper presents a precision feeding system with multilevel control. The three-level hierarchical control method was introduced: (1) CAN-based LAN was employed to unite the Center Controller and all ESFs within a gestation unit, ensuring reliable connections and consistent data; (2) WIFI-based LAN was utilized to establish a connection between an individual ESF and a PDA, optimizing the human–computer interaction experience; (3) the internet was leveraged for real-time sharing of production data between the Center Controller and the IoTP. In the context of the intensification and isolation of pig farms in China, our research enables the automatic adjustment of feed allowance during various stages of a sow's reproductive cycle. This approach may achieve a better balance between feed costs and productive performance, potentially contributing to addressing global food crises. Furthermore, our research dismantles information barriers within the farm, allowing breeders to concentrate on essential tasks such as breeding, vaccination, estrus detection, and more, rather than being burdened with data collection, compilation, consolidation, and distribution.

Author Contributions: J.X. (Jingjing Xia): conceptualization, methodology, software, supervision, and resources; J.X. (Jichen Xu): writing—original draft, writing—review and editing, conceptualization, methodology, investigation, data curation, software, formal analysis, and validation; Z.Z.: writing—review and editing, supervision, conceptualization, resources, and funding acquisition; E.L.: investigation, formal analysis, resources, conceptualization, and funding acquisition; F.W.: conceptualization, methodology, software, and validation; X.H.: data curation, investigation, methodology, validation, and software. Z.L.: data curation, supervision, and resources. All authors have read and agreed to the published version of the manuscript.

Funding: This research was funded by Guangzhou Key Research and Development Project (2023B03J 1363), Special Fund for Rural Revitalization Strategy of Guangdong (2023TS-3), Key Laboratory of Modern Agricultural Intelligent Equipment in South China, Ministry of Agriculture and Rural Affairs, China (HNZJ202209), Guangzhou Basic and Applied Research Project (2023A04J0752), Independent Research Project of Maoming Laboratory (2021ZZ003, 2021TDQD002), and Subproject of National Key Research and Development (2018YFD0701002-2).

Institutional Review Board Statement: Not applicable.

Informed Consent Statement: Not applicable.

Data Availability Statement: Data are contained within the article.

Acknowledgments: We are grateful for the administrative support from the South China Agricultural University and the generous help from the staff working in the pig unit.

Conflicts of Interest: The authors declare no conflict of interest.

References

1. Mustapa, M.A.C.; Kallas, Z. Towards More Sustainable Animal-Feed Alternatives: A Survey on Spanish Consumers' Willingness to Consume Animal Products Fed with Insects. *Sustain. Prod. Consum.* **2023**, *41*, 9–20. [CrossRef]
2. Van der Linden, A.; de Olde, E.M.; Mostert, P.F.; de Boer, I.J. A Review of European Models to Assess the Sustainability Performance of Livestock Production Systems. *Agric. Syst.* **2020**, *182*, 102842. [CrossRef]
3. Fields, B.; Matthews, N.; Bloemhof, S.; Fitzgerald, R.; Sosnicki, A. 103 Research Needs for the Development of the Pork Industry. *J. Anim. Sci.* **2019**, *97*, 58–59. [CrossRef]
4. El-Sabrout, K.; Aggag, S.; Mishra, B. Advanced Practical Strategies to Enhance Table Egg Production. *Scientifica* **2022**, *2022*, e1393392. [CrossRef]
5. Cwynar, P.; Stojkov, J.; Wlazlak, K. African Swine Fever Status in Europe. *Viruses* **2019**, *11*, 310. [CrossRef] [PubMed]
6. Hewitt, R.J.E.; Kemp, B.; Millet, S. Editorial: Addressing 21st Century Challenges in Pig Production. *Animal* **2022**, *16*, 100542. [CrossRef]
7. Garrido Sanchis, A.; Jin, L. Evaluation of the New Energy-Efficient Hot Bubble Pilot Plant (HBPP) for Water Sterilization from the Livestock Farming Industry. *Water Resour. Ind.* **2020**, *24*, 100135. [CrossRef]
8. Berckmans, D. General Introduction to Precision Livestock Farming. *Anim. Front.* **2017**, *7*, 6–11. [CrossRef]

9. Cavalett, O.; de Queiroz, J.F.; Ortega, E. Emergy Assessment of Integrated Production Systems of Grains, Pig and Fish in Small Farms in the South Brazil. *Ecol. Model.* **2006**, *193*, 205–224. [CrossRef]
10. Renaudeau, D.; Gilbert, H.; Noblet, J. Effect of Climatic Environment on Feed Efficiency in Swine. In *Feed Efficiency in Swine*; Patience, J.F., Ed.; Academic Publishers: Wageningen, The Netherlands, 2012; pp. 183–210. ISBN 978-90-8686-756-1.
11. Pomar, C.; Hauschild, L.; Zhang, G.-H.; Pomar, J.; Lovatto, P.A. Applying Precision Feeding Techniques in Growing-Finishing Pig Operations. *Rev. Bras. Zootec.* **2009**, *38*, 226–237. [CrossRef]
12. Gaillard, C.; Brossard, L.; Dourmad, J.-Y. Improvement of Feed and Nutrient Efficiency in Pig Production through Precision Feeding. *Anim. Feed. Sci. Technol.* **2020**, *268*, 114611. [CrossRef]
13. Thomas, L.L.; Goodband, R.D.; Tokach, M.D.; Woodworth, J.C.; DeRouchey, J.M.; Dritz, S.S. Effect of Parity and Stage of Gestation on Growth and Feed Efficiency of Gestating Sows. *J. Anim. Sci.* **2018**, *96*, 4327–4338. [CrossRef] [PubMed]
14. Thomas, L.L.; Goodband, R.D.; Tokach, M.D.; Dritz, S.S.; Woodworth, J.C.; DeRouchey, J.M. Partitioning Components of Maternal Growth to Determine Efficiency of Feed Use in Gestating Sows1,2. *J. Anim. Sci.* **2018**, *96*, 4313–4326. [CrossRef]
15. Kim, S.W.; Hurley, W.L.; Wu, G.; Ji, F. Ideal Amino Acid Balance for Sows during Gestation and Lactation1. *J. Anim. Sci.* **2009**, *87*, E123–E132. [CrossRef]
16. Carrión-López, M.J.; Madrid, J.; Martínez, S.; Hernández, F.; Orengo, J. Effects of the Feeding Level in Early Gestation on Body Reserves and the Productive and Reproductive Performance of Primiparous and Multiparous Sows. *Res. Vet. Sci.* **2022**, *148*, 42–51. [CrossRef] [PubMed]
17. Vignola, M. Sow Feeding Management during Lactation. In Proceedings of the London Swine Conference, Tools of the Trade, London, ON, Canada, 1–2 April 2009; pp. 107–117.
18. Solà-Oriol, D.; Gasa, J. Feeding Strategies in Pig Production: Sows and Their Piglets. *Anim. Feed. Sci. Technol.* **2017**, *233*, 34–52. [CrossRef]
19. Cromwell, G.L.; Hall, D.D.; Combs, G.; Hale, O.M.; Handlin, D.L.; Hitchcock, J.P.; Knabe, D.A.; Kornegay, E.T.; Lindemann, M.D.; Maxwell, C.V.; et al. S-145 Committee on Nutritional Systems for Swine to Increase Reproductive Efficiency Effects of Additional Feed during Late Gestation on Reproductive Performance of Sows: A Cooperative Study. *J. Anim. Sci.* **1989**, *67*, 3–14. [CrossRef]
20. Shelton, N.; Neill, C.; DeRouchey, J.; Tokach, M.; Goodband, R.; Nelssen, J.; Dritz, S. Effects of Increasing Feeding Level during Late Gestation on Sow and Litter Performance. *Kans. Agric. Exp. Stn. Res. Rep.* **2009**, 38–50. [CrossRef]
21. Gonçalves, M.A.D.; Gourley, K.M.; Dritz, S.S.; Tokach, M.D.; Bello, N.M.; DeRouchey, J.M.; Woodworth, J.C.; Goodband, R.D. Effects of Amino Acids and Energy Intake during Late Gestation of High-Performing Gilts and Sows on Litter and Reproductive Performance under Commercial Conditions1,2. *J. Anim. Sci.* **2016**, *94*, 1993–2003. [CrossRef]
22. Pomar, C.; Remus, A. Precision Pig Feeding: A Breakthrough toward Sustainability. *Anim. Front.* **2019**, *9*, 52–59. [CrossRef]
23. Collins, L.M.; Smith, L.M. Review: Smart Agri-Systems for the Pig Industry. *Animal* **2022**, *16*, 100518. [CrossRef] [PubMed]
24. Andretta, I.; Pomar, C.; Rivest, J.; Pomar, J.; Lovatto, P.A.; Radünz Neto, J. The Impact of Feeding Growing–Finishing Pigs with Daily Tailored Diets Using Precision Feeding Techniques on Animal Performance, Nutrient Utilization, and Body and Carcass Composition1. *J. Anim. Sci.* **2014**, *92*, 3925–3936. [CrossRef]
25. Trevisi, P.; Colombo, M.; Priori, D.; Fontanesi, L.; Galimberti, G.; Calò, G.; Motta, V.; Latorre, R.; Fanelli, F.; Mezzullo, M.; et al. Comparison of Three Patterns of Feed Supplementation with Live Saccharomyces Cerevisiae Yeast on Postweaning Diarrhea, Health Status, and Blood Metabolic Profile of Susceptible Weaning Pigs Orally Challenged with Escherichia Coli F4ac1. *J. Anim. Sci.* **2015**, *93*, 2225–2233. [CrossRef] [PubMed]
26. Stewart, V.; Buis, R.Q.; Christensen, B.; Hansen, L.L.; de Lange, C.F.M.; Mandell, I.B.; Huber, L.-A. The Effects of Precisely Meeting Estimated Daily Energy and Lysine Requirements for Gestating Sows over Three Consecutive Pregnancies on Sow Reproductive and Lactation Performance. *Transl. Anim. Sci.* **2021**, *5*, txab226. [CrossRef]
27. Gaillard, C.; Quiniou, N.; Gauthier, R.; Cloutier, L.; Dourmad, J.-Y. Evaluation of a Decision Support System for Precision Feeding of Gestating Sows. *J. Anim. Sci.* **2020**, *98*, skaa255. [CrossRef] [PubMed]
28. Gaillard, C.; Dourmad, J.-Y. Application of a Precision Feeding Strategy for Gestating Sows. *Anim. Feed. Sci. Technol.* **2022**, *287*, 115280. [CrossRef]
29. Buis, R.Q.; Wey, D.; de Lange, C.F.M. 266 Development of Precision Gestation Feeding Program Using Electronic Sow Feeders and Effects on Gilt Performance. *J. Anim. Sci.* **2016**, *94*, 125–126. [CrossRef]
30. Quiniou, N. Results of 15 Years of Precision Feeding of Hyper Prolific Gestating Sows. *Animals* **2021**, *11*, 2908. [CrossRef]
31. Iida, R.; Piñeiro, C.; Koketsu, Y. Behavior, Displacement and Pregnancy Loss in Pigs under an Electronic Sow Feeder. *JAS* **2017**, *9*, 43. [CrossRef]
32. Neethirajan, S. The Role of Sensors, Big Data and Machine Learning in Modern Animal Farming. *Sens. Bio-Sens. Res.* **2020**, *29*, 100367. [CrossRef]
33. Symeonaki, E.; Arvanitis, K.G.; Piromalis, D.; Tseles, D.; Balafoutis, A.T. Ontology-Based IoT Middleware Approach for Smart Livestock Farming toward Agriculture 4.0: A Case Study for Controlling Thermal Environment in a Pig Facility. *Agronomy* **2022**, *12*, 750. [CrossRef]
34. Su, J.-J.; Ding, S.-T.; Chung, H.-C. Establishing a Smart Farm-Scale Piggery Wastewater Treatment System with the Internet of Things (IoT) Applications. *Water* **2020**, *12*, 1654. [CrossRef]
35. Karunathilake, E.M.B.M.; Le, A.T.; Heo, S.; Chung, Y.S.; Mansoor, S. The Path to Smart Farming: Innovations and Opportunities in Precision Agriculture. *Agriculture* **2023**, *13*, 1593. [CrossRef]

36. Madushanki, A.R.; Halgamuge, M.N.; Wirasagoda, W.S.; Syed, A. Adoption of the Internet of Things (IoT) in Agriculture and Smart Farming towards Urban Greening: A Review. *IJACSA* **2019**, *10*, 11–28. [CrossRef]
37. Zeng, Z.; Zeng, F.; Han, X.; Elkhouchlaa, H.; Yu, Q.; Lü, E. Real-Time Monitoring of Environmental Parameters in a Commercial Gestating Sow House Using a ZigBee-Based Wireless Sensor Network. *Appl. Sci.* **2021**, *11*, 972. [CrossRef]
38. Lee, G.; Ogata, K.; Kawasue, K.; Sakamoto, S.; Ieiri, S. Identifying-and-Counting Based Monitoring Scheme for Pigs by Integrating BLE Tags and WBLCX Antennas. *Comput. Electron. Agric.* **2022**, *198*, 107070. [CrossRef]
39. Anil, L.; Anil, S.S.; Deen, J. Relationship between Postural Behaviour and Gestation Stall Dimensions in Relation to Sow Size. *Appl. Anim. Behav. Sci.* **2002**, *77*, 173–181. [CrossRef]
40. Zhou, Q.; Sun, Q.; Wang, G.; Zhou, B.; Lu, M.; Marchant, J.N.; Yang, X.; Zhao, R. Group Housing during Gestation Affects the Behaviour of Sows and the Physiological Indices of Offspring at Weaning. *Animal* **2014**, *8*, 1162–1169. [CrossRef]
41. Liu, Y.; Zhou, S.; Cao, Y.; Xia, J.; Wu, F.; Zeng, Z. Design and testing of intelligent feeder for pregnant sows. *J. Chin. Agric. Mech.* **2021**, *42*, 183. Available online: http://zgnjhxb.niam.com.cn/CN/10.13733/j.jcam.issn.2095-5553.2021.04.26 (accessed on 28 October 2023).
42. Xiong, B.; Yang, L.; Zheng, S.; Cao, P.; Pan, X.; Wu, G. Design and Test of Precise Blanking Control System for Lactating Sows. *Trans. Chin. Soc. Agric. Eng.* **2017**, *33*, 177–182. Available online: https://www.tcsae.org/cn/article/doi/10.11975/j.issn.1002-6819.2017.20.022 (accessed on 28 October 2023).
43. Chen, C.; Liu, X.; Liu, C.; Pan, Q. Development of the Precision Feeding System for Sows via a Rule-Based Expert System. *Int. J. Agric. Biol. Eng.* **2023**, *16*, 187–198. [CrossRef]
44. Thomas, L.L.; Gonçalves, M.A.; Vier, C.M.; Goodband, R.D.; Tokach, M.D.; Dritz, S.S.; Woodworth, J.C.; DeRouchey, J.M. Lessons Learned from Managing Electronic Sow Feeders and Collecting Weights of Gestating Sows Housed on a Large Commercial Farm. *JSHAP* **2018**, *26*, 270–275. [CrossRef] [PubMed]
45. De Rensis, F.; Kirkwood, R.N. Control of Estrus and Ovulation: Fertility to Timed Insemination of Gilts and Sows. *Theriogenology* **2016**, *86*, 1460–1466. [CrossRef] [PubMed]
46. Bai, J.H.; Qin, Y.S.; Zhang, S.L.; Xu, X.L.; Song, Y.Q.; Xiao, L.L.; Feng, T.; Tian, J.H.; Liu, Y. A Comparison of the Reproductive Performance in Primiparous Sows Following Two Timed Artificial Insemination Protocols. *Animal* **2021**, *15*, 100410. [CrossRef]

applied sciences

MDPI

Article

Comparison of the Spray Effects of Air Induction Nozzles and Flat Fan Nozzles Installed on Agricultural Drones

Seung-Hwa Yu [1], Yeongho Kang [2] and Chun-Gu Lee [1,*]

1 National Academy of Agricultural Science, Rural Development Administration, Jeonju 54875, Republic of Korea; yush6210@korea.kr
2 Jeollabuk-do Agricultural Research & Extension Services, Iksan 54591, Republic of Korea; dudgh0414@korea.kr
* Correspondence: llccgg@korea.kr

Abstract: Pest control is essential for increasing agricultural production. Agricultural drones with spraying systems for pest control have generated great interest among farmers. However, spraying systems installed on unmanned aerial vehicles, like any other sprayer, can cause damage to the environment due to drift of the agent. Air induction (AI) nozzles are known to produce less drift (e.g., larger spray drops) than other nozzles, but there is a lack of research analyzing their effectiveness in combination with drones. In this study, AI and flat fan nozzles were installed on drones to evaluate their spray and pest control performance. Aerial spraying was conducted on rice and soybeans to measure the coverage and penetration ratio and analyze the crop production as well as the crop damage due to pests and diseases. The drone flight was conducted at an altitude of 3 m and a velocity of 2 m/s. Spray droplets were collected using water-sensitive paper at two heights above the soil surface. The experiments showed that the crop coverage with the AI nozzle was 130% higher than that with the flat fan nozzle. The drift reduction of AI nozzles increased the coverage of spray droplets. But the difference in the penetration ratios, which is the ratio of agents to be delivered inside the crop, was not significant between the nozzles. Also, there was no significant difference in crop yield and pest control efficacy. Consequently, the performance of the AI nozzle did not show differences from that of the XR nozzle, except for coverage. However, the AI nozzle raised less drift, so it should be considered for use in aerial control.

Keywords: agricultural drones; air induction nozzle; aerial spraying; control efficacy; coverage; penetration ratio

Citation: Yu, S.-H.; Kang, Y.; Lee, C.-G. Comparison of the Spray Effects of Air Induction Nozzles and Flat Fan Nozzles Installed on Agricultural Drones. *Appl. Sci.* **2023**, *13*, 11552. https://doi.org/10.3390/app132011552

Academic Editors: José Miguel Molina Martínez, Ginés García-Mateos and Dolores Parras-Burgos

Received: 15 September 2023
Revised: 18 October 2023
Accepted: 20 October 2023
Published: 22 October 2023

1. Introduction

The increasing global population reached 8 billion in November 2022; it is projected to reach 9.7 billion by 2050 [1]. Thus, the demand for food has also increased with the population; therefore, food production will also need to increase by up to 70% [2]. To achieve this food production goal, pests must be controlled because crop yield would substantially decrease without pest control [3]. In particular, pests and insects are becoming increasingly common as a result of global warming, necessitating pest-prevention measures [4]. However, the number of farmers is continuously decreasing, causing a farm labor shortage [5]. In the past, the main method of control was pesticide spraying through knapsack sprayers and boom sprayers, which was laborious and exposed workers to agents. To solve this problem, many researchers have studied various sprayers for effective and efficient pest control. Moreover, recently, considerable attention has been focused on the use of agricultural spraying drones [6–9].

Agricultural spraying drones comprise a system that sprays plant protection products using a remote or automatic control tool [10–12]. Plant protection products are sprayed from unmanned aerial vehicles and delivered to crops using gravity and downwash airflow from the rotor. Agricultural drones facilitate convenient operations, high work efficiency for targeted application, and high worker safety; therefore, their use by farmers is gradually increasing [13–15]. In particular, agricultural drones are widely used in East Asia [16]. For example, China has 106,000 drones for pest control, and their work area is estimated to be 64 million hectares [17]. Xiongkui et al. [18] analyzed the current pest control status using drones in China, Japan, and South Korea and presented the challenges of aerial spraying. In Europe, a few studies of spray drones in mountainous terrain have been conducted in spite of regulation [19]. Several studies focused on improving the usability of agricultural drones have been also conducted [20,21].

With the growth of the agricultural spraying drone market, there is a huge concern about the drift of pesticides and drop bounce from agricultural drone spray systems [22]. In Europe, aerial spraying is generally not permitted due to the risk of drifts [23]. In South Korea, aerial spraying has also become a problem due to revised pesticide use standards in 2019. The quantity of drift is generally associated with the ratio of fine spray droplets [24]. Depending on the size of the spray droplets, more than 70% of the volume can be drifted [25].

Many factors including flight altitude, additives, and downwash can affect the spraying performance of agricultural drones [26–28]. Various studies have been conducted to analyze droplet downwash from drones [29]. Yang et al. [30] conducted a study to build a CFD model to analyze the effects of downwinds and crosswinds on the behavior of sprayed droplets. Zhan et al. [31] researched influence of the downwash airflow of UAVs and found that the strength of the airflow is positively correlated with deposition and penetration and negatively correlated with uniformity and drift. The type of nozzle determines the diameter of the sprayed droplets and the spray angle and shape, thereby substantially impacting spraying performance [32,33]. Chen et al. [25] used four flat fan nozzles with different orifice sizes to analyze the effect of droplet size on droplet deposition and drift. They verified that the deposition, penetration of droplets, and drift distribution were affected by the droplet size. Guo et al. [34] used machine learning methods to analyze the droplet diameter and distribution pattern of the nozzles. Among the various nozzles, flat fan nozzles are widely used for unmanned aerial sprayers [35] because of a high spray efficiency [36]. In contrast to flat fan nozzles, AI nozzles produce large droplets by including air bubbles inside the droplets and reduce the drift of droplets [37–39]. Dafsari et al. [40] developed a new AI nozzle for aerial spraying, tested it, and found that it worked as designed and had a smaller spray diameter than a conventional AI nozzle. However, there is a lack of research that actually uses drones equipped with AI nozzles to conduct control and analyze their effectiveness. Therefore, research is needed to actually perform aerial spraying with AI nozzles and compare performance differences with conventional nozzles when using drones.

When comparing the performance of two nozzles, the effect of pest reduction when pest control is conducted using agricultural drones must also be investigated [41]. The control efficacy is one of the most important criteria of the spraying system. However, most previous studies have only examined the deposition of spray droplets using collectors to evaluate the physical spraying performance of agricultural drones. Furthermore, even when pest control is conducted using the same system, the penetration rate of plant protection products may vary depending on the size and shape of the crops, affecting pest control efficacy [25,42,43]. However, only a few studies have been conducted to compare the differences in the pest control efficacy of crops when using agricultural drones.

The main objective of this study was to compare the spraying performance of a flat fan nozzle and an AI nozzle when using a drone. To compare spraying performance, the deposition, penetration rate, and control efficacy have to be analyzed using water-sensitive paper. In order to compare the different results between crops, we planned to perform the experiment on soybeans and rice. Crops have to be collected after pest control to compare control efficacy by investigating the yield and damage caused by pests.

2. Materials and Methods

2.1. Experimental Site

The experiments were conducted at a farmland in Gimje, South Korea (Figure 1). The experimental field was where rice and soybeans were cultivated, and the areas for rice and soybean cultivation were 4000 m^2, respectively. Rice and soybeans are the main food crops in South Korea, and, recently, soybean cultivation in paddy fields has been increasing, so both were selected as target crops. The rice variety was Sindongjin, which was planted on 2 June with a planting density of 15 plants/m^2. The soybean variety was Daechan, which was planted on 17 May with a planting density of 8.3 plants/m^2. The spraying was conducted around 9 am to minimize the influence of wind. The weather conditions during the experiment were measured using a kestrel 5500 weather meter (KestrelMeters, Boothwyn, PA, USA). The temperature during the experiment was 20.9~25.3 °C, and the wind speed was 1~2 m/s. There were no significant obstacles that affected pest control around the experimental site.

Figure 1. View of experimental site. On the left, the field with rice cultivar 'Sindongjin', and right, Soyabean cultivar 'Daechan'.

2.2. Equipment

The agricultural drone used in the experiment was a multicopter (SG-10p, Hankook-samgong, Seoul, Repubic of Korea) with eight rotors (Figure 2). An agricultural drone was equipped with a flight controller (K++, JIYI, Shanghai, China). Maximum thrust of rotor is 5.1 kg. The detailed specifications of the agricultural spraying drone are listed in Table 1. The pump installed on the drone had a rated pressure of 0.8 MPa and a rated flow rate of 5.5 L/min. During the experiment, the chemical solution tank was filled with 10 L of the plant protection product solution before the flight. Only the two front nozzles were used to eliminate the overlapping effect of spraying the front and back nozzles. The nozzles were installed in the pipe located just below the rotor. The horizontal distance between the nozzles was 158 cm and the spray width of the drone was 4 m. The drone was operated by a professional using a remote control (T12 Data link Remote Controller, Skydroid, Quanzhou, China).

Figure 2. Agricultural spraying drone with quad rotor (SG-10P).

Table 1. Specifications of agricultural spraying drone.

Parameter	Value
Dimension (mm)	2075 (W) × 2075 (L) × 700 (H)
Total weight (kg)	14.5
Flight controller	K++
Maximum thrust (kg/rotor)	5.1
Maximum takeoff weight (kg)	24.9
Rated pressure (MPa)	0.8
Rated flow (L/min)	5.5
Capacity of solution tank (L)	10
Spray width(m)	4
Number of used nozzles	2

Conventional flat fan nozzles (XR110015VS, Teejet Technologies, Glendale Heights, IL, USA) and AI nozzles were used with drones. The XR110015VS has a flat fan-shaped spray form, and the droplet size is classified as fine on a general spraying pressure. In the AI nozzle, the sprayed droplet diameter is enlarged via air induction. The AI nozzle is less affected by wind compared with other nozzles. To develop a new AI nozzle for aerial spraying, a research on nozzle design factors was conducted and a new AI nozzle was developed based on it [40]. The developed AI nozzle has a spray rate of 0.6 LPM, an air to liquid mass ratio (ALR) of 0.0005, and a spray angle of 110 degrees. The volume median diameter (VMD) was measured based on the standardized method to identify the characteristics of the droplets sprayed from the two nozzles [44]. The VMD is an important factor in determining the amount of drift because it determines how far the droplets travel horizontally in the wind. If the VMD of the spray droplets is smaller, spray droplets can move farther away depending upon the wind speed and air inversion characteristics at the time of spraying [45]. The measurement results showed that the VMDs of the flat fan and AI nozzles were 189 and 450 μm, respectively. The specifications of the nozzles are listed in Table 2.

Table 2. Characteristics of flat fan and AI nozzles.

Nozzle	Pressure (MPa)	Spray Angle	VMD (μm)
XR110015	0.28	110	189
AI	0.28	110	450

2.3. Evaluation of Spray Coverage and Penetration

To analyze the effect of nozzle type on coverage and penetration, the agricultural spraying drone was equipped with a flat fan nozzle and an AI nozzle and sprayed with water on 6 September and 5 October 2021. The drone flew at a height of 3 m above the ground and a speed of 2 m/s during the experiment. The flight distance of single spraying was 40 m. The pressure of pump was set to 0.28 Mpa based on the nozzle recommendation [44]. The flow rate of the pump at this pressure was 1.2 L/min. The sprayed droplets were collected using water-sensitive paper (76 mm × 52 mm, Teejet Technologies, Glendale Heights, IL, USA) [46]. Since the water-sensitive paper is sensitive to moisture, the humidity was controlled below 80% by draining the water in the rice field the day before the experiment. To minimize the effects of humidity, the paper was put out immediately prior to spraying and collected a few seconds after the drone sprayed the plant protection product. Three sampling points were symmetrically arranged along the perpendicular direction of the agricultural drone flight route with three repetitions (Figure 3). The interval distance between sampling points was 2 m. The total length of the sampling points was designed for the effective spray width. The distance between the flight center line was 10 m and, while the drone was returning, the spraying stopped. The water-sensitive paper was attached to the angle-control label stand (Figure 4). The water-sensitive paper was installed at two levels: the canopy (h_1 = 80 cm) and middle point (h_2 = 40 cm) of the crop at each sampling point. The collectors were installed at three points on the flight path. After the experiment, each water-sensitive paper was sealed in a zipper bag to block contact with moisture and delivered to the laboratory. The images of the water-sensitive paper were obtained using a red, green, and blue camera and analyzed via binarization.

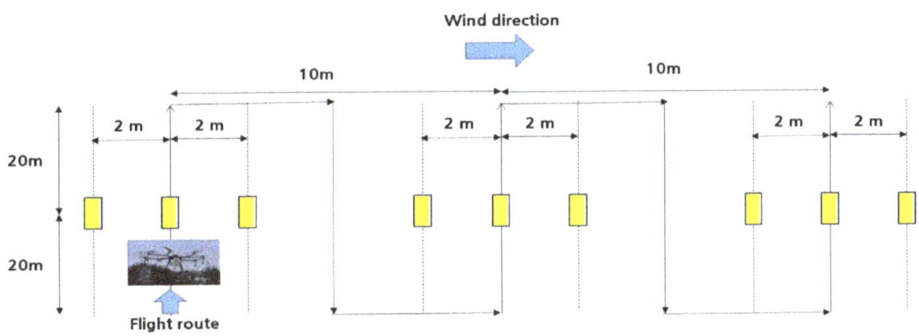

Figure 3. Layout of flight path and collectors. The distance between groups of collectors is 10 m and the drones fly perpendicular to the wind direction.

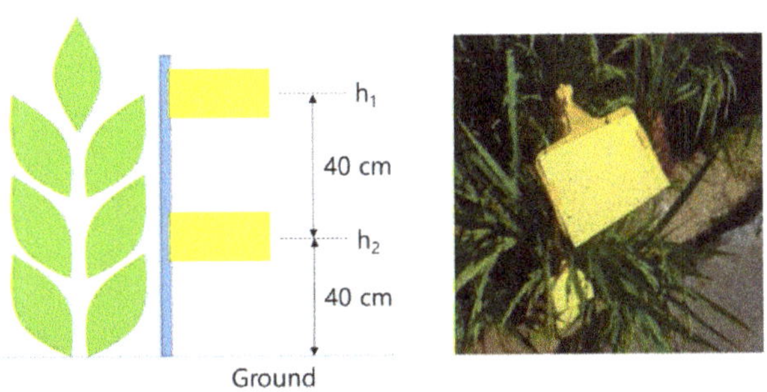

Figure 4. (**left**) Arrangement of water-sensitive paper at two height levels (40, 80 cm) and (**right**) appearance of angle-control label stand.

The coverage and penetration ratio (PR) were measured to evaluate the nozzle spray performance because the amount of agent delivered to the surface or inside the crop affects control effectiveness. Coverage refers to the ratio of the droplet deposition area on the collector to the entire collector area. The amount of deposition can be compared based on coverage (Equation (1)) [47]. The collector located at the canopy was used to analyze the coverage. The PR was measured to identify the ratio of agents to be delivered inside the crop. PR is important, especially for leafy plants, because it affects the delivery of the agent [48]. The PR was calculated using the ratio of droplet areas attached to the canopy and middle point of crops (Equation (2)). R (Version 4.2.2, The R Foundation, Vienna, Austria), a statistical analysis software, was employed to analyze the data. A two-way analysis of variance was conducted to analyze sensitive factors affecting the spray performance.

$$\mathrm{C} = \frac{A_D}{A_W} \tag{1}$$

$$\mathrm{PR} = \frac{\mathrm{C}_2}{\mathrm{C}_1} \tag{2}$$

where C is coverage, A_D is total area of droplets deposited, A_W is total area of water-sensitive paper, PR is penetration ratio, C_1 is coverage of water-sensitive paper at h_1, and C_2 is coverage of water-sensitive paper at h_2.

2.4. Efficacy Analysis

The plant protection product was sprayed three times on rice and soybeans to compare the pest control efficacy based on the nozzle type in July and August 2021. Insecticides and fungicides are shown in Table 3. The plant protection product was diluted with water in a ratio of 1:8. The amount of plant protection product sprayed in each pest control operation for each plot was 0.44 L.

Table 3. Ingredients of plant protection products.

Spraying	Date	Plant Protection Product	Dosage
		Rice	
1	16 July	20% Ferimzone + 1.5% Fluxapyroxad + 8% Etofenprox + 3% Metaflumizone	0.44 L
2	5 August	6% Azoxystrobin + 5% Validamycin A + 10% Etofenprox	0.44 L
3	12 August	8% Etofenprox + 3% Metaflumizone + 30% Ferimzone + 3% Thifluzamide	0.44 L
		Soybean	
1	5 July	8% Etofenprox + 3% Metaflumizone	0.44 L
2	5 August	6% Azoxystrobin + 5% Validamycin A + 10% Etofenprox	0.44 L
3	12 August	8% Etofenprox + 3% Metaflumizone + 13.6% Boscalid + 8% Pyraclostrobin	0.44 L

The rice and soybean fields were divided into three areas: the area of pest control using an AI nozzle, the area of pest control using a flat fan nozzle, and the nontreated area (Figure 5). The total area of 4000 m^2 for each crop was divided into areas of 1600, 1600, and 800 m^2. To reduce experimental error, the seed variety, soil condition, and cultivation method were the same between treatments. The drone worked across the plot and flew perpendicular to the boundary to minimize the impact on neighboring areas. Based on the treatment, the yield and damaged crop ratio caused by pests and diseases were investigated. In the case of rice, we targeted rice blast and rice stem borer and sheath blight, and in the case of soybeans, we targeted anthracnose, litura, and clavatus. For sheath blight, rice stem borer, anthracnose, and litura, the percentage of affected stems were examined. For rice blast, the percentage of affected ears were examined. For anthracnose and clavatus, the percentage of affected soybeans were examined. Once the plant reached full maturity, a harvest was conducted for pest inspection and production investigation. The three subplots per treatment, area of 3 m × 1.2 m, were selected and investigated by collecting crops for yield and damaged crop ratio data. Duncan's test was conducted to analyze the difference between the experimental results.

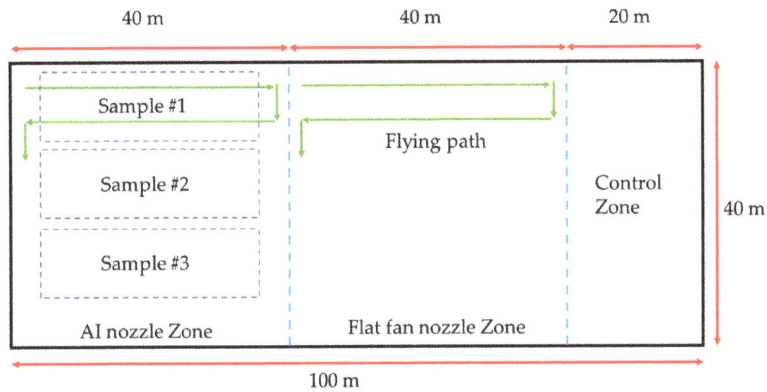

Figure 5. Layout of field for experiment treatment, sampling zone, and flying path.

3. Results and Discussion

3.1. Coverage and Penetration Ratio of Spray

The analysis of the coverage of droplet deposition by plants and nozzles showed that the coverage of the experiment with the AI nozzle (5.05 and 3.35%) was 2.5 times higher than that of the experiment with the flat fan nozzle (2.11 and 1.44%) in both plants (Figure 6). When the same amount of spray is applied using both nozzles, an increase in deposition indicates a decrease in drift. This means that in drone aerial sprayers, just like in ground-based sprayers, the use of AI nozzles reduces drift compared with the use of flat fan nozzles. The larger diameter of the droplets sprayed from the AI nozzle resulted in less drift, consistent with previous research [25]. However, Hunter et al. [49] showed that coverage decreases as the size of the droplet increases. This appears to be caused by downwash from the drone affecting the behavior of the spray droplets. For more accurate analysis, a downwash analysis of the drone should also be performed. According to two-way analysis of variance, there is significant difference only in the effect of the nozzle type ($p < 0.05$) (Table 4). This means that it is useful to use AI nozzles to deliver more spray droplets to the crop during spraying operations. Although the average deposition of rice was higher than that of soybeans, there was no significant difference by plant [50,51]. The deposition of droplets is affected by the height of the crop, size of the spray drops, and wind speed, but in this study the crops were of similar height, so there does not seem to be a difference in coverage [52].

Table 4. Two-way analysis of variance of droplet deposition.

Factor	Degree of Freedom	Sum of Squares	F	*p*-Value
Plant	1	6.15	0.764	0.3904
Nozzle	1	43.9	5.450	0.0279
Plant × Nozzle	1	1.92	0.238	0.6300

The C_2 of the AI nozzle was 1.68% in rice and 1.12% in soybean. And the C_2 of the flat fan nozzle was 0.27% in rice and 0.73% in soybean. The PR averages of the AI nozzle were (27.4 and 33.3%) and those of the flat fan nozzle were (19.0 and 31.5%), respectively (Figure 7). To analyze the effect of plant and nozzle on the PR, a two-way analysis of variance was performed, and it was found that there was no significant difference in the treatment (Table 5). According to previous research, it was expected that larger size spray droplets have a greater PR [25]. However, in this study, spray droplets applied to soybeans using the flat fan nozzle resulted in higher-than-expected C_2. After checking the experimental data, the C_2 value was measured to be higher in the downwind direction in two out of three iterations. This may have been caused by a temporary strong wind at the time of the experiment, and we believe that it needs to be improved through additional experiments in the future.

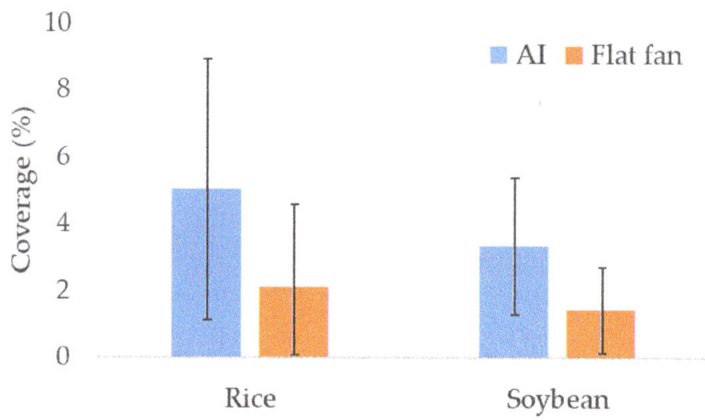

Figure 6. Coverage of droplet deposition with error bar representing standard deviation by plants and nozzles.

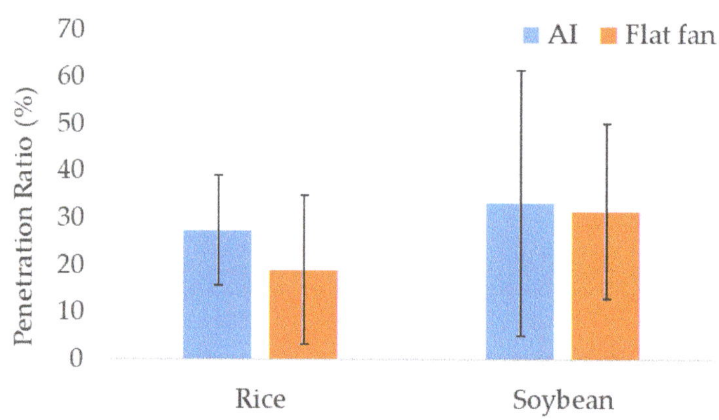

Figure 7. Penetration ratio of droplet deposition with error bar representing standard deviation by plants and nozzles.

Table 5. Two-way analysis of variance of penetration ratio.

Factor	Degree of Freedom	Sum of Squares	F	*p*-Value
Plant	1	1104	2.668	0.117
Nozzle	1	398	0.962	0.337
Plant × Nozzle	1	2	0.005	0.947

3.2. *Control Efficacy*

The crop productions at sprayed areas were higher than that of non-sprayed areas for both crops (Table 6). To compare the difference caused by the nozzle, Duncan's test was conducted to determine whether there was a difference in crop production. As a result, no significant difference was found between the experimental groups using AI and flat fan nozzles for both crops ($p < 0.05$). Based on the coverage measurements in Section 3.1, the amount of spray droplets delivered to the crop canopy was higher with the AI nozzle, but there was no difference in the PR, so there was no difference in yield. And it is considered that this is the result of trying to minimize the impact of various factors on crop yields, but not being able to fully control them. Also, the increased size of the spray droplets resulted in increased deposition with less drift, but the decreased surface area seems to have reduced the effectiveness. Further research is needed to determine the optimal spray droplet size.

Table 6. Crop production rate by treatment.

	Rice (g/m^2)	Soybean (g/m^2)
AI nozzle	465a	324a
Flat fan nozzle	449a	309a
Control	322b	240b

a,b: the same letter note indicates no significant difference at the $p = 0.05$ significant level, different letters indicate significant differences at $p = 0.05$ significant level.

To compare direct pest control efficacy, the damage ratio of rice caused by pests and diseases was investigated (Figure 8). The investigation results showed that in the case of AI nozzle treatment, the damage ratios of rice blast, rice stem borer, and sheath blight were 39.3%, 1.0%, and 6.5%, respectively; in the case of flat fan nozzle treatment, these values were 42.0%, 2.9%, and 8.0%, respectively; and for the control group, these values were 71.1%, 7.6%, and 35.0%, respectively. The damage ratios caused by three pests and diseases were reduced by 6.4%, 65.5%, and 18.7% when using AI nozzles compared with those when using flat fan nozzles. Therefore, the AI nozzle is shown to have better control efficacy, giving some indication of smaller drop drift using the conventional nozzles.

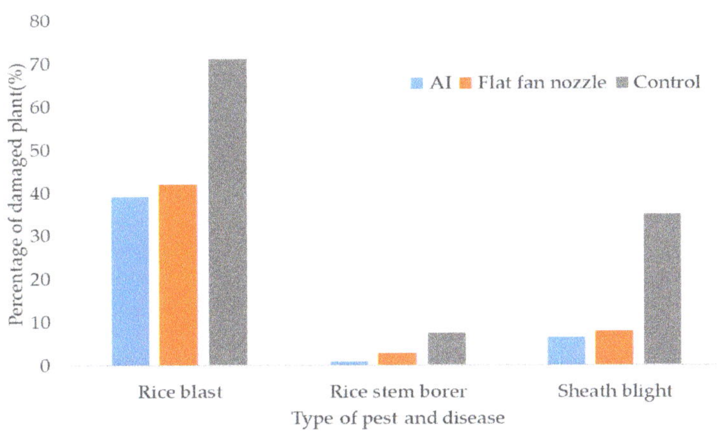

Figure 8. Control efficacy of rice after spraying by pests and diseases.

The damage ratios of soybeans caused by pests and diseases were also investigated (Table 7). The pest that causes damage to the leaves was anthracnose and the insect was litura. And the pest that damages grain was anthracnose and the insect was clavatus. It was shown that the damage caused by pests outweighed the damage caused by diseases and that the leaf damage was more severe than grain damage. For more detailed analysis, Duncan's test was conducted to determine whether there was a difference in the damage ratios. However, no significant difference was found between the experimental groups using AI and flat fan nozzles for both crops ($p < 0.05$). Similar to the results of the production investigation, there was no significant difference in the PR values between treatments, so it seems that there is no difference in control efficacy. When analyzing the effects of aerial control in the future, it is considered more accurate to examine PR values rather than canopy coverage. Wang et al. [53] conducted a study using four different types of sprayers and found that control efficacy varied depending on the type of sprayer. Further analysis is needed to determine if the no significant difference is caused by a small difference from the nozzle or other factors (e.g., wind speed).

Table 7. Control efficacy of soybean after spraying.

	Damaged Leaf Ratio		Damaged Grain Ratio	
	Disease (%)	Pest (%)	Disease (%)	Pest (%)
AI nozzle	0.5a	11.8a	2.2a	3.6a
Flat fan nozzle	0.7a	15.9a	2.4a	4.7a
Control	2.3b	49.7b	8.0b	18.4b

a,b: the same letter note indicates no significant difference at the $p = 0.05$ significant level, different letters indicate significant differences at $p = 0.05$ significant level.

4. Conclusions

The number of farmers using spraying drones is increasing owing to its various advantages. However, spray drift has been a challenge in aerial spraying systems. To address this issue, an AI nozzle with an effective drift reduction capacity was used. However, there are few research works that installed AI nozzles to drones and analyzed their effectiveness. In this study, AI and flat fan nozzles were installed on drones to evaluate their spray performance. Moreover, the pest control performance was also evaluated. The deposition and penetration ratios were measured using a flat fan nozzle and an AI nozzle. Pest control efficacy was analyzed based on the type of target crops (rice and soybean). The crops used were rice and soybean. The study results are listed as follows:

1. For both crops, coverage was more than 2.3 times higher when using AI nozzles compared with that when using flat fan nozzles. The larger size droplets of the AI nozzle apparently reduced drift, which was attributed to increased deposition of the sprayed droplets. A two-way analysis of variance showed that the effect of the nozzle on the coverage was significant ($p < 0.05$). For the effect of the plant on coverages, there was no significant difference.
2. The penetration ratios of both nozzles were analyzed. A two-way analysis of variance showed no significant difference due to the treatment method. It seems that weather conditions may have affected the results of the experiment. Previous studies of vertically growing maize, which has a similar morphology to rice, have yielded different results than this study, and additional analysis is required.
3. The crop production rate was investigated. Duncan's test showed no significant difference between the experimental groups. Environmental and genetic factors seem to have influenced production. Also, it seemed that the decreased surface area reduced the effectiveness.
4. For the control efficacy, Duncan's test showed no significant difference between the experimental groups. As the results of the production investigation, similar PR values between treatments made no difference in control efficacy.
5. When measuring aerial spray performance, penetration rates must be measured at the same time.

The analysis of the aforementioned results showed that using AI nozzles instead of flat fan nozzles for aerial pest control increased the droplet deposition of the sprayed droplets. However, there was no significant difference in production and control efficacy. Thus, this study showed that aerial control with AI nozzles delivers more spray droplets to the field and the effect on control efficacy needs further study.

In this study, a single flight condition (2 m/s forward speed at 3 m height) was used to compare the effect of nozzle type on control performance. In the future, it is necessary to conduct experiments under various flight conditions and/or to compare UAV loadings over time to drift. However, since it is difficult to control various conditions in the field, the primary analysis is performed through Computational Fluid Dynamic (CFD) simulations and supplemented with field experiments.

Author Contributions: Conceptualization, C.-G.L. and S.-H.Y.; methodology, C.-G.L. and S.-H.Y.; software, C.-G.L.; validation, C.-G.L., Y.K. and S.-H.Y.; formal analysis, C.-G.L. and Y.K.; investigation, C.-G.L. and Y.K.; resources, S.-H.Y.; data curation, C.-G.L.; writing—original draft preparation, C.-G.L.; writing—review and editing, C.-G.L. and S.-H.Y.; visualization, C.-G.L.; supervision, S.-H.Y.; project administration, S.-H.Y.; funding acquisition, S.-H.Y. All authors have read and agreed to the published version of the manuscript.

Funding: This work was carried out with the support of "Cooperative Research Program for Agriculture Science and Technology Development (Project No. PJ01557501)" Rural Development Administration, Republic of Korea.

Institutional Review Board Statement: Not applicable.

Informed Consent Statement: Not applicable.

Data Availability Statement: Not applicable.

Conflicts of Interest: The authors declare no conflict of interest.

References

1. UN. *World Population Prospects: Summary of Results*; UN DESA/POP/2022/TR/NO. 3; UN: New York, NY, USA, 2022.
2. Hunter, M.C.; Smith, R.G.; Schipanski, M.E.; Atwood, L.W.; Mortensen, D.A. Agriculture in 2050: Recalibrating Targets for Sustainable Intensification. *BioScience* **2017**, *67*, 386–391. [CrossRef]
3. Oerke, E.C. Crop losses to pests. *J. Agric. Sci.* **2005**, *144*, 31–43. [CrossRef]
4. Skendžić, S.; Zovko, M.; Živković, I.P.; Lešić, V.; Lemić, D. The Impact of Climate Change on Agricultural Insect Pests. *Insects* **2021**, *12*, 440. [CrossRef]
5. FAO. *World Food and Agriculture–Statistical Yearbook 2022*; FAO: Rome, Italy, 2022.
6. Giles, D.K.; Akesson, N.B.; Yates, W.E. Pesticide Application Technology: Research and Development and the Growth of the Industry. *Trans. ASABE* **2008**, *51*, 397–403. [CrossRef]
7. OECD. *State of the Knowledge—Literature Review on Unmanned Aerial Spray Systems in Agriculture*; OECD: Paris, France, 2021.
8. Hanif, A.S.; Han, X.; Yu, S.-H. Independent Control Spraying System for UAV-Based Precise Variable Sprayer: A Review. *Drones* **2022**, *6*, 383. [CrossRef]
9. Koo, Y. Nozzle Pressure Response Characteristics of Variable Rate System for Unmanned Aerial Applications. *J. Biosyst. Eng.* **2019**, *44*, 1–11. [CrossRef]
10. Chen, P.; Douzals, J.P.; Lan, Y.; Cotteux, E.; Delpuech, X.; Pouxviel, G.; Zhan, Y. Characteristics of unmanned aerial spraying systems and related spray drift: A review. *Front. Plant Sci.* **2022**, *13*, 870956. [CrossRef]
11. Kim, J.; Kim, S.; Ju, C.; Son, H.I. Unmanned Aerial Vehicles in Agriculture: A Review of Perspective of Platform, Control, and Applications. *IEEE Access* **2019**, *7*, 105100–105115. [CrossRef]
12. *DIS 23117-1*; Agricultural and Forestry Machinery—Unmanned Aerial Spraying Systems—Environmental Requirements. ISO: Geneva, Switzerland, 2021.
13. Chen, H.; Lan, Y.; Fritz, B.K.; Hoffmann, W.C.; Liu, S. Review of agricultural spraying technologies for plant protection using unmanned aerial vehicle (UAV). *Int. J. Agric. Biol. Eng.* **2021**, *14*, 38–49. [CrossRef]
14. Tsouros, D.C.; Bibi, S.; Sarigiannidis, P.G. A Review on UAV-Based Applications for Precision Agriculture. *Information* **2019**, *10*, 349. [CrossRef]
15. Morales-Rodríguez, P.A.; Cano Cano, E.; Villena, J.; López-Perales, J.A. A Comparison between Conventional Sprayers and New UAV Sprayers: A Study Case of Vineyards and Olives in Extremadura (Spain). *Agronomy* **2022**, *12*, 1307. [CrossRef]
16. Wang, L.; Huang, X.; Li, W.; Yan, K.; Han, Y.; Zhang, Y.; Pawlowski, L.; Lan, Y. Progress in Agricultural Unmanned Aerial Vehicles (UAVs) Applied in China and Prospects for Poland. *Agriculture* **2022**, *12*, 397. [CrossRef]
17. Zhang, Y.; Huang, X.; Lan, Y.; Wang, L.; Lu, X.; Yan, K.; Deng, J.; Zeng, W. Development and Prospect of UAV-Based Aerial Electrostatic Spray Technology in China. *Appl. Sci.* **2021**, *11*, 4071. [CrossRef]
18. Xiongkui, H.; Bonds, J.; Herbst, A.; Langenakens, J. Recent development of unmanned aerial vehicle for plant protection in East Asia. *Int. J. Agric. Biol. Eng.* **2017**, *10*, 18–30.
19. Sarri, D.; Martelloni, L.; Rimediotti, M.; Lisci, R.; Lombardo, S.; Vieri, M. Testing a multi-rotor unmanned aerial vehicle for spray application in high slope terraced vineyard. *J. Agric. Eng.* **2019**, *50*, 38–47. [CrossRef]

20. Hussain, S.; Cheema, M.J.; Arshad, M.; Chatta, a.; Latif, M.; Ashraf, S.; Siddique, S. Spray uniformity testing of unmanned aerial spraying system for precise agro-chemical applications. *Pak. J. Agric. Sci.* **2019**, *56*, 897–903.
21. Lee, C.-G.; Yu, S.-H.; Rhee, J.-Y. Effects of Unmanned Aerial Spray System Flight Altitude and Collector Height on Spray Deposition Measured Using a Food Dye Tracer. *Agriculture* **2023**, *13*, 96. [CrossRef]
22. Kirk, I.W. Aerial Spray Drift from Different Formulations of Glyphosate. *Trans. ASABE* **2000**, *43*, 555–559. [CrossRef]
23. *DIRECTIVE 2009/128/EC*; Establishing a Framework for Community Action to Achieve the Sustainable Use of Pesticides. EU: Brussel, Belgium, 2009; pp. 71–86.
24. Hilz, E.; Vermeer, A.W.P. Spray drift review: The extent to which a formulation can contribute to spray drift reduction. *Crop. Prot.* **2013**, *44*, 75–83. [CrossRef]
25. Chen, S.; Lan, Y.; Zhou, Z.; Ouyang, F.; Wang, G.; Huang, X.; Deng, X.; Cheng, S. Effect of Droplet Size Parameters on Droplet Deposition and Drift of Aerial Spraying by Using Plant Protection UAV. *Agronomy* **2020**, *10*, 195. [CrossRef]
26. Zhang, X.-Q.; Song, X.-P.; Liang, Y.-J.; Qin, Z.-Q.; Zhang, B.-Q.; Wei, J.-J.; Li, Y.-R.; Wu, J.-M. Effects of Spray Parameters of Drone on the Droplet Deposition in Sugarcane Canopy. *Sugar Tech* **2020**, *22*, 583–588. [CrossRef]
27. Shouji, C.; Yu, S.-H.; Kang, Y.-H.; Choi, Y.; Dafsari, R.A.; Lee, J. Experimental Analysis of the Downwash Airflow Created by a Single Rotor Blade in Agricultural Drones. *J. Biosyst. Eng.* **2021**, *46*, 317–331. [CrossRef]
28. Wang, X.; He, X.; Song, J.; Wang, Z.; Wang, C.; Wang, S.; Wu, R.; Meng, Y. Drift potential of UAV with adjuvants in aerial applications. *Int. J. Agric. Biol. Eng.* **2018**, *11*, 54–58. [CrossRef]
29. Wang, L.; Xu, M.; Hou, Q.; Wang, Z.; Lan, Y.; Wang, S. Numerical verification on influence of multi-feature parameters to the downwash airflow field and operation effect of a six-rotor agricultural UAV in flight. *Comput. Electron. Agric.* **2021**, *190*, 106425. [CrossRef]
30. Yang, F.; Xue, X.; Cai, C.; Sun, Z.; Zhou, Q. Numerical Simulation and Analysis on Spray Drift Movement of Multirotor Plant Protection Unmanned Aerial Vehicle. *Energies* **2018**, *11*, 2399. [CrossRef]
31. Zhan, Y.; Chen, P.; Xu, W.; Chen, S.; Han, Y.; Lan, Y.; Wang, G. Influence of the downwash airflow distribution characteristics of a plant protection UAV on spray deposit distribution. *Biosyst. Eng.* **2022**, *216*, 32–45. [CrossRef]
32. Yu, S.-H.; Yun, Y.-T.; Choi, Y.; Dafsari, R.A.; Lee, J. Effect of Injection Angle on Drift Potential Reduction in Pesticide Injection Nozzle Spray Applied in Domestic Agricultural Drones. *J. Biosyst. Eng.* **2021**, *46*, 129–138. [CrossRef]
33. Kirk, I.W. Measurement and Prediction of Atomization Parameters from Fixed-Wing Aircraft Spray Nozzles. *Trans. ASABE* **2007**, *50*, 693–703. [CrossRef]
34. Guo, H.; Zhou, J.; Liu, F.; He, Y.; Huang, H.; Wang, H. Application of Machine Learning Method to Quantitatively Evaluate the Droplet Size and Deposition Distribution of the UAV Spray Nozzle. *Appl. Sci.* **2020**, *10*, 1759. [CrossRef]
35. Cavalaris, C.; Karamoutis, C.; Markinos, A. Efficacy of cotton harvest aids applications with unmanned aerial vehicles (UAV) and ground-based field sprayers—A case study comparison. *Smart Agric. Technol.* **2022**, *2*, 100047. [CrossRef]
36. Technologies, T. *Catalog 51A-M*; TeeJet: Glendale Heights, IL, USA, 2014.
37. Post, S.L. Drift of Droplets from Air-Induction Nozzles. *Trans. ASABE* **2019**, *62*, 1683–1687. [CrossRef]
38. Matthews, G.A. Developments in application technology. *Environmentalist* **2008**, *28*, 19–24. [CrossRef]
39. Guler, H.; Zhu, H.; Ozkan, H.E.; Derksen, R.C.; Yu, Y.; Krause, C.R. Spray Characteristics and Drift Reduction Potential with Air Induction and Conventional Flat-Fan Nozzles. *Trans. ASABE* **2007**, *50*, 745–754. [CrossRef]
40. Dafsari, R.A.; Khaleghi, M.; Yu, S.-H.; Choi, Y.; Lee, J. Design and Performance Evaluation of Air Induction (AI) Nozzles to Reduce Drift Potential for Agricultural Unmanned Aerial Vehicles (UAVs). *J. Biosyst. Eng.* **2021**, *46*, 462–473. [CrossRef]
41. Lou, Z.; Xin, F.; Han, X.; Lan, Y.; Duan, T.; Fu, W. Effect of Unmanned Aerial Vehicle Flight Height on Droplet Distribution, Drift and Control of Cotton Aphids and Spider Mites. *Agronomy* **2018**, *8*, 187. [CrossRef]
42. Zhu, H.; Dorner, J.W.; Rowland, D.L.; Derksen, R.C.; Ozkan, H.E. Spray Penetration into Peanut Canopies with Hydraulic Nozzle Tips. *Biosyst. Eng.* **2004**, *87*, 275–283. [CrossRef]
43. Zhu, H.; Brazee, R.D.; Fox, R.D.; Derksen, R.C.; Ozkan, H.E. Development of a Canopy Opener to Improve Spray Deposition and Coverage Inside Soybean Canopies: Part 1. Mathematical Models to Assist Opener Development. *Trans. ASABE* **2008**, *51*, 1905–1912. [CrossRef]
44. *ANSI/ASAE S572.3*; Spray Nozzle Classification by Droplet Spectra. ASABE: St. Joseph, MI, USA, 2020.
45. Hofman, V.; Solseng, E. *Spray Equipment and Calibration*; North Dakota State University: Fargo, ND, USA, 2004.
46. Turner, C.R.; Huntington, K.A. The use of a water sensitive dye for the detection and assessment of small spray droplets. *J. Agric. Eng. Res.* **1970**, *15*, 385–387. [CrossRef]
47. Cerruto, E.; Manetto, G.; Longo, D.; Failla, S.; Papa, R. A model to estimate the spray deposit by simulated water sensitive papers. *Crop. Prot.* **2019**, *124*, 104861. [CrossRef]
48. Negrisoli, M.M.; Raetano, C.G.; Souza, D.M.d.; e Souza, F.M.S.; Bernardes, L.M.; Del Bem Junior, L.; Rodrigues, D.M.; Sartori, M.M.P. Performance of new flat fan nozzle design in spray deposition, penetration and control of soybean rust. *Eur. J. Plant Pathol.* **2019**, *155*, 755–767. [CrossRef]
49. Hunter, J.E.; Gannon, T.W.; Richardson, R.J.; Yelverton, F.H.; Leon, R.G. Coverage and drift potential associated with nozzle and speed selection for herbicide applications using an unmanned aerial sprayer. *Weed Technol.* **2020**, *34*, 235–240. [CrossRef]
50. Zhang, P.; Deng, L.; Lyu, Q.; He, S.; Yi, S.; Liu, Y.; Yu, Y.; Pan, H. Effects of citrus tree-shape and spraying height of small unmanned aerial vehicle on droplet distribution. *Int. J. Agric. Biol. Eng.* **2016**, *9*, 45–52. [CrossRef]

51. Smith, D.B.; Askew, S.D.; Morris, W.H.; Shaw, D.R.; Boyette, M. Drolet Size and Leaf Morphology Effects on Pesticide Spray Deposition. *Trans. ASAE* **2000**, *43*, 255–259. [CrossRef]
52. Zheng, Y.; Yang, S.; Zhao, C.; Chen, L.; Lan, Y.; Yu, T. Modelling operation parameters of UAV on spray effects at different growth stages of corns. *Int. J. Agric. Biol. Eng.* **2017**, *10*, 57–66.
53. Wang, G.; Lan, Y.; Yuan, H.; Qi, H.; Chen, P.; Ouyang, F.; Han, Y. Comparison of Spray Deposition, Control Efficacy on Wheat Aphids and Working Efficiency in the Wheat Field of the Unmanned Aerial Vehicle with Boom Sprayer and Two Conventional Knapsack Sprayers. *Appl. Sci.* **2019**, *9*, 218. [CrossRef]

applied sciences

MDPI

Article

Determination of the Dependences of the Nutritional Value of Corn Silage and Photoluminescent Properties

Dmitriy Y. Pavkin, Mikhail V. Belyakov, Evgeniy A. Nikitin *, Igor Y. Efremenkov and Ilya A. Golyshkov

Federal Scientific Agroengineering Center VIM, 109428 Moscow, Russia; dimqaqa@mail.ru (D.Y.P.); bmw20100@mail.ru (M.V.B.); efremenkovigor55@mail.ru (I.Y.E.); golyshkovila@gmail.com (I.A.G.)
* Correspondence: evgeniy.nicks@yandex.ru; Tel.: +7-9251-186-506

Abstract: This article examines existing optical methods for the diagnostics of food and feed products used in agriculture to determine their nutritional value or detect maximum permissible indicators. Among the most common feeds used for cattle, corn silage is considered. Its nutritional value depends on many external factors that need to be taken into account when formulating feeding rations. This article is dedicated to assessing the prospects of using visible-range photoluminescence for determining dry matter content, total protein content, and NDF (neutral detergent fiber) using a portable device in field conditions without lengthy sample preparation. This research aims to develop a laboratory device and establish the theoretical foundations for determining the nutritional value of agricultural feeds using photoluminescence. The study revealed that the most indicative range for measuring nutritional corn silage is to use excitation via radiation with a wavelength of about 362 nm. At the same time, the luminescent radiation flux must be measured in a range of 440–620 nm. Moreover, R^2 values greater than 0.8 were achieved in correlation after constructing luminescence relationships only for the determination of dry matter content/moisture, total protein content, and NDF. This indicates the potential use of the proposed method for determining these parameters.

Keywords: corn silage; photoluminescence; dry matter; protein; NDF; regression models

Citation: Pavkin, D.Y.; Belyakov, M.V.; Nikitin, E.A.; Efremenkov, I.Y.; Golyshkov, I.A. Determination of the Dependences of the Nutritional Value of Corn Silage and Photoluminescent Properties. *Appl. Sci.* **2023**, *13*, 10444. https://doi.org/10.3390/app131810444

Academic Editors: José Miguel Molina Martínez, Ginés García-Mateos and Dolores Parras-Burgos

Received: 28 August 2023
Revised: 13 September 2023
Accepted: 14 September 2023
Published: 18 September 2023

1. Introduction

Modern livestock complexes for the maintenance of large cattle specialized in milk and beef production are enterprises with a high level of automation of technological processes. However, some of these processes cannot be executed with 100% efficiency without human intervention [1]. One such process is the feeding of animals, which involves a complex of preparatory measures for calculating and justifying the composition of feed components and the energy value of the mixed diet [2].

In general, when forming the feeding ration for animals, the farmer adheres to a strictly rational approach from the perspective of milk or beef productivity, as well as the preservation of animal health, ensuring productive longevity [3]. Achieving high levels of animal efficiency in terms of milk and beef production is determined by the feedstock of the animals, the balance of the diet, and the energy value of the feed [4].

The nutritional or energy value of feeds used for animal feeding is typically determined through preliminary analysis, where a key indicator is the dry matter content. This indicator determines the amount of nutrients that the animal will receive by consuming a specific component of the feed mixture [5].

The dry matter content in plant feeds for agricultural animals is measured using gravimetric, dielcometric, and optical methods.

The gravimetric method, most common in countries with low investment attractiveness in the agricultural sector, particularly in terms of technological provision, involves the sampling of the analyzed feed, weighing, and then recording the initial mass indicator of the taken feed portion. Subsequently, the feed portion undergoes drying by placing it in a dehydrator or a household drying chamber using special containers. It is processed until the mass of the processed feed portion ceases to decrease (indicating complete drying of the sample). By calculating the proportion, the percentage of lost moisture is determined, and conclusions are drawn about the level of dry matter/moisture of the investigated feed.

This described method is most frequently used by farmers to determine dry matter indicators on the farm without laboratory services. However, it does not provide entirely accurate results due to the fact that complete drying of the feed to 100% is impossible when interacting with the surrounding environment. Part of the feed may adhere to the used utensils and absorb moisture from the surroundings. The significant disadvantages of this method include it being a fire hazard, as well as the labor-intensive sample preparation (up to 30 min per sample) [1,6].

Among commercial devices, the dielcometric method has gained the widest prevalence, which is based on the correlation between the dielectric permittivity of a material and its moisture content. Instruments that operate based on the dielcometric method are designed as a rod, one end of which is placed into the tested sample, while the measuring part is located at the other end. The moisture assessment results can be obtained within 20 s. However, a significant drawback of the method is that the tested sample must be tightly compressed, which is why the method is most effective for determining the moisture content of wood, feed compacted in a trench, or gathered into briquettes using special presses. Otherwise, the device introduces a high level of error [1,7].

Optical methods represent the most promising direction, with broad developmental potential compared to the methods mentioned above. Instruments based on optics are more costly in terms of development and manufacturing. However, the potential of the hardware aspect offers significantly wider possibilities in terms of the number of determinable parameters. For example, devices released to the market, such as Aurora NIR (Mayville, WI, USA) and Dinamica Generale (Poggio Rusco, Italy), enable not only the determination of dry matter content but also protein, ash, and fiber. The principle on which the operation of such devices is based is near-infrared spectroscopy [1,8].

However, concerning the non-contact determination of ash, ADF, and NDF cellulose concentrations, as well as fats, it is expedient to conduct research into the effectiveness of applying an optical method in the visible range from 380 to 780 nm [9].

According to the conducted literature review, the photoluminescence of the visible range demonstrates its effectiveness in terms of detecting okadaic acid (OA), a marine biotoxin produced by microalgae that poses a significant threat to mariculture, seafood safety, and human health. The results of processing experimental data showed that the reliability of detection using luminescence is confirmed by a coefficient of variation of 2.54% when employing alternative methods [10].

Luminescence is employed to identify the adulteration of butter by detecting non-dairy fats through the dependence of the intensity ratios of components in the extended spectrum of luminescence on palm oil content when excited at a wavelength of λ = 266 nm [11]. Additionally, it is used for the quantitative determination of fat content in milk [12].

The broader diagnostic functionality of plant products, including those that have undergone extensive processing, is provided by luminescent nanoprobes, enabling the diagnosis of heavy metal content, pesticides, and the presence of veterinary drugs, microbes, and mycotoxins [13].

Photoluminescence is a useful technique for the non-destructive and quick evaluation of cereals and other starchy products. Visible light peaking at around λ = 460 nm is observed from cereals, such as rice, wheat, barley, millet, flour, corn starch, and peanut, under the illumination of ultraviolet light at λ = 365 nm [14].

The photoluminescence method demonstrates its effectiveness in detecting riboflavin and its derivative flavin mononucleotide in cellulose materials [15], which also suggests the potential for identifying important vitamins in livestock feed.

Moreover, photoluminescence has been actively utilized in the industrial processing of food products and the preparation of semi-finished goods. For instance, in the range of 550 nm, myoglobin in poultry meat actively fluoresces, allowing the degree of poultry cooking to be determined through its detection [16].

Similarly, for the detection of vitamin B2 (riboflavin) in cellulose materials, its concentration is determined in milk at peak intensity values of $\lambda = 545$ nm, by measuring the configuration of backscattering using ultraviolet LEDs, violet LEDs, or a blue LED as the excitation light [17].

In confirmation of the efficiency of detecting qualitative feed indicators and ensuring safety prior to direct animal consumption, photoluminescence enables the visualization of serum albumin within the range of 480 to 535 nm [18].

The most commonly used mathematical method for processing optical signals obtained from photodiodes is correlation–regression analysis, often accompanied by the pre-training of the hardware algorithm. In this process, preliminary results obtained through chemical analysis are utilized as the reference parameter for the corresponding sample.

In the scope of this present study, we examined the prospect of utilizing the photoluminescence method as a tool for determining the dry matter content in livestock feed. To test the key hypotheses, a prototype device equipped with LEDs and photodiodes calibrated to a specific range of luminescence was developed for corn silage and compound feed.

The goal of the presented research is to develop a laboratory apparatus and establish theoretical foundations for determining the nutritional value of agricultural feeds using photoluminescence.

2. Materials and Methods

Study of Spectral Characteristics of Feed Mix Components

The physical phenomenon serving as a basis for the offered method consists of measuring the image reflection ability of a forage mix surface layer.

At the preliminary laboratory study stage, it was necessary to determine spectral characteristics of components of feeding mixes as exemplified by a cattle feed ration. This type of ration can contain natural-origin forages (grass green mass, corn silage, alfalfa haylage) and concentrated mixed fodder (consisting of grain mash, corn, and barley as rape meal and sunflower meal).

The study aimed to determine the optical properties and establish correlation spectral dependencies of the most commonly used components in cattle feed mixtures. The investigated materials included corn silage. The exploration of optical properties was carried out using a spectrofluorimeter CM 2203 (Lumex, St. Petersburg, Russia) (Figure 1), through which the most indicative ranges of photoluminescence for corn silage were determined.

Based on the conducted measurements, it was revealed that the excitation of photoluminescence in corn silage occurs from 220 nm to 550 nm. The most representative range of photoluminescence for corn silage was found to be from 350 nm to 365 nm, where the strongest light signal absorption occurred. This allowed for the use of light-emitting diodes and photodiodes tuned to the corresponding frequency in the portable device prototype.

Preliminary measurements from the spectrofluorimeter, which determined the most representative measurement range of 350–365 nm, shaped the concept of constructing a portable optical device. As there is a high probability of external light exposure within this range, the decision was made to eliminate external light exposure by creating a light-permeable casing with a retractable case.

For the utmost representativeness of optical measurement results, the internal part containing the optical module was colored in a matte black shade to eliminate reflections.

The external appearance of the portable optical device for determining the photoluminescence of agricultural feeds is depicted in Figure 2.

Figure 1. Spectrofluorimeter CM 2203 for determining the indicative ranges of photoluminescence in agricultural feed.

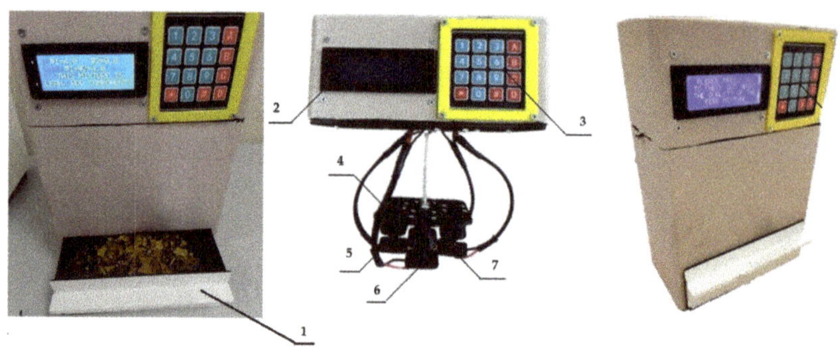

Figure 2. Portable optical device for determining the nutritional value of corn silage through photoluminescence. 1—feed bed, 2—display, 3—interface control unit, 4—console, 5—photodiodes, 6—light-emitting diode, 7—adjusting screw.

Since the luminescent emission from silage has extremely low intensity, signal amplifiers are necessary for its detection alongside the photodiode. Operational amplifiers AD820ANZ (Analog Devices, Shanghai, China) are used as amplifiers. Once the signal passes through the operational amplifier, it is processed via the microcontroller ATmega328P (Atmel Corporation, Shenzhen, China) and subsequently displayed on the LCD 2004 (Winstar Display, Taiwan, China). The laboratory analyzer is powered by batteries with a nominal capacity of 2200 mAh. Device control is carried out through the keyboard (Figure 2, position 3).

The key indicator that determines the quantity of nutrients in feed is the dry matter content or its inverse value (humidity). Based on these values, farmers determine the agrotechnical deadlines for plant processing or the timing for starting feed harvesting. Additionally, it involves modeling a multi-component diet, considering the balance between total dry matter consumption and moisture level. This is especially important for livestock groups during peak lactation periods.

The algorithm for determining the dry matter content using a portable optical device based on photoluminescence is presented in Figure 3.

Figure 3. Algorithm for determining nutritional value indicators using an optical device.

The main hypothesis of the presented study is to determine the nutritional value of agricultural feed using an optical method that does not require preliminary sample preparation. This will ensure the construction of a device for conducting measurements in the field. To do this, we measured photoluminescence fluxes using a laboratory sample according to the following equations.

Sampling of corn silage from the storage was carried out using a manual sampler; then, vacuum sealing of the material was carried out. Up to 59 measurements were carried out to construct correlation dependencies for each indicator, while the values with the largest spread from the average were entered in Table 1.

Table 1. Photo of the voltage and parameters of the silo at different humidities.

U, mV	20.27	15.88	24.93	22.61	23.83	32.38	34.21	40.08
Humidity w, %	74.22	71.59	66.02	60.01	59.08	54.07	47.27	44.98
Starch C, %	27.41	27.21	28.78	34.5	33.66	33.81	34.7	35.1
Protein P_t, %	7.35	7.61	7.23	7.04	7.24	6.78	6.80	6.67
ADF C_{ADF}, %	28.05	28.99	24.62	22.82	22.24	21.53	21.00	20.50
NDF C_{NDF}, %	47.06	47.50	45.32	40.52	38.96	40.01	38.66	38.02
Ash A, %	5.69	5.20	4.71	3.98	3.99	3.76	3.55	3.31
Raw fat F_t, %	3.60	3.16	3.16	3.26	3.00	3.04	2.99	2.90

The integral value of photoluminescence flux is determined using Formula (1).

$$\Phi = \int_{\lambda_1}^{\lambda_2} \varphi_l(\lambda) d\lambda \quad (1)$$

$\varphi_l(\lambda)$—photoluminescence spectra;
λ_1–λ_2—limits of the operating spectral range of photoluminescence.
In this case, photonic signals are determined according to Formula (2).

$$U = U_g - U_d \quad (2)$$

U_g—total voltage on the photodiode;
U_d—dark voltage.
Formula (3) is expressed through the photonic signal.

$$U = S_U \cdot \Phi \quad (3)$$

S_U—voltage sensitivity of the photodiode.

Further, by using the provided mathematical data processing algorithm and the developed original photoluminescence device, measurements of corn silage were carried out with the construction of correlation–spectral dependencies concerning the dry matter content.

The research was conducted in June 2023 on the basis of the Federal Agroengineering Center VIM, while the samples of corn silage were grown on the basis of the agricultural firm JSC Zelenogradskoye (Moscow region) and laid for storage in 2022. Sampling was carried out using a special sampler, while the packaging of the corn silo sample was vacuum- and light-tight.

3. Results

Among the three previously identified excitation maxima in plant organisms (362 nm, 424 nm, 485 nm) (PHP23), the silage absorption spectrum shows the most pronounced peak at 362 nm. The peak at 424 nm is more than twice as small and shifted to the short-wavelength region. The peak at 485 nm is not evident in the spectrum. Peaks in the long-wavelength region above 550 nm are instrument noise and do not induce luminescence, as experimentally verified by the authors. Furthermore, as evident from Figure 4, the excitation and, consequently, photoluminescence are very weak, necessitating work with integral parameters.

The spectral excitation characteristic of the silage is presented in Figure 4.

Earlier, we found that in plant feeds with maximum excitation (362 nm, 424 nm, 485 nm) [19], in the excitation (absorption) spectrum of the silos, the maximum 362 nm is most pronounced. The maximum of 424 nm is more than twice as small and shifted to the shortwave region. The maximum of 485 nm does not appear in the spectrum. Bursts in the long-wavelength region above 550 nm are the noise of the device and do not excite luminescence, which was verified experimentally by the authors. Also, from Figure 4, it can be seen that the excitation and, consequently, photoluminescence are very weak; therefore, it is necessary to work with integral parameters.

In Table 1, we entered data characterizing the largest variation in the measurement process in the amount of eight values for each indicator. At the same time, the experiment assumed a repetition of measurements up to 59 times for each sample.

Based on the array of obtained values, dependencies of the photo signal on silage moisture content were constructed, along with their approximations, as presented in Figure 5.

The dependence $U(w)$ can be statistically and reliably approximated using a linear function with a coefficient of determination R^2 = 0.8522.

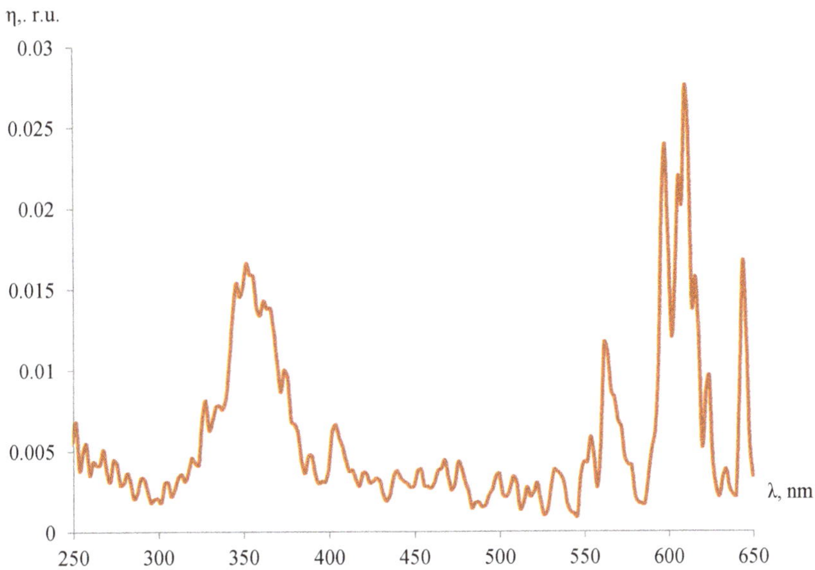

Figure 4. Spectral characteristic of silo excitation.

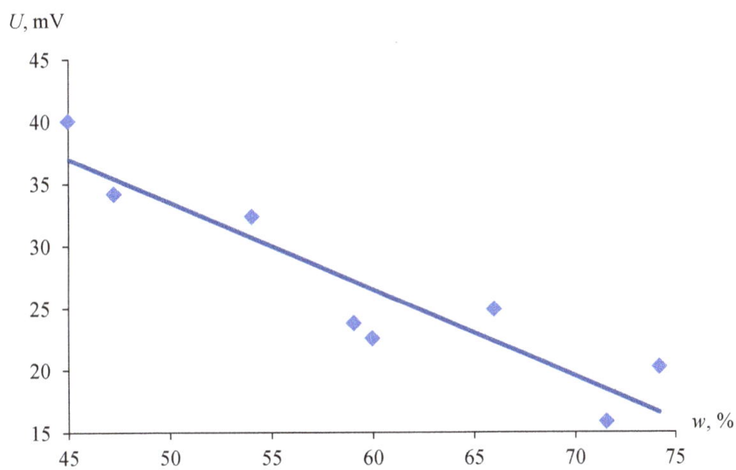

Figure 5. Dependence of photon voltage on silage moisture and its linear approximation.

For silage with varying moisture content, the change in the photoluminescence intensity is caused by the quenching of fluorescence due to changes in the concentration of mechanically bound (free) water in the near-surface tissues. The quenching of fluorescence is caused by collisions of excited atoms with their surrounding unexcited particles. As a result of the collisions, the excitation energy transfers into the kinetic energy of the colliding particles or into the excitation energy of the partner. The efficiency of the quenching depends on the collision frequency with the excited atom and the probability of quenching during collisions. Quenching becomes pronounced at high concentrations when the free path length of particles is small and the collision frequency is high. The quenching effect is particularly strong in condensed media. The quenching of fluorescence as a result of the collisions of the excited particles is accompanied by a reduction in the average lifetime of the excited state. The nature of the particles colliding with excited atoms significantly influences the efficiency of quenching.

Static quenching is caused by the formation of a non-luminescent product as a result of the interaction between the fluorophore and the quencher (water). Since static quenching is related to the formation of non-luminescent complexes Z between the fluorophore molecule L and the quencher molecule Q, the quenching process can be described using Formula (4).

$$L + Q = Z \tag{4}$$

If the absorption of the fluorophore and the complex is the same, then the following expression can be written (Formula (5)).

$$\frac{\eta}{\eta_Q} = 1 + \beta_y Q \tag{5}$$

η—luminescence yield in the absence of quencher;
η_Q—luminescence yield in the presence of quencher;
β_y—stability constant of the non-luminescent complex;
Q—concentration of quencher molecules.

In this case, the value of Q is proportional to the moisture content of the silage, w.

In addition to determining the dry matter content in corn silage, this research aimed to test the hypothesis regarding the determination of other nutritional value indicators using photoluminescence. These indicators also hold significant importance in the management of feeding for cattle and other animals, such as protein content, ADF and NDF cellulose content, ash content, etc.

The investigated sample of corn silage on the developed device was also subjected to the determination of starch concentration using a chemical method. However, in the process of searching for correlation dependencies between the intensity of photoluminescence and starch content in corn silage, no relationships were identified (Figure 6).

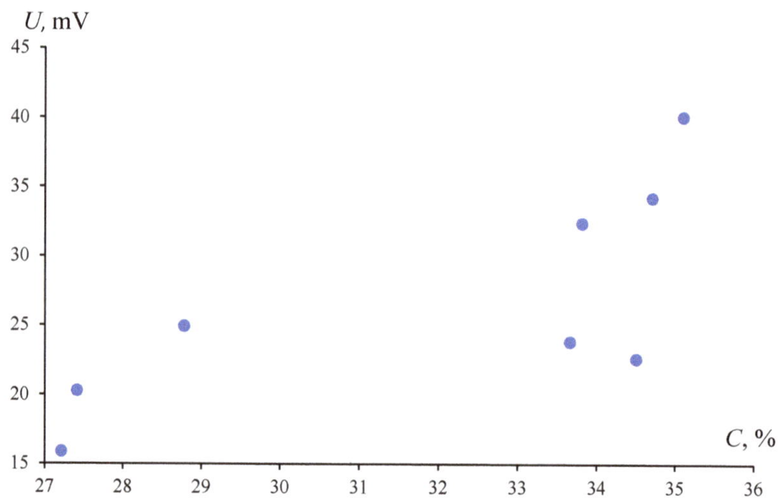

Figure 6. Dependence of photon voltage on starch content in corn silage and its linear approximation.

The relationship between photon voltage and starch content in the silage cannot be statistically and reliably approximated due to the fact that the value of $R^2 < 0.8$.

However, for the determination of protein concentration in corn silage, photoluminescence proves to be a promising method, as evidenced by the identified dependencies with $R^2 > 0.8$ (Figure 7).

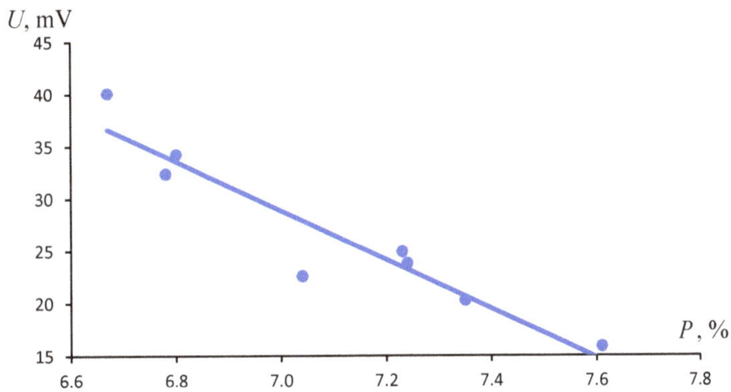

Figure 7. The dependence of the photon voltage on the protein content in the silo and its linear approximation.

The dependence of $U(P)$ can also be statistically significantly approximated using a linear function with a coefficient of determination $R^2 = 0.8945$, where P is the protein content.

Furthermore, during the course of the research, we investigated the correlation of photoluminescent signals with the content of ADF in corn silage (Figure 8).

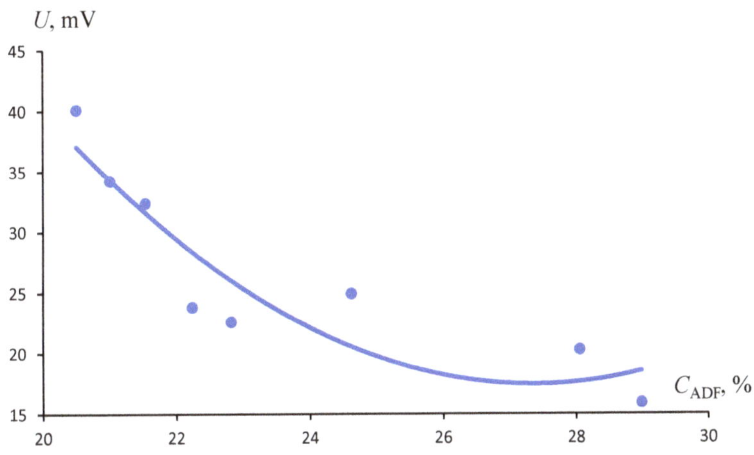

Figure 8. The dependence of the photon voltage on the ADF content in the silo and its approximation using a second-order polynomial.

The relationship of U to C_{ADF} is statistically significant and is confirmed by the fact that $R^2 > 0.8$; however, the values can only be approximated by a second-order polynomial, as $R^2 = 0.8346$ (Figure 9).

During the laboratory research, we also analyzed the content of ash, taking into account the variation in photon voltage in the corn silage (Figure 10).

As a result of processing the experimental data, the value of $R^2 < 0.8$ indicates the absence of the possibility to statistically and reliably approximate the relationship between the photon voltage and the ash content in corn silage.

Furthermore, during the literature analysis, sources referring to the potential determination of fat content in various materials were explored. However, concerning the feeding of large ruminants and the formulation of rations, it is crucial to assess the total fat content. Through photoluminescence analysis and statistical data processing, no correlation relationships were detected, as evidenced by the graph (Figure 11).

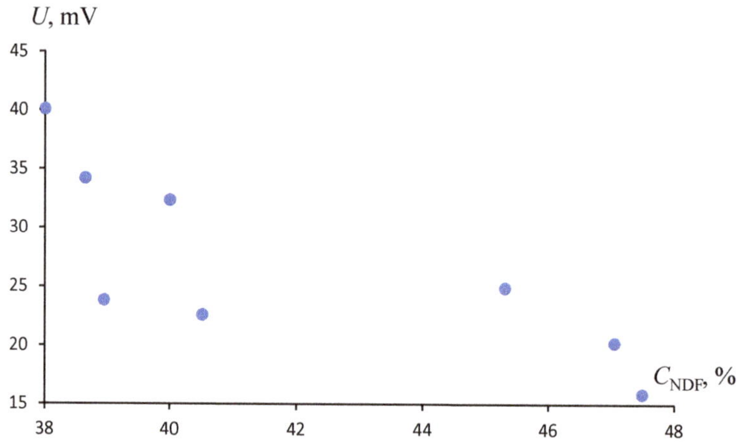

Figure 9. Dependence of the photon voltage of corn silage on the NDF content.

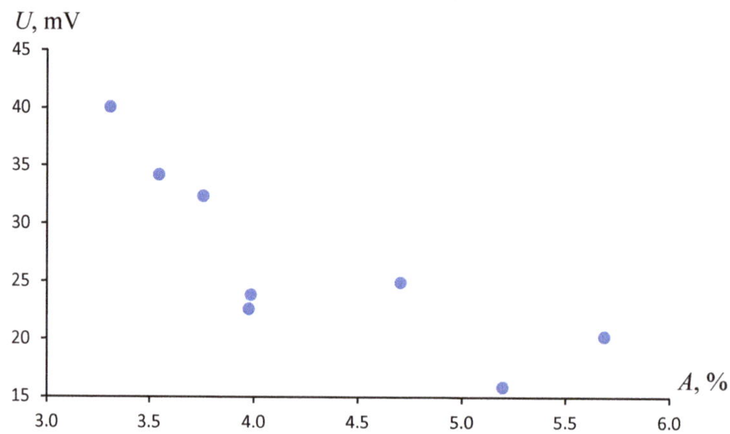

Figure 10. Dependence of photon voltage on the ash content in corn silage.

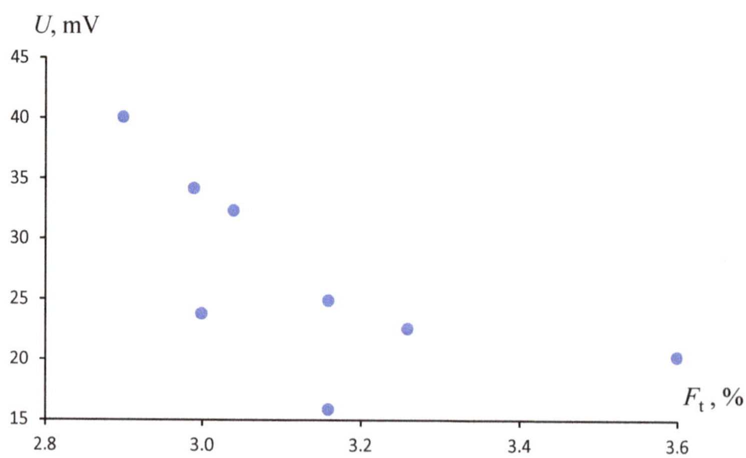

Figure 11. Dependence of the photon voltage on the total fat content in the corn silo.

The relationship between photon voltage and raw fat content in silage cannot be statistically and reliably approximated ($R^2 < 0.8$).

Based on the obtained results for three silage parameters (moisture content, protein content, and ADF content), calibration equations for the developed laboratory instrument can be derived.

For dry matter/moisture content, see Equation (6):

$$w = -1.22U + 92.35 \quad (6)$$

For total protein content, see Equation (7):

$$P_t = -0.038U + 8.11 \quad (7)$$

For ADF content in corn silage, see Equation (8):

$$C_{ADF} = 0.019U^2 - 1.42U + 46.90 \quad (8)$$

4. Discussion

We previously revealed that the foundation of the automation process and the elimination of human factors in agriculture are typically built upon optical technologies. These technologies enable the automatic positioning of manipulators in milking robots using LiDAR or TOF cameras, measurements of milking intensity in cows using LED sensors, and the assessment of milk quality through the interpretation of color characteristics [20–23].

Feeding processes are equally important factors that shape the profitability of the industry. For farmers, it is crucial to make quick managerial decisions regarding adjustments to feeding rations or crop harvesting. The proposed solution for the optical express analysis of feed crops, exemplified by corn silage, allows farmers to determine the nutritional value of the feed used in-field conditions without the involvement of specialized laboratories.

In the explored literature and commercial device brochures, the use of NIR analysis methods for the nutritional value of feeds in agriculture is noted. Our presented research contributes to testing the hypothesis of using photoluminescence to determine nutritional value, providing corresponding correlation dependencies of photoluminescence concerning total protein content, moisture, ADF, and other indicators [24–26].

A similar spectral method, but based on reflectivity data, was used to detect Fusarium pepper disease [27]. Unfortunately, the creation of a device implementing the method was not reported.

The device proposed in this study, unlike analogs, does not need to analyze the structure of volatile organic compounds [28] and does not require the construction of an image [29,30].

We found that among the existing studies, the most common are manuscripts that reveal the essence of spectroscopy in the near-infrared range (700–1400 nm) as a tool for determining the nutritional value of agricultural feed. At the same time, the essence of the method is to register the reflecting signals associated with the vibrational movements of the molecules of the surface layer of the biomaterial that is being analyzed. Also, the NIR method requires the use of an expensive spectrometer for recording the reflecting signal and hardware with software for processing spectral data.

We proposed a method related to the effect of photoluminescence in the visible range and confirmed our hypothesis with the results of the experimental studies. At the same time, the practical value of photoluminescence as a tool for determining the nutritional value of feed is the low cost of electronic components (diode and photodiode) compared to a spectrometer and a light source for implementing this task in the near-infrared range.

5. Conclusions

1. To measure the nutritional value of corn silage, it is advisable to use excitation via radiation with a wavelength of about 362 nm. At the same time, the luminescent radiation flux must be measured in the range of 440–620 nm.

2. Based on the preliminary measurements, a component base was selected for the original photoluminescence device, where the photodiodes and light-emitting diodes were tuned to the indicative photoluminescence range of corn silage.
3. Through the development and fabrication of an experimental photoluminescence device, measurements of corn silage were performed to determine its dry matter content, total protein content, ash, ADF, and NDF using the physical method of photoluminescence. Dependencies were identified that allowed for the assertion that photoluminescence can be used to determine the dry matter content/moisture, total protein content, and ADF content in corn silage. These results are supported by statistically significant indicators ($R^2 < 0.8$) that can be approximated.
4. In order to conduct further research and improve the results, we plan to modernize the laboratory sample of the device, which will allow for measuring nutritional values with greater reliability in narrow ranges of visible radiation.

6. Patents

1. Patent No. RU 2775170 C1, IPC G06T 7/60, G06T 7/90, A01K 5/02. A system for assessing the quality of animal feed/Pavkin D.Y. Nikitin E.A. Kiryushin I.A./Patent holder: Federal Scientific Agroengineering Center VIM. Application: 2021129058, dated 5 October 2021.
2. Patent No. RU 2781751 C1, IPC G01N 21/25. Portable spectral meter of feed quality indicators/Pavkin D.Y. Lednev V.N. Nikitin E.A. Pershin S.M. Grishin M.Y. Sdvighenskiy P.A./Patent holder: Federal Scientific Agroengineering Center VIM. Application: 2021129056, dated 5 October 2021.
3. Patent No. RU 2021663154, Program for automated recognition of feed mixture components for farm animals/Pavkin D.Y. Kiryushin I.A. Nikitin E.A. Vladimirov F.E. Yurochka S.S./Patent holder: Federal Scientific Agroengineering Center VIM. Application: 2021662352, dated 9 August 2021.

Author Contributions: Conceptualization, M.V.B.; methodology, E.A.N.; software, I.Y.E.; validation, D.Y.P. and I.A.G.; formal analysis, D.Y.P.; investigation, E.A.N.; resources, D.Y.P. and E.A.N.; data curation, I.A.G. and I.Y.E.; writing—original draft preparation, M.V.B. and E.A.N.; writing—review and editing, E.A.N. and M.V.B.; visualization, I.Y.E.; supervision, D.Y.P.; project administration, E.A.N.; funding acquisition, D.Y.P. All authors have read and agreed to the published version of the manuscript.

Funding: This work was supported by a grant from the Ministry of Science and Higher Education of the Russian Federation for large scientific projects in priority areas of scientific and technological development (grant number 075-15-2020-774).

Institutional Review Board Statement: The animal study protocol was approved by the Ethics Committee of Federal Scientific Agroengineering Center VIM (protocol code 321 and date of approval 15 July 2022) for studies involving animals.

Data Availability Statement: The data presented in this study are available on request from the corresponding author. The data are not publicly available due to the terms of the contract under which the study was funded.

Conflicts of Interest: The authors declare no conflict of interest.

References

1. Pavkin, D.Y.; Nikitin, E.A.; Shilin, D.V.; Belyakov, M.V.; Golyshkov, I.A.; Mikhailichenko, S.; Chepurina, E. Development Results of a Cross-Platform Positioning System for a Robotics Feed System at a Dairy Cattle Complex. *Agriculture* **2023**, *13*, 1422. [CrossRef]
2. Belyakov, M.V. Photoluminescent Sensor of Scarification Efficiency of Fodder Plants' Seeds. *Sensors* **2023**, *23*, 106. [CrossRef]
3. Nikitin, E.A.; Pavkin, D.Y.; Izmailov, A.Y.; Aksenov, A.G. Assessing the homogeneity of forage mixtures using an RGB camera as exemplified by cattle rations. *Appl. Sci.* **2022**, *12*, 3230. [CrossRef]
4. Belyakov, M.V.; Pavkin, D.Y.; Nikitin, E.A.; Efremenkov, I.Y. Substantiation of the choice of spectral ranges of photoluminescent control of the composition and nutritional value of feed. *Mach. Equip. Rural. Area* **2023**, *2*, 31–36. [CrossRef]

5. Belyakov, M.V.; Nikitin, E.A.; Efremenkov, I.Y. Efficiency of the Photoluminescent Method for Monitoring the Homogeneity of Feed Mixtures in Animal Husbandry. *Agric. Mach. Technol.* **2022**, *16*, 55–61. [CrossRef]
6. Buelvas, R.M.; Adamchuk, V.I.; Leksono, E.; Tikasz, P.; Lefsrud, M.; Holoszkiewicz, J. Biomass estimation from canopy measurements for leafy vegetables based on ultrasonic and laser sensors. *Comput. Electron. Agric.* **2019**, *164*, 104896. [CrossRef]
7. Giménez-Gallego, J.; González-Teruel, J.D.; Soto-Valles, F.; Jiménez-Buendía, M.; Navarro-Hellín, H.; Torres-Sánchez, R. Intelligent thermal image-based sensor for affordable measurement of crop canopy temperature. *Comput. Electron. Agric.* **2021**, *188*, 106319. [CrossRef]
8. Michelucci, U.; Venturini, F. Multi-Task Learning for Multi-Dimensional Regression: Application to Luminescence Sensing. *Appl. Sci.* **2019**, *9*, 4748. [CrossRef]
9. Qin, Y.; Li, J.; Kuang, J.; Shen, S.; Zhou, X.; Zhao, X.; Huang, B.; Han, B. Okadaic Acid Detection through a Rapid and Sensitive Amplified Luminescent Proximity Homogeneous Assay. *Toxins* **2023**, *15*, 501. [CrossRef]
10. Chaikov, L.L.; Kirichenko, M.N.; Krivokhizha, S.V.; Kupov, M.R.; Lobanov, A.N.; Sverbil, P.P.; Nagaev, E.I.; Shkirin, A.V. Two Convenient Methods for Detection of Non-Dairy Fats in Butter by Dynamic Light Scattering and Luminescence Spectroscopy. *Appl. Sci.* **2023**, *13*, 8563. [CrossRef]
11. Yuan, H.; Li, Y.; Lv, J.; An, Y.; Guan, D.; Liu, J.; Tu, C.; Wang, X.; Zhou, H. Recent Advances in Fluorescent Nanoprobes for Food Safety Detection. *Molecules* **2023**, *28*, 5604. [CrossRef] [PubMed]
12. Iyer, S.N.; Behary, N.; Nierstrasz, V. Study of photoluminescence property on cellulosic fabric using multifunctional biomaterials riboflavin and its derivative Flavin mononucleotide. *Sci. Rep.* **2019**, *9*, 8696. [CrossRef] [PubMed]
13. Liu, S.; Liu, C.; Ren, Y.; Kuang, S.; Zhou, C.; Cai, H.; Yao, Z. Fluorescence spectra of nutrients in chicken and skin under baking conditions. *Optik* **2020**, *218*, 164795. [CrossRef]
14. Katsumata, T.; Aizawa, H.; Komuro, S.; Ito, S.; Matsumoto, T. Evaluation of photoluminescence from milk with various vitamin B2 concentrations. *Int. Dairy J.* **2020**, *109*, 104744. [CrossRef]
15. Kim, S.-B.; Kamiya, G.; Furuta, T.; Kitada, N.; Maki, S.A. Coelenterazine Indicators for the Specific Imaging of Human and Bovine Serum Albumins. *Sensors* **2023**, *23*, 6020. [CrossRef] [PubMed]
16. Moskovsky, M.N.; Shogenov, Y.H.; Lavrov, A.V.; Gulyaev, A.A.; Efremenkov, I.Y.; Pyatchenkov, D.S.; Belyakov, M.V. Spectral Photoluminescent Parameters of Barley Seeds (*Hordéum vulgáre*) Infected with Fusarium ssp. *Photochem. Photobiol.* **2023**, *99*, 29–34. [CrossRef]
17. Xu, H.Y.; Jiang, G.Y.; Yu, M.; Luo, T. A Color Image Watermarking Based on Tensor Analysis. *IEEE Access* **2018**, *6*, 51500–51514. [CrossRef]
18. Sun, X.; Zhao, L. RGB Pixel Brightness Characteristics of Linked Color Imaging in Early Gastric Cancer: A Pilot Study. *Gastroenterol. Res. Pract.* **2020**, *2020*, 2105874. [CrossRef]
19. Lukac, R.; Plataniotis, K.N. Color filter arrays: Design and performance analysis. *IEEE Trans. Consum. Electron.* **2005**, *51*, 1260–1267. [CrossRef]
20. Burmistrov, D.E.; Ignatenko, D.N.; Lednev, V.N.; Gudkov, S.V.; Pavkin, D.Y.; Khakimov, A.R.; Nikitin, E.A.; Lobachevsky, Y.P.; Zvyagin, A.V. Application of optical quality control technologies in the dairy industry: An overview. *Photonics* **2021**, *8*, 551. [CrossRef]
21. Halachmi, I.; Ben Meir, Y.; Miron, J.; Maltz, E. Feeding behavior improves prediction of dairy cow voluntary feed intake but cannot serve as the sole indicator. *Animal* **2016**, *10*, 1501–1506. [CrossRef] [PubMed]
22. Ding, K.; Xiao, L.; Weng, G. Active contours driven by region-scalable fitting and optimized Laplacian of Gaussian energy for image segmentation. *Signal Process.* **2017**, *134*, 224–233. [CrossRef]
23. Jin, R.; Weng, G. Active contours driven by adaptive functions and fuzzy c-means energy for fast image segmentation. *Signal Process.* **2019**, *163*, 1–10. [CrossRef]
24. Sun, H.; Xing, Z.Z.; Zhang, Z.Y.; Ma, X.Y.; Long, Y.W.; Liu, N.; Li, M.Z. Visualization Analysis of Crop Spectral Index Based on RGB-NIR Image Matching. *Spectrosc. Spectr. Anal.* **2019**, *39*, 3493–3500. [CrossRef]
25. Manokhina, A.A.; Dorokhov, A.S.; Kobozeva, T.P.; Fomina, T.N.; Starovoitov, V.I. Jerusalem Artichoke as a Strategic Crop for Solving Food Problems. *Agronomy* **2022**, *12*, 465. [CrossRef]
26. Alshehawy, A.M.; Mansour, D.-E.A.; Ghali, M.; Lehtonen, M.; Darwish, M.M.F. Photoluminescence Spectroscopy Measurements for Effective Condition Assessment of Transformer Insulating Oil. *Processes* **2021**, *9*, 732. [CrossRef]
27. Karadağ, K.; Tenekeci, M.E.; Taşaltın, R.; Bilgili, A. Detection of pepper fusarium disease using machine learning algorithms based on spectral reflectance. *Sustain. Comput. Inform. Syst.* **2020**, *28*, 100299. [CrossRef]
28. Sanjay, M.; Kalpana, B. Early Mass Diagnosis of Fusarium Wilt in Banana Cultivations using an E-Nose Integrated Autonomous Rover System. *Int. J. Appl. Sci. Biotechnol.* **2017**, *5*, 261–266. [CrossRef]
29. Zhang, D.; Wang, D.; Du, S.; Huang, L.; Zhao, H.; Liang, D.; Gu, C.; Yang, X. A Rapidly Diagnosis and Application System of Fusarium Head Blight Based on Smartphone. In Proceedings of the 8th International Conference on Agro-Geoinformatics (Agro-Geoinformatics), Istanbul, Turkey, 16–19 July 2019; pp. 1–5. [CrossRef]
30. Zhang, D.; Wang, Z.; Jin, N.; Gu, C.; Chen, Y.; Huang, Y. Evaluation of Efficacy of Fungicides for Control of Wheat Fusarium Head Blight Based on Digital Imaging. *IEEE Access* **2020**, *8*, 109876–109890. [CrossRef]

applied sciences

Article

A Miniaturized and Low-Cost Near-Infrared Spectroscopy Measurement System for Alfalfa Quality Control

Candela Melendreras [1], Ana Soldado [1,*], José M. Costa-Fernández [1], Alberto López [2] and Francisco Ferrero [2,*]

[1] Department of Physical and Analytical Chemistry, University of Oviedo, 33006 Oviedo, Spain; melendrerascandela@uniovi.es (C.M.); jcostafe@uniovi.es (J.M.C.-F.)
[2] Department of Electrical, Electronic, Computers and Systems Engineering, University of Oviedo, 33204 Gijón, Spain; uo181549@uniovi.es
* Correspondence: soldadoana@uniovi.es (A.S.); ferrero@uniovi.es (F.F.); Tel.: +34-985-103-583 (A.S.); +34-985-182-552 (F.F.)

Abstract: Food safety and quality are the first steps in the food chain. This work reports a miniaturized, low-cost, and easy-to-use near-infrared spectroscopy (NIRS) measurement system for alfalfa quality control. This is a significant challenge for dairy farm technicians and producers who need rapid and reliable knowledge of the forage quality on their farms. In most cases, the instrumentation suitable for these specifications is expensive and difficult to operate. The core of the proposed NIR spectroscopy measurement system is Texas Instruments' NIRscan Nano evaluation module (EVM) spectrometer. This module has a large sensing area and high resolution, suitable for forage samples. To evaluate the feasibility of the prototype for analyzing agrifood samples, different ways of presenting the sample, intact or ground, were tested. The final objective of the research is not just to check the efficiency of the proposed system. It is also to determine the characteristics of the measurement system, and how to improve them for alfalfa quality control.

Keywords: agrifood quality control; digital micromirror device (DMD); forage; near-infrared spectroscopy (NIRS)

Citation: Melendreras, C.; Soldado, A.; Costa-Fernández, J.M.; López, A.; Ferrero, F. A Miniaturized and Low-Cost Near-Infrared Spectroscopy Measurement System for Alfalfa Quality Control. *Appl. Sci.* **2023**, *13*, 9290. https://doi.org/10.3390/app13169290

Academic Editors: José Miguel Molina Martínez, Ginés García-Mateos and Dolores Parras-Burgos

Received: 6 July 2023
Revised: 11 August 2023
Accepted: 14 August 2023
Published: 16 August 2023

1. Introduction

The nutritional value of animal feed is essential to quality, safe feed consumption, and animal welfare [1]. In addition to this fact, and due to the great variability in the raw materials used to feed animals, it is necessary to develop strategies focused on the tight control of animal feed products. These should be combined with the research and development of new, simple, economical, and robust methods for monitoring quality and safety parameters [2,3].

Forage is one of the main feed products in animal husbandry and must, therefore, be subject to safety and quality controls. Among the most important quality parameters for forages, the following three can be highlighted [4]. The fiber content is mainly provided by the fodder cell wall, and represents its carbohydrate. The mineral content (ash) gives information about possible contamination with soil, and supplies micronutrients to the diet. It also provides information on the quality of the forage. The third parameter is the protein content, which is extremely important in animal production.

Several important books on NIR spectroscopy are currently available, but some of them are not up to date. Ozaki et al. [5] report a new state-of-the-art textbook on NIR spectroscopy, covering its principles, spectral analysis and data treatments, instrumentation, and applications. In [6], Huck et al. review the fundamental principles of NIRS, and its applicability, regulatory issues, advantages, and limitations in natural product research.

Near-infrared spectroscopy (NIRS) techniques have always been valued and used in food analysis and quality control, due to the speed of analysis, the simplicity of sampling, the non-invasive nature of the techniques, and the possibility of their being implemented in the production line. In this review [7], Shenk et al. introduce scientific and technical reports using the NIRS to evaluate food, agriculture, and forest products.

In addition to the advantages of this technique, the possibility of developing miniaturized NIR systems that are easy to use, and specialized in the quality control of the raw forages used in animal feed, makes it possible to increase quality control (sampling). Thus, Cherney et al. [8] evaluate several hand-held NIR instruments for the precision and accuracy of the currently available calibrations for dry-matter and forage nutritive value. Crocombe [9] outlines the technologies used in portable spectroscopy, and Beć et al. [10] discuss the characteristics of miniaturized NIR sensors in comparison to benchtop laboratory spectrometers.

The use of easy-to-use portable NIRS instrumentation minimizes time losses, because nonspecialized personnel are able to analyse samples on site, and a real-time response is achieved as soon as the analysis is carried out. These characteristics are some of those required in precision farming [11].

Several miniaturized spectrophotometers have been developed and marketed in the last decade, some of which are extremely small, light, and inexpensive. By allowing measurements in the field, at the point of delivery, production, sale, purchase, and use, these spectrometers transform NIRS technology. Table 1 lists some commercial miniaturized NIRSs, and presents their main characteristics, as found on the websites of the manufacturers.

Table 1. Main specifications of some commercial miniaturized NIRSs.

Manufacturer	Model	Technology	Spectral Resolution (nm)	Spectral Range (nm)	Size (mm)	Prize (USD)
Texas Instruments	NIRscan Nano EVM	Grating–MEMS DMD	10	900–1700	58×62 × 36	1000
VIAVI Solutions	MicroNIR 1700	LVF–Linear	12–20	950–1650	50 × 45	14,500
Si-Ware Systems	NeoSpectra	MEMS–FT	8–16	1250–1700	178 × 91 × 62	6950
Consumer Physics	SCiO	X	X	740–1070	67.7 × 40.02 × 18.8	3395
Ocean Optics	Flame NIR	Grating	~10.0	970–1700	89.1 × 63.3 × 34.4	9926
SouthNest Technology	nanoFTIR NIR	MEMS Michelson	6	1000–2600	143 × 49 × 28	X
Spectral Engines	NIRONE S1.7	Fabry–Pérot	13–17	1350–1550	25 × 25 × 17.5	8940

X: No information is provided on the manufacturer's website.

This work evaluates the feasibility of a miniaturized and low-cost NIRS measurement system for alfalfa quality control. The core of this system is a NIR spectrometer based on the Texas Instruments NIRscan Nano evaluation module (EVM). It has a large sensing area and high resolution to analyse forage samples. To evaluate the feasibility of the prototype, different ways of presenting the sample, intact (raw) or ground, were studied. This equipment has already been tested for use in liquid sample analysis [12], with a specific cuvette for liquid samples. For this purpose, alfalfa samples were analysed. Alfalfa is one of the main forages used to feed animals, due to its high biomass production, and protein and fiber contents. In summary, this work contributes:

- to confirming the efficiency of the proposed NIRS measurement system;
- to identifying the qualities of the equipment; and
- to looking for aspects to improve and implement in the future.

The remainder of this paper is organized as follows: Section 2 reports the materials and methods for analysing dairy farm forage quality. Section 3 presents the results and discussion. Finally, Section 4 contains conclusions and future works.

2. Materials and Methods

2.1. NIRscan Nano Evaluation Module

The core of the optoelectronic measurement system is the Texas Instruments NIRscan Nano evaluation module (USD 1000 approx.) [13]. The block diagram of this module for reflectance measurements is shown in Figure 1. A slit collects and concentrate diffuse reflections, by illuminating samples at an angle that eliminates specular reflections. Light passing through the slit is collimated, low-pass filtered, and then dispersed into its constituent wavelengths via a reflective grating. Each wavelength is represented by a separate image created by the focusing lens of the digital micromirror device (DMD).

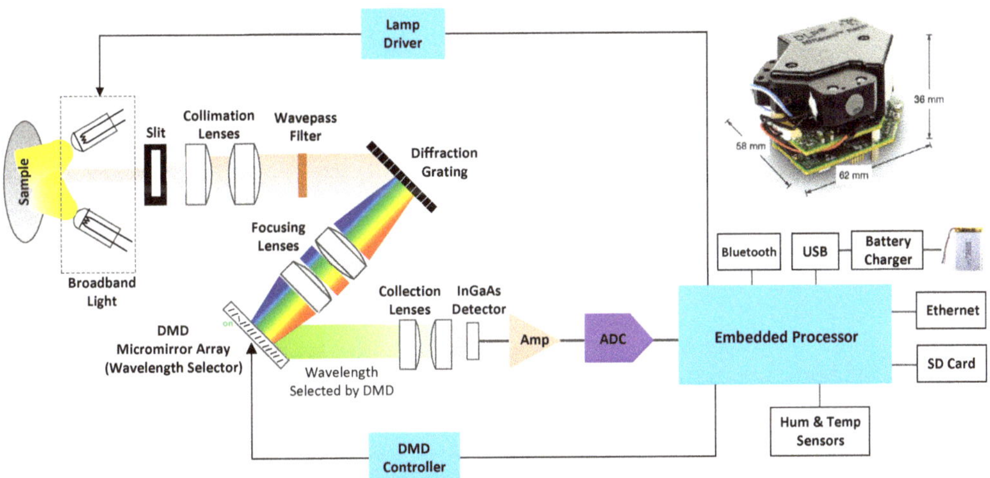

Figure 1. Block diagram and image (upper right corner) of the Texas Instruments DLP NIRscan Nano EVM.

The DMD is controlled by the embedded processor, which turns only certain mirrors on and off at certain times. The width of the DMD columns selected as "on" determines the amount of light directed to the photodetector, as well as the resolution of the system. The DMD columns selected as "off" divert unselected wavelengths away from the photodetector's optical path, to prevent interference. By doing this, they can achieve high signal-to-noise ratios (SNRs). An array detector cannot take advantage of the adaptive scanning techniques that can be performed using this type of architecture. Light energy is collected and concentrated, by the collection lenses, onto the single-point InGaAs photodetector. Analog-to-digital converters (ADCs) convert photodetector signals to digital values through transimpedance amplifiers.

2.2. NIRS Measurement System

The instrument sample window can be enlarged if raw samples with larger particle sizes are analyzed. This will produce reproducible spectra. However, this is difficult when measuring samples directly, as the sample window is very small (10 mm × 10 mm). This problem was overcome through the attaching of a semicircular sample holder to the spectrometer, as shown in Figure 2a. To ensure homogenous results for the sample, 10 measurements are made at each holder position. Figure 2b shows the microcontroller (LOLIN ESP32, USD 10) attached to the spectrometer module that drives the servomotor

(MG90S, USD 10). This servo motor is powered by a 3.7 V to DC–DC converter (Pololu U1V0F5, USD 6). When the start button is pressed, the servo motor rotates, and stops at different positions to take measurements. After rotating 180°, the holder returns to its original position, and waits for another measurement to take place. When the load button is pressed, the servo motor rotates 90° to an intermediate position, making loading and unloading easier. When neither of these two actions occur, the microcontroller enters sleep mode.

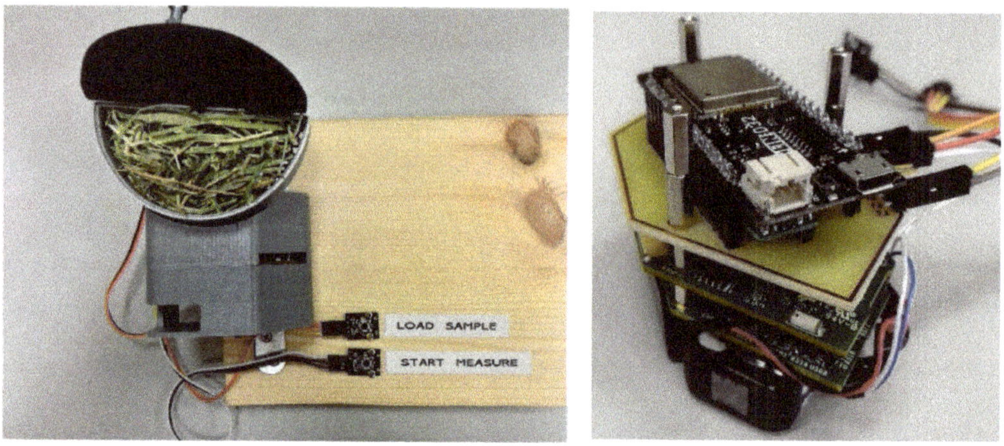

(a) (b)

Figure 2. Measurement system with (**a**) a whole sample, and (**b**) a microcontroller board attached over the spectrometer module.

2.3. Forage Samples

For this study, a calibration set of 57 samples of hay or dehydrated alfalfa collected in the north of Spain was utilized. Approximately 200 g of alfalfa was collected during the sampling procedure, to be analyzed using miniaturized NIRs. Prior to the collection of the raw scans, the samples were homogenized. After that, the alfalfa samples were milled, using a domestic spice mill (which is cheap and easy to use), and re-scanned in their ground form. This type of mill does not allow mesh-size setting. The variability associated with this factor appears in the collected spectra, and also in the chemometric results. This is because all spectroscopic information is considered when developing calibrations.

When the NIR analysis was complete, the quality of the alfalfa samples, based on animal feed requirements, was characterized, according to their nutritive value parameters. This was done using laboratory reference procedures. A Van Soest analysis [14] was performed on the neutral detergent fiber (NDF), a gravimetric analysis was conducted on the mineral content (MC), and a Kjeldahl analysis was conducted on the crude protein (CP). The CP and NDF are two of the critical parameters for classifying alfalfa quality. CP values of 19% or higher, combined with crude fiber values of 29% or lower, are related to good forage quality [15]. Table 2 summarizes the statistics for the nutritive parameters of all the samples involved in this study. It includes the range and variability in each analyzed parameter. Appendix A includes the individual values for each alfalfa sample included in this study.

Table 2. Statistics for the nutritive value of the alfalfa samples (N = 57).

Parameter (%)	Mean	Min.	Max.	SD
NDF	40.70	29.24	61.01	5.50
MC	10.64	8.47	13.71	0.91
CP	14.48	7.19	17.01	1.70

NDF: neutral detergent fiber; MC: mineral content; CP: crude protein.

2.4. Spectral Acquisition

Spectral acquisition requires a scan configuration (see Figure 3). Texas Instruments provides two different scan configurations within the "Factory" settings: "Column" and "Hadamard".

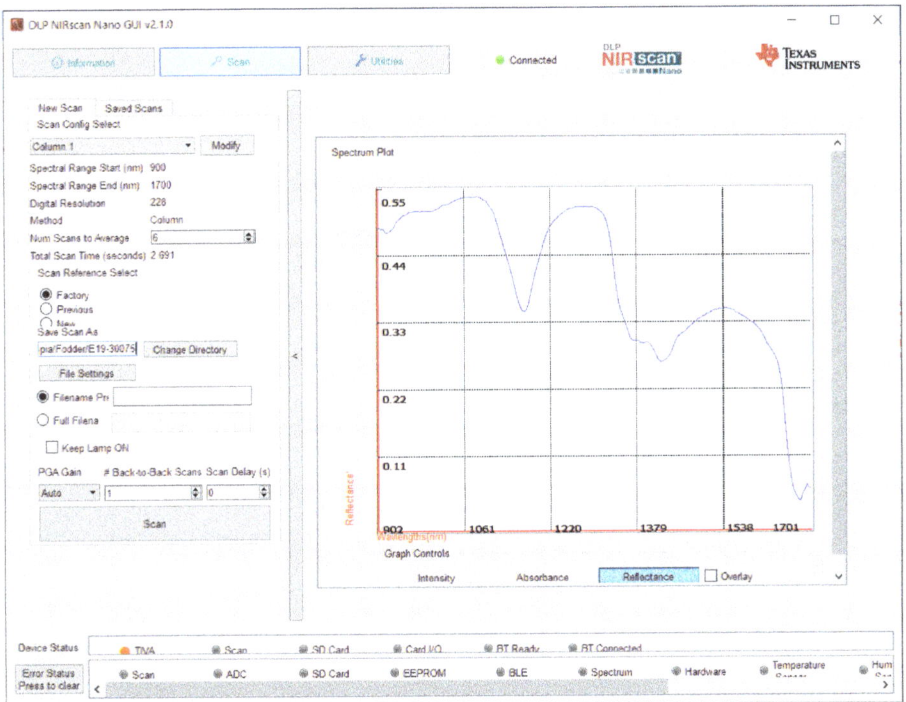

Figure 3. NIRscan Nano GUI scan screen (reflectance signal).

Column scanning selects wavelengths one at a time. With Hadamard scanning, several wavelengths are multiplexed at a time, and the wavelengths are decoded individually. Hadamard scanning collects light, and offers a higher SNR, but it depends heavily on the spectrum being measured, and on the measurement system [15]. Column analysis is more effective for analyzing forage samples, because it is more accurate than other methods of analysis, due to its reproducibility.

Before scan collection, one to five sections can be configured, depending on the type of method (Column or Hadamard), the spectral range (the starting wavelength and ending wavelength), the digital resolution (the wavelength points captured within the defined spectral range) the exposure time (between 0.635 and 60.960 milliseconds), and the number of scans per sample (in this work, 10 scans at 10 different points in the sample).

In this work, all samples were scanned in reflectance mode, using the measurement system shown in Figure 2.

In order to increase the sampling window and improve the spectroscopic information for each sample, after the alfalfa samples were homogenized, each was divided into three subsamples and scanned using miniaturized NIRs. Each spectrum is an average of 10 spectra, in a wavelength range between 901 and 1700 nm, with a non-linear path wavelength between 2.9 and 3.9 nm. A total of 30 scans were collected for each alfalfa sample. The precision of the collected spectrum for each sample or signal reproducibility was evaluated for the raw and ground alfalfa. The statistic of the root-mean-square error (*RMSE*) was calculated. Using (1), it is possible to calculate the *RMSE* for each sample spectrum. Lower *RMSE* values are related to reproducible and repeatable spectra.

$$RMSE = 10^6 \times \sqrt{\frac{\sum D^2}{n}}\,; D = y_a - y_b \tag{1}$$

y_a = log 1/R to λ for the average spectrum resulting from averaging a number of scans, and R is reflectance.
y_b = log 1/R to λ for the average spectrum resulting from averaging b number of scans
n = number of spectral data

2.5. Spectral Data Processing

To establish a calibration model to quantify alfalfa nutritive parameters, different chemometric strategies were assayed. The software Unscrambler X version 10.4 was employed to find the linear correlation between the NIRS spectra and nutritive parameters. This software takes NIRS spectra and transforms them into a matrix with X and Y variables, defined as the wavelength and reflectance. To detect potential spectral outliers, a principal component analysis was performed on the calibration set, before regression models were constructed using partial least squares (PLSs) [16]. An optimal number of PLSs factors is determined using the Unscrambler X software version 10.4 package, based on the minimum residual variance and 20 segments. Different spectrum mathematical pretreatment strategies were used for NDF, MC, and CP quantification. These approaches were performed using both the full range of equipment, from 900 to 1700 nm, and a reduced one, from 900 to 1600 nm.

To establish a successful model, a combination of pretreatments, including scatter correction with the standard normal variate (SNV), and the first and second Savitzky–Golay derivatives (SG), were tested in this study. Six possible pre-treatments were developed for three parameters, studying two possible wavelength ranges, for both the raw and milled samples. A total of 72 mathematical pre-treatments were assayed.

For the evaluation and selection of the most suitable chemometric model, the following statistics were calculated: the coefficient of determination for calibration (R^2, see Equation (2)), and the standard error of calibration (*SEC*, see Equation (3)) [17].

$$R^2 = 1 - \frac{\sum_{i=1}^{n}(y_i - \hat{y}_i)^2}{\sum_{i=1}^{n}(y_i - \bar{y})^2} \tag{2}$$

y_i are the reference values obtained in the laboratory, $\hat{y}_i$ is the prediction of the model, and n represents the number of samples used in the calibration set.

$$SEC = \sqrt{\frac{1}{n}\sum_{i=1}^{n}(y_i - \hat{y}_i)^2} \tag{3}$$

This parameter provides an average of the typical uncertainty for future sample prediction based on the $\hat{y}_i$ and y_i values for the ith sample.

For each parameter evaluated, the best mathematical pre-treatment was selected for the raw and ground NIR sample analysis. The criteria for the selection of these pre-treatments were based on the lowest values of calibration standard error (*SEC*), as well as the values closest to one for the calibration determination coefficients (R^2) [18].

3. Results and Discussion

For an NIRS procedure to be understood, the characterization of the spectrum data obtained using the NIRscan Nano prototype is essential. Figure 4 shows the raw and SNV spectra, plus the first Savitzky–Golay derivative spectra of the alfalfa samples included in the calibration procedure. Within the NIR wavelength range of the NIRscan Nano prototype, we can identify some characteristic bands of forages. According to the bibliography, those bands observed at 1166 nm are related to the protein content of the samples [18], and those observed at around 1350 nm are related to the fiber content [19]. Additionally, around 1400 nm, there is a band that can give information about the moisture content because, at that wavelength, OH bond overtone vibrations are observed [20]. In Figures 4 and 5, a rectangle highlights all the cited wavelengths, and the referenced respective parameters. Hence, in Figure 4b, which is an extended area between 1650 and 1700 nm of Figure 4a, we can observe that some of the collected spectra show a noisy signal at the end of the collected spectra. This noisy wavelength range, as shown in Figure 4c, can be minimized after scattering correction (SNV) is applied, along with other mathematical pretreatments, such as derivatives.

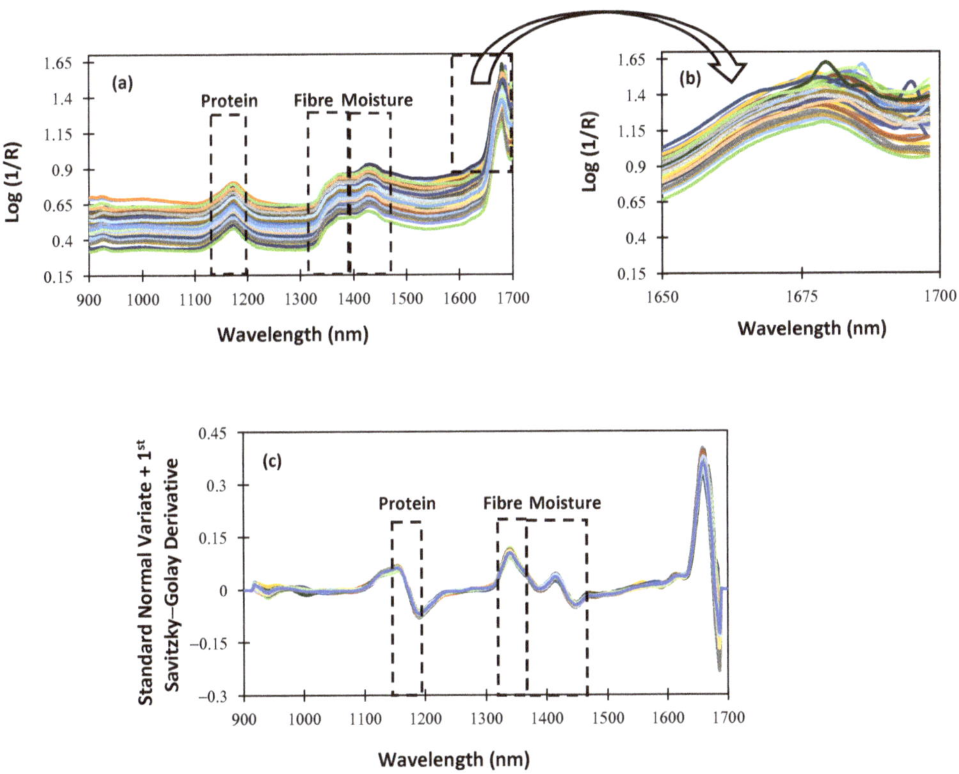

Figure 4. Spectra of the whole samples: (**a**) raw spectra; (**b**) extended area from 1650 to 1700 nm; (**c**) standard normal variate + first Savitzky–Golay derivative pretreatment.

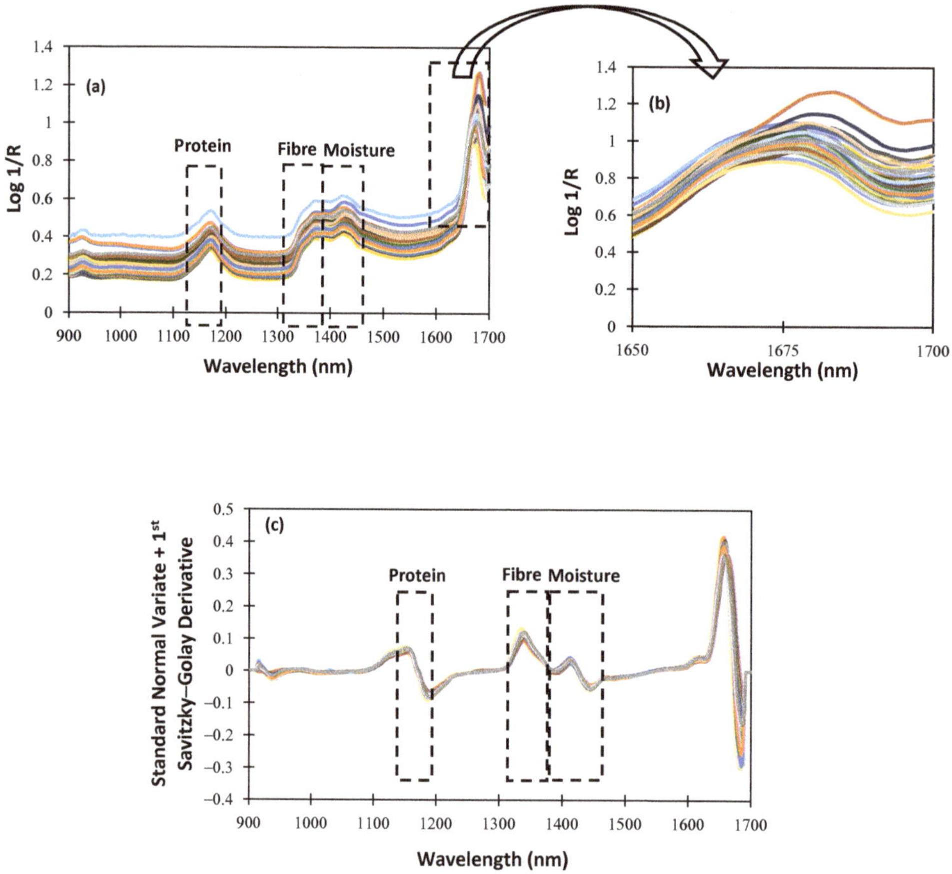

Figure 5. Spectra of the ground samples: (**a**) raw spectra; (**b**) extended area from 1650 to 1700 nm; (**c**) standard normal variate + first Savitzky–Golay derivative pretreatment.

To evaluate the effect of the sample pretreatment on the spectral data set, ground samples were scanned using NIRscan Nano. Figure 5a shows the spectral data set. As can be seen, no differences in the representative bands are observed. Moreover, the extended wavelength range (1650–1700 nm; see Figures 4b and 5b) shows that, after milling the sample pre-treatment, the noisy wavelength range disappears. These data confirm that the spectral quality depends on the sampling procedure (raw or ground). This distorted wavelength range is due to the huge and non-homogeneous particle size of the alfalfa samples. It is worthy of mention that, after the application of mathematical pretreatments, no differences were observed in the collected spectra of the alfalfa samples (see Figures 4c and 5c).

As observed in Figures 4 and 5, between 1600 and 1700 nm, the absorbance increases, and the SNR is lower than in other ranges. It is because the intensity measured at the detector is proportional to the number of DMD mirrors positioned to reflect the incident illumination towards it. As the number of pixels changes, the measured intensity is affected as well, resulting in an increase in noise levels.

Once the spectra were evaluated, the precision of the subsampling procedure for each scanned sample (raw and ground alfalfa) was evaluated [21]. Five samples were randomly selected from the 57 analysed. The RMS value was calculated for both the intact and ground samples with the two ranges proposed (901–1700 nm and 901–1600 nm). The results are shown in Table 3.

Table 3. The root-mean-square error (*RMS*) values for the paired subsamples of the same scanned sample.

	901–1700 nm		901–1600 nm	
Sample	Raw	Ground	Raw	Ground
1	60,529	27,209	43,283	14,560
2	58,378	23,853	33,386	10,886
3	54,190	20,057	34,085	7029
4	54,854	30,655	36,262	17,678
5	48,472	33,325	34,239	21,939

Once the values have been calculated, two clear trends can be observed. As expected, the RMS values obtained for the intact samples (raw) are higher than those obtained for the ground ones. These results could be due to heterogeneity in the raw alfalfa. The difference was also significant when the entire spectrum was compared or the last 100 nm was suppressed. Table 3 shows that the RMS values ranging from 900 to 1600 nm are lower than the full range of values. These results highlight the influence of the sampling procedure on the spectrum data precision.

After characterizing the spectral signal, the next step was to develop a calibration model. To attempt calibration, it is necessary to build a data matrix including nutritive values (the NDF, MC, and CP) and spectral data. After that, prior to the carrying out of calibrations, as mentioned in the Material and Methods section, different mathematical pre-treatments were applied, for the three parameters, to the raw and ground samples, both for the full range and the reduced range. Partial-least-square regression is used to establish the correlation between the spectra and the assayed parameters.

Table 4 summarizes the NIRS models' calibration statistics to quantify the NDF. As can be seen in Table 4, the R^2 values are higher and the SEC values are lower in the chemometric models developed with the reduced wavelength range than in those developed using the full one. In relation to the variability in the results depending on the mathematical pretreatment, it is important to note that the SNV plus the second Savitzky–Golay derivative reached the best calibration statistics for the raw and ground samples. Previous authors [22], after evaluating different commercial portable NIRS instruments to analyze ground forages, obtained R^2 values for the NDF ranging between 0.95 and 0.71, depending on the instrumentation employed. Regarding the SEC values, their results were between 2.85 and 1.21. It is not possible to obtain SEC values lower than 1, because the standard error of the laboratory (SEL) for this parameter is higher than 1.3 [22].

Table 4. Calibration statistics of NIR multivariate models for neutral detergent fiber quantification.

Raw alfalfa				
Wavelength range:	901–1600 nm		901–1700 nm	
Mathematical pretreatment	R^2	*SEC*	R^2	*SEC*
1 4 4 SG	0.898	1.554	0.883	1.670
2 4 4 SG	0.784	2.184	0.786	2.289
SNV 1 4 4SG	0.911	1.392	0.791	2.187
1 4 4 SG SNV	0.840	1.910	0.145	1.398
SNV 2 4 4SG	0.955	1.066	0.514	3.155
2 4 4 SG SNV	0.726	2.558	0.540	3.238

Table 4. *Cont.*

Raw alfalfa				
Ground alfalfa				
Wavelength range:	901–1600 nm		901–1700 nm	
Mathematical pretreatment	R^2	*SEC*	R^2	*SEC*
1 4 4 SG	0.756	2.749	0.598	2.830
2 4 4 SG	0.842	2.258	0.623	3.371
SNV 1 4 4SG	0.761	2.694	0.043	5.383
1 4 4 SG SNV	0.796	2.421	0.510	3.946
SNV 2 4 4SG	0.892	1.861	0.540	3.803
2 4 4 SG SNV	0.730	2.860	0.524	3.321

N1N2N3: derivative order, number of left smoothing points, number of right smoothing points; SG: Savitzky–Golay derivative; SNV: standard normal variate; R^2: coefficient of determination of calibration; *SEC*: standard error of calibration.

Table 5 summarizes the calibration statistics for the CP. Most mathematical treatments reach R^2 values lower than 0.5 for raw alfalfa samples. This could be related to the heterogeneity in alfalfa forage, with two clearly different parts, the leaf and the stem. The leaf is the part of the plant that contains a protein fraction. However, it is important to remark that using the reduced range, the spectrum mathematical pretreatment of the SNV for the scatter correction, and the second Savitzky–Golay derivative (the same mathematical pretreatment as for the NDF), R^2 values of 0.885, with an SEC of 0.377, were achieved. A typical SEL for reference CP analysis is around 0.210 [22].

Table 5. Calibration statistics of NIR multivariate models for crude protein quantification.

Raw alfalfa				
Wavelength range:	901–1600 nm		901–1700 nm	
Mathematical pretreatment	R^2	*SEC*	R^2	*SEC*
1 4 4 SG	0.742	0.510	0.884	0.428
2 4 4 SG	0.262	0.911	0.608	1.262
SNV 1 4 4SG	0.307	1.314	0.156	1.524
1 4 4 SG SNV	0.678	0.842	0.257	1.378
SNV 2 4 4SG	0.885	0.377	0.345	0.855
2 4 4 SG SNV	0.328	0.812	0.318	1.368
Ground alfalfa				
Wavelength range:	901–1600 nm		901–1700 nm	
Mathematical pretreatment	R^2	*SEC*	R^2	*SEC*
1 4 4 SG	0.671	0.986	0.706	0.927
2 4 4 SG	0.906	0.530	0.290	1.014
SNV 1 4 4SG	0.773	0.816	0.790	0.650
1 4 4 SG SNV	0.734	0.882	0.723	0.651
SNV 2 4 4SG	0.862	0.660	0.216	1.145
2 4 4 SG SNV	0.820	0.746	0.179	1.433

N1N2N3: derivative order, number of left smoothing points, number of right smoothing points; SG: Savitzky–Golay derivative; SNV: standard normal variate; R^2: coefficient of determination of calibration; *SEC*: standard error of calibration.

Based on the ground samples, and a reduced range (901–1600 nm), the developed models showed statistics around 0.7 or higher, with SEC values between 0.530 and 0.986. Considering these results, it is worth mentioning that, even though the homogeneity in the ground samples gives better calibration statistics, NIRscan Nano reached acceptable values when scanning the raw samples.

Feeding animals with minerals is a common practice; however, if there is an abnormal mineral content, there is a high probability of contamination with soil, which is not desirable in animal feeding systems. To quantify the MC in alfalfa forages, 24 different calibration models have been developed, assaying different mathematical pretreatments of spectrum data. The statistics of the proposed PLS models are shown in Table 6. As stated before, the reduced range gave better calibration statistics than the full one. Comparing mathematical pretreatments, the scatter correction applied after the derivatization procedure increased the R^2 values, and reduced the SEC. The highest R^2 and the lowest SEC values were 0.861/0.219 and 0.867/0.318 for the raw and ground samples, respectively.

Table 6. The calibration statistics of the mineral content multivariate models.

		Raw alfalfa		
Wavelength range:	901–1600 nm		901–1700 nm	
Mathematical pretreatment	R^2	*SEC*	R^2	*SEC*
1 4 4 SG	0.524	0.503	0.211	0.572
2 4 4 SG	0.619	0.492	0.129	0.861
SNV 1 4 4SG	0.783	0.374	0.734	0.464
1 4 4 SG SNV	0.502	0.579	0.312	0.491
SNV 2 4 4SG	0.675	0.409	0.679	0.434
2 4 4 SG SNV	0.861	0.219	0.687	0.444
		Ground alfalfa		
Wavelength range:	901–1600 nm		901–1700 nm	
Mathematical pretreatment	R^2	*SEC*	R^2	*SEC*
1 4 4 SG	0.650	0.530	0.652	0.506
2 4 4 SG	0.770	0.435	0.243	0.819
SNV 1 4 4SG	0.570	0.586	0.670	0.519
1 4 4 SG SNV	0.867	0.318	0.347	0.723
SNV 2 4 4SG	0.604	0.566	0.591	0.579
2 4 4 SG SNV	0.781	0.424	0.301	0.625

N1N2N3: derivative order, number of left smoothing points, number of right smoothing points; SG: Savitzky–Golay derivative; SNV: standard normal variate; R^2: coefficient of determination of calibration; SEC: standard error of calibration.

These NDF, CP, and MC calibration model statistics, obtained using the NIRS measurement system, are like those acquired via commercial portable instruments, using a wavelength range like that evaluated in this work [22,23]. The SEC values are in accordance with laboratory results, and the effect of the sampling procedure has been studied comparatively in this work. As a summary of the obtained results, Table 7 selects the best models obtained for each sampling procedure (raw or ground alfalfa) and parameter. As can be seen, the second derivative is the best of the assayed pretreatments that provide satisfactory results for the nutritive value quantification.

Table 7. Statistical analysis of alfalfa nutritive values (N = 57).

Parameter	Sampling	Mathematical Pretreatment	Range (nm)	R^2	*SEC*
NDF	Raw	SNV 2 4 4 SG	900–1600	0.955	1.066
	Ground	SNV 2 4 4 SG	900–1600	0.892	1.861
CP	Raw	SNV 2 4 4 SG	900–1600	0.885	0.377
	Ground	2 4 4 SG	900–1600	0.906	0.530
MC	Raw	2 4 4 SG SNV	900–1600	0.861	0.219
	Ground	1 4 4 SG SNV	900–1600	0.867	0.318

N1N2N3: derivative order, number of left smoothing points, number of right smoothing points; SG: Savitzky–Golay Derivative; SNV: standard normal variate; R^2: coefficient of determination for calibration, SEC: standard error of calibration; NDF: neutral detergent fiber; MC: mineral content; CP: crude protein.

4. Conclusions and Future Work

In this work, heterogeneous forage (alfalfa) has been selected as a model to evaluate the precision of instrumental measures (spectra collected), and the effects of sampling presentation (raw or ground), on calibration statistics. The results have revealed that homogeneous forage samples (those milled) allow us to achieve better calibration models than those scanned in their raw form (heterogeneous). However, for all sampling procedures, it has been possible to obtain satisfactory calibration to quantify the nutritive parameters.

Through the proposed instrumentation, users can evaluate the forage quality, increase sampling without incurring costs, and obtain results in real time. This is done by avoiding delays related to carrying samples from the farm to the laboratory. Furthermore, this instrument does not require specialized training.

In the future, with the use of internet of things (IoT) tools, data can be sent to the cloud for processing. In this way, they would be accessible from any device. Thanks to storage and processing in the cloud, data are accessible from any site with internet access. This allows the use of data in decisions.

Author Contributions: Conceptualization, A.S., C.M. and J.M.C.-F.; methodology, A.S. and C.M.; validation, C.M. and A.L.; formal analysis, J.M.C.-F.; investigation, A.L.; resources, F.F., J.M.C.-F. and A.L.; data curation, F.F.; writing—original draft preparation, C.M. and F.F.; writing—review and editing, C.M., A.S., A.L. and F.F.; visualization, A.L. and F.F.; supervision, A.S.; funding acquisition, J.M.C.-F. All authors have read and agreed to the published version of the manuscript.

Funding: This research was funded by the Spanish Ministry of Science and Innovation (PID2020-117282RBI00 and MCI-20-PID2019-109698GB-I00) and by Principado de Asturias GRUPIN IDI/2021/000081.

Institutional Review Board Statement: Not applicable.

Informed Consent Statement: Not applicable.

Data Availability Statement: Not applicable.

Acknowledgments: We would like to thank Jairo Tuñón Díaz for his help scanning the samples.

Conflicts of Interest: The authors declare no conflict of interest.

Appendix A

Alfalfa Sample Code	Mineral Content	Crude Protein	Neutral Detergent Fiber
1	11.41	13.69	48.42
2	13.71	12.12	52.30
3	10.80	14.68	40.53
4	10.46	15.06	39.69
5	12.04	14.05	42.10
6	9.96	13.59	43.37
7	10.83	7.19	61.01
8	9.55	13.18	43.82
9	10.65	14.20	40.02
10	11.09	15.17	38.52
11	9.92	16.53	32.17
12	12.16	17.01	38.92
13	9.44	13.35	46.84
14	11.26	15.98	34.66
15	9.98	16.27	36.25
16	9.58	13.88	43.68
17	10.61	13.78	43.17
18	11.52	15.90	37.41
19	10.24	14.49	42.38
20	10.54	14.22	41.76

Alfalfa Sample Code	Mineral Content	Crude Protein	Neutral Detergent Fiber
21	10.86	13.28	43.66
22	11.42	16.28	37.79
23	10.78	16.83	39.99
24	10.88	14.01	42.53
25	10.61	14.56	40.62
26	10.27	14.35	42.58
27	10.42	12.74	47.38
28	11.41	15.89	37.76
29	9.41	10.53	51.14
30	10.47	14.23	42.64
31	10.77	16.86	35.55
32	11.32	14.17	42.64
33	10.04	14.64	29.24
34	11.14	13.76	43.73
35	10.49	15.50	37.09
36	10.87	14.23	40.99
37	10.51	14.96	40.71
38	9.39	13.69	39.99
39	11.57	14.67	44.27
40	11.47	13.28	36.94
41	10.08	13.95	45.79
42	11.33	14.28	39.69
43	11.45	16.35	35.44
44	11.48	14.91	39.50
45	8.49	16.43	44.22
46	9.89	16.69	40.03
47	9.22	13.60	45.70
48	8.47	14.17	31.49
49	10.95	16.65	31.44
50	9.40	13.87	35.89
51	11.15	15.85	36.56
52	11.86	16.42	33.51
53	10.56	14.50	39.19
54	10.79	14.69	36.60
55	10.46	14.40	36.64
56	10.35	10.82	47.84
57	10.67	15.13	36.26
Average	10.64	14.48	40.70
Maximum	13.71	17.01	61.01
Minimum	8.47	7.19	29.24
Standard Deviation	0.91	1.70	5.50

References

1. Food and Agricultural Organization of the United Nations. Available online: https://www.google.com.hk/url?sa=t&rct=j&q=&esrc=s&source=web&cd=&cad=rja&uact=8&ved=2ahUKEwiuhcH6iNuAAxW34DgGHanIAWMQFnoECB0QAQ&url=https%3A%2F%2Fwww.fao.org%2Fabout%2Fen%2F&usg=AOvVaw2ikccPM_zDoInRJYBZakYo&opi=89978449 (accessed on 4 August 2023).
2. FAO. *The Feed Analysis Laboratory: Establishment and Quality Control*; FAO: Rome, Italy, 2013; Available online: https://www.fao.org/3/i3535e/i3535e.pdf (accessed on 4 August 2023).
3. Prache, S.; Adamiec, C.T.; Astruc, E.; Baéza-Campone, P.E.; Bouillot, A.; Clinquart, C.; Feidt, E.; Fourat, J.; Gautron, A.; Girard, L.; et al. Review: Quality of animal-source foods. *Animal* **2022**, *16*, 100376. [CrossRef] [PubMed]
4. Katoch, R. Forage Quality Indices. In *Techniques in Forage Quality Analysis*; Springer: Singapore, 2023.
5. Ozaki, Y.; Huck, C.W.; Tsuchikawa, S.; Engelsen, S.B. (Eds.) *Near-Infrared Spectroscopy*; Springer: Singapore, 2021.
6. Huck, C.W.; Beć, K.B.; Grabska, J. Near infrared spectroscopy in natural product research. In *Encyclopedia of Analytical Chemistry: Applications, Theory and Instrumentation*; Meyers, R.A., Ed.; John Wiley & Sons: Hoboken, NJ, USA, 2020; pp. 1–29.
7. Shenk, J.S.; Workman, J.J.; Westerhaus, M.O. Application of NIR spectroscopy to agricultural products. In *Handbook of Near Infrared Analysis*; Burns, D.A., Ciurczak, E.W., Eds.; CRC Press: Boca Raton, FL, USA, 2008; pp. 347–387.
8. Cherney, J.H.; Digman, M.F.; Cherney, D.J. Handheld NIRS for forage evaluation. *Comput. Electron. Agric.* **2021**, *190*, 106469. [CrossRef]
9. Crocombe, R.A. Portable spectroscopy. *App. Spectr.* **2018**, *72*, 1701–1751. [CrossRef] [PubMed]
10. Beć, K.B.; Grabska, J.; Huck, C.W. Miniaturized NIR Spectroscopy in Food Analysis and Quality Control: Promises, Challenges, and Perspectives. *Foods* **2022**, *11*, 1465. [CrossRef] [PubMed]

11. Mizik, T. How can precision farming work on a small scale? A systematic literature review. *Precis. Agric.* **2023**, *24*, 384–406. [CrossRef]
12. Melendreras, C.; Soldado, A.; Costa-Fernández, J.M.; López, A.; Valledor, M.; Campo, J.C.; Ferrero, F.J. An affordable NIR spectroscopic system for fraud detection in olive oil. *Sensors* **2023**, *23*, 1728. [CrossRef] [PubMed]
13. Texas Instruments. DLP NIRscan™ Nano EVM User´s Guide. 2017. Available online: https://www.ti.com/lit/pdf/dlpu030 (accessed on 2 January 2023).
14. Van-Soest, P.J.; Robertson, J.B.; Lewis, B.A. Methods for dietary fiber, neutral detergent fiber, and nonstarch polysaccharides in relation to animal nutrition. *J. Dairy Sci.* **1991**, *74*, 3583–3597. [CrossRef] [PubMed]
15. Thompson, M.; Wood, R. Using uncertainty functions to predict and specify the performance of analytical methods. *Accred. Qual. Assur.* **2006**, *10*, 471–478. [CrossRef]
16. Xu, J.; Liu, H.; Lin, C.; Sun, Q. SNR analysis and Hadamard mask modification of DMD Hadamard transform near-infrared spectrometer. *Opt. Commun.* **2017**, *383*, 250–254. [CrossRef]
17. Lacefield, G.D. Alfalfa Hay Quality Makes the Difference. In *Agriculture and Natural Resources*; University of Kentucky UKnowledge, Agriculture and Natural Resources Publications: Lexington, KY, USA, 1988; p. 32.
18. Rego, G.; Ferrero, F.; Valledor, M.; Campo, J.C.; Forcada, S.; Royo, L.J.; Soldado, A. A portable IoT NIR spectroscopic system to analyze the quality of dairy farm forage. *Comput. Electron. Agric.* **2020**, *175*, 105578. [CrossRef]
19. Coelho, M.H.; Hembry, F.G. Laboratory methods of forage evaluation. Near infrared reflectance analysis. *Livest. Prod. Day Rep. La. Agric. Exp. Stn. Anim. Sci. Dep.* **1982**, *3*, 195–199.
20. Andrés, S.; Murray, I.; Calleja, A.; Giráldez, F.J. Nutritive Evaluation of Forages by near Infrared Reflectance Spectroscopy. *J. Near Infrared Spectrosc.* **2005**, *13*, 301–311. [CrossRef]
21. Martínez, M.L.; Garrido-Varo, A.; Pedro, E.D.; Sánchez, L. Effect of sample heterogeneity on near infrared meat analysis: The use of the RMS statistic. *J. Near Infrared Spectrosc.* **1998**, *6*, A313–A320. [CrossRef]
22. Berzaghi, P.; Cherney, J.H.; Casler, M.D. Prediction performance of portable near infrared reflectance instruments using preprocessed dried, ground forage samples. *Comput. Electron. Agric.* **2021**, *182*, 106013. [CrossRef]
23. Gorla, G.; Taiana, A.; Boqué, R.; Bani, P.; Gachiuta, O.; Giussani, B. Unravelling error sources in miniaturized NIR spectroscopic measurements: The case study of forages. *Anal. Chim. Acta* **2022**, *1211*, 339900. [PubMed]

applied sciences

MDPI

Article

Real-Time Deployment of MobileNetV3 Model in Edge Computing Devices Using RGB Color Images for Varietal Classification of Chickpea

Dhritiman Saha [1,2,*], Meetkumar Pareshbhai Mangukia [1] and Annamalai Manickavasagan [1]

1 School of Engineering, University of Guelph, Guelph, ON N1G 2W1, Canada; meetmangukia1998@gmail.com (M.P.M.); mannamal@uoguelph.ca (A.M.)
2 ICAR-Central Institute of Post-Harvest Engineering and Technology (CIPHET), Ludhiana 141 004, India
* Correspondence: dsaha@uoguelph.ca

Abstract: Chickpeas are one of the most widely consumed pulses globally because of their high protein content. The morphological features of chickpea seeds, such as colour and texture, are observable and play a major role in classifying different chickpea varieties. This process is often carried out by human experts, and is time-consuming, inaccurate, and expensive. The objective of the study was to design an automated chickpea classifier using an RGB-colour-image-based model for considering the morphological features of chickpea seed. As part of the data acquisition process, five hundred and fifty images were collected per variety for four varieties of chickpea (CDC-Alma, CDC-Consul, CDC-Cory, and CDC-Orion) using an industrial RGB camera and a mobile phone camera. Three CNN-based models such as NasNet-A (mobile), MobileNetV3 (small), and EfficientNetB0 were evaluated using a transfer-learning-based approach. The classification accuracy was 97%, 99%, and 98% for NasNet-A (mobile), MobileNetV3 (small), and EfficientNetB0 models, respectively. The MobileNetV3 model was used for further deployment on an Android mobile and Raspberry Pi 4 devices based on its higher accuracy and light-weight architecture. The classification accuracy for the four chickpea varieties was 100% while the MobileNetV3 model was deployed on both Android mobile and Raspberry Pi 4 platforms.

Keywords: chickpea; convolutional neural network; transfer learning; classification

Citation: Saha, D.; Mangukia, M.P.; Manickavasagan, A. Real-Time Deployment of MobileNetV3 Model in Edge Computing Devices Using RGB Color Images for Varietal Classification of Chickpea. *Appl. Sci.* **2023**, *13*, 7804. https://doi.org/10.3390/app13137804

Academic Editors: José Miguel Molina Martínez, Ginés García-Mateos and Dolores Parras-Burgos

Received: 29 May 2023
Revised: 27 June 2023
Accepted: 29 June 2023
Published: 2 July 2023

1. Introduction

Pulses constitute one of the most significant crops in the leguminous family. About 90 million metric tonnes of pulses are produced worldwide [1]. Common beans, lentils, chickpeas, dry peas, cowpeas, mung beans, urad beans, and pigeon peas are the main pulses farmed worldwide. Pulses are significant sources of protein, phosphorus, iron, calcium, and other essential minerals, and help millions of individuals in impoverished nations meet their nutritional needs [2,3]. Pulses have recently acquired popularity on a global scale as an alternative to animal-based sources of protein [4]. With a production of 15 million metric tonnes and a third-place ranking among pulses after beans and peas, chickpeas are one of the high protein pulses grown in more than 57 nations [1]. Depending on the cultivar, agronomic practices, and climatic factors, the protein content of chickpeas varies from 17% to 24% [5]. Additionally, chickpeas are a good source of energy and include vitamins, minerals, fibre, and phytochemicals. According to Wood and Grusak, 2007 [6], regular consumption of chickpeas lowers the risk factors for diabetes, cancer, and cardiovascular disease.

For processors, customers, and other stakeholders, the quality of chickpeas is crucial in influencing their preference [7]. The various factors considered in evaluating the quality of chickpeas are 100-seed weight, ash content, colour, cooking time, cooked chickpea seed stiffness, moisture content, protein content, seed size distribution, starch content, total dietary fibre, water absorption, and others [8]. It is difficult to assess the quality of a sample that contains mixed chickpea varieties, so maintaining varietal purity is the first crucial step in determining chickpea quality. Prior to a deeper examination of their visible or internal properties, the identification or classification of chickpea types assumes relevance. The most used imaging method for classifying agricultural products by varietal is RGB imaging [9]. Around the world, different chickpea cultivars are produced in various agroclimatic zones. The nutrient content, physical characteristics, and economic worth of each variety vary. Different chickpea varieties can be identified by their physical characteristics, such as colour and texture, which are visible and aid in classification. However, this classification process is time-consuming, expensive, and frequently performed by human professionals. Due to the intricacy of the task and the abundance of visual and environmental components, research into the creation of tools for the automation of these occupations is still underway. The need for accurate variety identification systems in precision agriculture is also growing as a result of the ramifications for the economy, ecology, and society [10].

Deep learning models are being used more frequently, which has led to significant improvements, notably in classification tasks [11–15]. Deep learning has been used in recent research for agricultural crop classification due to the increased accuracy and hardware accessibility. In addition, the rapid advancement of open-source hardware in recent years has promoted the creation and use of low-cost agricultural surveillance tools with image processing and AI capabilities. Single-board computers (SBCs) in many configurations, like the Raspberry Pi (RPi), have spread quickly across many different applications like food manufacturing [16] and surveillance monitoring [17]. Osroosh et al., 2017 [16] detailed the design and construction of a monitoring system that employs thermal and RGB images, built on a Raspberry Pi 3, that is able to operate in difficult real-world scenarios. Raspberry Pi was utilised by Hsu et al., 2018 [18] to build a low-cost monitoring network for agricultural applications that is intended for wide adoption. A monitoring environment with many devices and interfaces was developed by Morais et al., 2019 [19], enabling communication in low-cost agricultural equipment.

In light of this backdrop, the current work provides the results of training using transfer learning in architectures that have just been investigated for agricultural applications, as well as their implementation on affordable hardware like the android mobile and Raspberry Pi 4 microcomputer. In order to obtain findings that help in the classification of chickpeas, one of the objectives of this research is to incorporate CNN models in a low-cost, low-power device that can process information in real time. Therefore, four common Canadian chickpea varieties were utilised and an attempt was made to identify them utilising computer vision and machine-learning-based methodologies in order to automate the process of chickpea classification. The work's goal is broken down into two subcategories. The first stage involves creating the classification model using computer vision and machine-learning-based approaches. The deployment phase, which involves deploying the trained machine-learning model on Android mobile and Raspberry Pi 4 devices, and evaluating its performance, takes place in the second stage.

2. Materials and Methods

In this study, the Crop Development Centre (CDC), University of Saskatchewan, Saskatoon, Canada, provided four popular Canadian chickpea varieties (CDC-Alma, CDC-Consul, CDC-Cory, and CDC-Orion) harvested in 2020. To remove unwanted foreign particles, the seeds were sieved. The seeds were also hand-sorted to remove broken, mouldy, and discoloured seeds, as well as washed to remove any leftover dust particles. The initial moisture content of each variety of cleaned chickpeas was evaluated after 24 h in a 105 °C hot air oven [20]. To achieve equal moisture distribution among the samples, all chickpea varieties were reconditioned to 12.5 ± 0.5% wet basis by adding measured distilled water and well mixing it, followed by seven days of storage in airtight low-density polyethylene (LDPE) bags at 5 °C.

The proposed approach in this study is presented as a block diagram in Figure 1. It combines data acquisition, pre-processing, model training, and deployment. The images used for the experiment were collected using a smartphone camera and an industrial camera in a natural lightning environment. Various data pre-processing techniques were used to augment the existing dataset. For training purposes, transfer learning was used to create and train machine-learning models. The model evaluation was performed on a separate test dataset. Thereafter, the optimal machine-learning model was deployed on an Android device and a Raspberry Pi 4 to check its real-time performance.

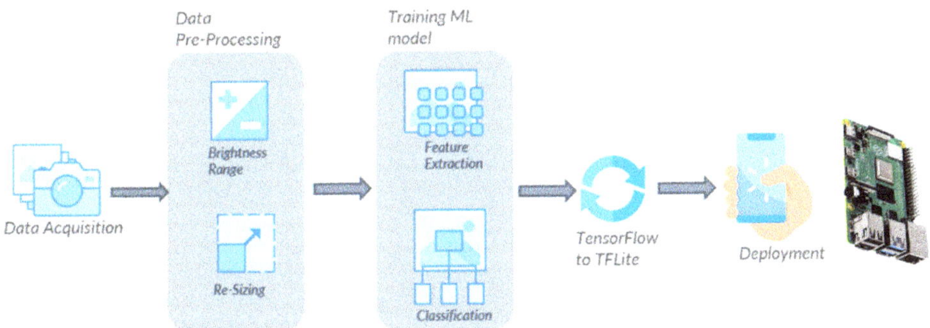

Figure 1. Flow diagram of the experimental process.

2.1. Data Acquisition

To build robust classification models, the RGB images of the chickpea sampled were captured using two different cameras (smartphone camera and industrial camera) to accommodate potential variations due to imaging systems. The smartphone (Samsung Galaxy A6 Plus) used comes with a dual camera, where the primary camera has 16 MP, followed by 5 MP for the secondary camera with an f/1.7 aperture opening. First, the chickpeas were placed in a tray in a manner such that it was fully filled with chickpeas. In order to collect images through a smartphone, the smartphone camera was set on top of the chickpea tray in such a way that it could capture the maximum area of the tray without including the borders of the tray. The images were 3456 × 4608 in resolution and a total of 150 images were collected for each variety. As for the industrial camera, a GigE colour industrial camera (Model: DFK 33GX273; Specifications: 1.6 MP, 60 FPS; The Imaging Source, USA) was used. The chickpea seeds were placed over a conveyor belt and the speed of the belt was fixed at 1 m/min to avoid image distortion. The industrial camera was set up on top of a conveyor belt and the video of the chickpea variety was captured. Thereafter, the video was processed and each frame of the video was extracted as an image. The extracted images were used to build our chickpea dataset with different cameras. Thereafter, a total of 400 images were collected for each variety from captured video with 1440 × 1080 resolution. Thereafter, all the images were combined (smartphone and industrial camera) and the final dataset consisted of 550 images per variety. The reason behind using two different sources of dataset (i.e., industrial camera and smartphone) is to increase the variability in the dataset which helps in better generalisation by the model. In other words, it will increase the robustness of our trained model since each source of data is exposed to different environmental conditions and, by combining the data, the model becomes more tolerant to changes in lighting, noise, and other factors, producing more reliable and accurate results. Further, this variability will make the model better generalise on datasets acquired with different camera resolution and conditions. Figure 2 displays the collected images for the four chickpea varieties.

Figure 2. Images of four chickpea varieties.

2.2. Data Augmentation

Image data augmentation is a way of artificially increasing the amount and diversity of a training dataset by realistically transforming images currently in the training dataset. Data augmentation aids in the building of more efficient models, which improves the model's generalisability, and prevents overfitting and the problem of the network memorising specific information of the training images [21,22]. Image alteration methods used in this work for real-time augmentation include rotation, horizontal and vertical translation, shear, and zoom. The training dataset is made up of both the original images and the augmented images produced through alterations.

Different pre-processing techniques were applied using the *ImageDataGenerator* function defined in TensorFlow. Table 1 shows the list of different data augmentation pre-processing techniques and their values. The rotation range rotates the image clockwise by a given number of degrees. A horizontal flip flips the image horizontally. The width shift range shifts the image left or right by considering the given number as a percentage (i.e., 0.2 = 20%). The height shift range shifts the image up or down by considering the given number as a percentage. Shear range distorts the image by considering the given value as an angle. Zoom range zooms the image by considering the given number as a percentage.

Table 1. Different data augmentation parameters along with their values.

Parameter	Value
Rotation Range	40
Horizontal Flip	True
Width Shift Range	0.2
Height Shift Range	0.2
Shear Range	0.2
Zoom Range	0.2

2.3. Feature Extraction

The present work uses a CNN-based model for extracting features from the input images. In our study, three different models named NasNet-Mobile, MobleNetV3-Small, and EfficientNetB0 were selected. The architecture of these models is fast and efficient, designed for frequent or real-time predictions on mobile devices and single-board computers [23,24]. All three models are designed using neural architecture search (NAS) to achieve optimal architecture for on-device machine learning. NAS is the process of automatically searching for the optimal deep neural network (DNN) architecture using three major components: search space, optimisation methods, and candidate evolution methods. The main idea of search space is to define the optimal architecture of the model by considering the input dataset [25]. Since it requires prior knowledge of the dataset, it is not ideal for novel domains. Another problem with search space is its limitations in exploring the available architecture, as some of the excluded architecture might be a better choice. Optimisation methods help search space determine the best possible architecture by considering the effectiveness of the selected architecture. The last component, candidate evolution, is designed for comparing the results produced by optimisation methods and helps search space choose the best possible architecture [26].

2.3.1. NasNet-A (Mobile)

The first model used in our study was NasNet-A (mobile). As the name suggests, the architecture of this model was developed using NAS. Researchers redesigned a search space (the first component of NAS) by including controller recurrent neural network (RNN) in order to obtain the best architecture for the CIFRA 10 dataset. Thereafter, the same architecture (NasNet architecture) was taken, and stacking of the copies of the developed CNN layers on each other was performed and applied to the ImageNet dataset. As a result, the new architecture was able to achieve 82.7% top-1 and 96.2% top-5 accuracy on the ImageNet dataset. Table 2 showcases the high-level representation of the used NasNet-A (mobile) architecture. As can be observed from the table, the architecture consists of reduction cells and normal cells. Reduction cells produce a feature map of their inputs by reducing their height and weight, whereas normal cells produce a feature map with the same dimensions as their inputs [27].

Table 2. NasNet-A (mobile) model architecture.

Stage	Operator	Resolution	Channels	Layers
1	Conv2D, 3 × 3	224 × 224	3	1
2	Reduction Cell	111 × 111	32	2
3	Normal Cell	28 × 28	44	4
4	Reduction Cell	28 × 28	88	1
5	Normal Cell	14 × 14	88	4
6	Reduction Cell	14 × 14	176	1
7	Normal Cell	7 × 7	176	4

2.3.2. MobileNetV3 (Small)

The second model chosen was MobileNetV3 (small) because of its high efficiency for on-device machine learning. In order to develop the cell for MobileNetV3, researchers took MobileNetV2′s cell (consisting of residual and linear bottlenecks) and added squeeze-and-excite to the residual layers. Moreover, upgraded layers (with modified switch non-linearities), squeeze, and residual have sigmoid functions instead of sigmoid to get better results. After designing this cell, NAS was applied to obtain the best possible network architecture. Additionally, researchers also applied the NetAdapt algorithm to the developed model to fine-tune each layer. The model had 67.5% top-1 accuracy with 16.5 ms average latency on the ImageNet dataset. Table 3 indicates the high-level architecture of the model [28].

Table 3. MobileNetV3 (small) model architecture.

Stage	Operator	Resolution	Channels	Layers
1	Conv2d, 3 × 3	224 × 224	3	2
2	Bottleneck, 3 × 3	112 × 112	16	2
3	Bottleneck, 3 × 3	56 × 56	16	2
4	Bottleneck, 3 × 3	28 × 28	24	1
5	Bottleneck, 3 × 3	28 × 28	24	2
6	Bottleneck, 3 × 3	14 × 14	40	1
7	Bottleneck, 3 × 3	14 × 14	40	1
8	Bottleneck, 3 × 3	14 × 14	40	1
9	Bottleneck, 3 × 3	14 × 14	48	1
10	Bottleneck, 3 × 3	14 × 14	48	2
11	Bottleneck, 3 × 3	7 × 7	96	1
12	Bottleneck, 3 × 3	7 × 7	96	1
13	Conv2d, 1 × 1	7 × 7	96	1
14	Pool	7 × 7	576	1
15	Conv2D, 1 × 1, NBH	1 × 1	576	1
16	Conv2D, 1 × 1, NBH	1 × 1	1024	1

2.3.3. EfficientNetB0

The third model selected for the study was EfficientNetB0 because of its state-of-the-art accuracy, small size, and speed compared to other ConvNets. In order to develop the EfficientNet, researchers used NAS on a new mobile-sized convolutional network and came up with EfficieNetB0, which had 77.1% top-1 accuracy and 93.3% top-5 accuracy on the ImageNet dataset. In order to make EfficientNetB0 more efficient, researchers proposed a new compound scaling method that uniformly scales all dimensions (depth, width, and resolution) of any given model and used that method on EfficientNetB0 to produce better versions of it (i.e., EfficientNetB1, EfficientNetB2, EfficientNetB3, EfficientNetB4, EfficientNetB5, EfficientNetB6, and EfficientNetB7). For our study, we used EfficentNetB0, which consists of a mobile inverted bottleneck MBConv cell and squeeze-and-excitation, as shown in Table 4 [29].

Table 4. EfficientNetB0 architecture.

Stage	Operator	Resolution	Channels	Layers
1	Conv2d, 3 × 3	224 × 224	32	1
2	MBConv1, k 3 × 3	112 × 112	16	1
3	MBConv6, k 3 × 3	112 × 112	24	2
4	MBConv6, k 5 × 5	56 × 56	40	2
5	MBConv6, k 3 × 3	28 × 28	80	3
6	MBConv6, k 5 × 5	14 × 14	112	3
7	MBConv6, k 5 × 5	14 × 14	192	4
8	MBConv6, k 3 × 3	7 × 7	320	1
9	Conv2D, 1 × 1	7 × 7	1280	1

2.4. *Training of CNN Architectures*

The chickpea image dataset was organised into four labelled files, each with 550 original images. The dataset was split into three parts: training, validation, and testing. The training was performed on 80% of the original images, with the remaining 10% used for validation and the other 10% for testing. As a result, the 80:10:10 ratio was consistent across all varieties and models. The validation set was used to validate training performance, whereas the test set was used to validate classifier performance. The images for training, validation, and testing were chosen by initialising the GPU random seeds to ensure that the model networks were trained, validated, and tested on the same dataset. All three model architectures utilised in this work were trained five times using five different random seeds and the findings reported in this study are for a single random seed (3) shared by all model architectures. The different hyperparameters utilised during network training, such as momentum, number of epochs, and optimiser (stochastic gradient descent), were modified to properly train the CNN for image classification. Momentum is employed during network training to accelerate the learning rate. An epoch is the number of times the network traverses the complete training dataset. It is important to note that the number of epochs that a model should be trained for depends on a number of factors, including the size and complexity of the dataset, the complexity of the model, and the desired level of accuracy. The optimiser is used to update the network's learnable parameters during training in order to reduce the loss function [30]. The settings of these hyperparameters (Table 5) were chosen based on available literature and then kept constant to allow for fair comparisons between networks [31]. The learning rate specifies how frequently the weights in the layers are updated during training, whereas the batch size specifies the number of images used to train the network in each epoch. The minimum and maximum ranges of the hyperparameters were determined through several trials and based on existing related literature. Since the training was performed on pre-trained networks with defined weights, the learning rate was kept low at 0.005. A very low learning rate may result in prolonged training time without convergence, while a very high learning rate may result in poor learning of complexity from the training dataset [32]. Similarly, choosing the right batch size is critical for network training because a small batch size will converge faster than a large batch size. Furthermore, a bigger batch size will achieve optimum minima, which a very tiny batch will find difficult to achieve. For the training, 'TensorFlow Lite Model Maker' was used that simplified the process of training a TensorFlow Lite model using custom dataset. It uses transfer learning to reduce the quantity of required training data and training time, resulting in fewer epochs of model training. In this study, we have added two layers, a dropout layer and a dense layer, after each pre-trained CNN. The dropout rate used was 20% in the dropout layer and the dense layer had four units, or neurons, in order to classify four different varieties of chickpea with the SoftMax activation function. The study was conducted on a Google Colab configured with an NVIDIA Tesla K80 GPU and 12 GB of RAM.

Table 5. Values of hyperparameters used in machine-learning models.

Parameters	Values
Max Epochs	5
Mini Batch Size	32
Optimiser	Stochastic Gradient Descent (SGD)
Initial Learning Rate	0.005
Momentum	0.9
Loss Function	Categorical Cross Entropy

2.5. Model Evaluation

In this study, the model's performance was assessed using accuracy, precision, sensitivity, and specificity obtained from the confusion matrix of the models [33].

$$\text{Accuracy} = \frac{\text{TP} + \text{TN}}{\text{TP} + \text{TN} + \text{FP} + \text{FN}} \tag{1}$$

$$\text{Precision} = \frac{\text{TP}}{\text{TP} + \text{FP}} \tag{2}$$

$$\text{Specificity} = \frac{\text{TN}}{\text{TN} + \text{FP}} \tag{3}$$

$$\text{Sensitivity} = \frac{\text{TP}}{\text{TP} + \text{FN}} \tag{4}$$

where TP, TN, FP, and FN represent true positive, true negative, false positive, and false negative, respectively. True positive is a consequence in which the model predicts the positive class accurately and true negative indicates accurate prediction of the negative class. False positive indicates the incorrect prediction of the positive class by the model and false negative indicates the incorrect prediction of the negative class.

In addition to the previously mentioned metrics used to evaluate classifier performance, the models were also assessed based on the area under the receiver operating characteristic (ROC) curve, known as AUC. AUC is a significant scalar value that provides an overall assessment of a classifier's performance. The ROC curve plots the true positive rate (TPR) against the false positive rate (FPR), with TPR on the y-axis and FPR on the x-axis. AUC measures the classifier's ability to distinguish between classes and represents the degree of separability. It ranges from 0.5 to 1.0, where the minimum value corresponds to a random classifier and the maximum value indicates a perfect classifier. AUC has the advantageous properties of threshold and scale invariance, making it a robust metric. This implies that AUC is not influenced by the chosen threshold or the scale of probabilities. Computationally, AUC is determined by aggregating trapezoid areas beneath the ROC curve. AUC values between 0.7 and 0.8 are considered acceptable, 0.8 to 0.9 as excellent, and greater than 0.9 as outstanding [33].

The classification time measure would utilise the model's average time to predict an image class. This was accomplished by employing a timer at the beginning and end of the evaluation procedure, and the classification time was calculated using the following formula [34]:

$$\text{Classification time} = \frac{Test\ evaluation\ time}{Number\ of\ steps \times Batch\ size} \quad (5)$$

The Python batch dataset format was applied to the test dataset. The batch size hyperparameter was used to group the images in this format. A batch in this experiment consisted of 32 images.

2.6. Confusion Matrix

The confusion matrix of the test set was used to assess the models' performances. The confusion matrix results give the quantitative and predictive values for chickpea varietal categorisation. The anticipated class/output class is on the X-axis of the matrix shown in Figure 3, while the true class/target class is on the Y-axis. The diagonal cells in the matrix reflect correctly categorised observations, showing that the anticipated and actual classes are the same, whereas the remaining observations have been misclassified [33].

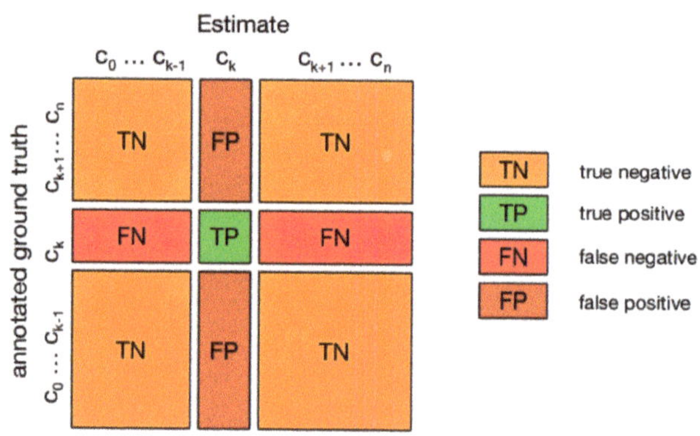

Figure 3. Confusion matrix for multi–class classification.

2.7. Deployment Platforms

Among the three studied machine-learning models, it is planned to deploy the model with optimal performance on an Android device and a Raspberry Pi 4 device to verify its functionality in the real world.

2.7.1. Mobile Application Development

To deploy trained machine learning on mobile devices, the models were converted to TensorFlow Lite termed interpreters in the format of ".tflite" files using the TensorFlow Lite Task Library API. TensorFlow Lite is a collection of tools built for edge, embedded, and mobile devices, allowing for on-device machine learning. The benefits of edge machine learning include real-time latency (no data offloading), privacy, robustness, connectivity, a smaller model size, and efficiency (costs of computation and energy in watts per FPS). It supports Linux, Android, iOS, and the MCU. The TensorFlow Lite converter and TensorFlow Lite interpreter are two components for the deployment of TensorFlow models on mobile devices [35]. To begin, the Keras model that is created using the TensorFlow 2.0 library was exported to pb (protocol buffer) models. Second, the PB models were converted to TensorFlow Lite models using the TensorFlow Lite Converter. Finally, the TensorFlow Lite interpreter was set to execute the TensorFlow Lite models on smartphones and take advantage of smartphone hardware resources to boost detection performance even further [36]. To facilitate the use of TensorFlow Lite models on mobile phones, the mobile application was developed for Android OS using the Java programming language and the TensorFlow Lite library. The application takes live camera image streams as input and processes them individually. Each image was processed by a Keras-based Tensorflow Lite model and the model produced a list of confidence scores, indicating the probability of the image belonging to a particular class. The application would take the highest confidence score index and display the image with a label with that index. A screenshot of the application is given in Figure 4.

Figure 4. Deployed model in Android device (smartphone) showing the prediction accuracy for Orion chickpea variety.

2.7.2. Raspberry Pi 4

Raspberry Pi is a low-cost, credit-card-sized single-board computer that was developed in 2012 by the Raspberry Pi Foundation in the UK. Raspberry Pi has its own operating system, previously called Raspbian, based on Linux. The Raspberry Pi includes 26 GPIO (General Purpose Input/Output) pins, allowing users to connect a larger variety of external hardware devices. Furthermore, it supports practically all of the peripherals offered by Arduino. This board accepts code in practically any language, including C, C++, Python, and Java. The Raspberry Pi has a faster processor than the Arduino and other microcontroller modules [37]. It can function as a portable computer. The details of the Raspberry Pi 4 are given in Table 6.

The Tensorflow Lite model was deployed on the Raspberry Pi 4 through a script written in Python that provides basic information, including the real-time prediction of chickpea variety and the percentage of prediction accuracy. The script takes real-time images as input and feeds them one by one to the TFLite model to process them. The model provides a set of confidence scores, which represent the probability of the image belonging to different categories. Then, the script will select the index corresponding to the highest confidence score and present the image along with the label associated with that index. The experimental arrangement for model deployment using the Raspberry Pi 4 is given in Figure 5.

Table 6. Technical attributes of the single-board computer Raspberry Pi 4.

Processor	Broadcom BCM2711, Quad core Cortex-A72 (ARM v8) 64-bit SoC @ 1.5GHz
RAM	8GB LPDDR4-3200 SDRAM
Bluetooth	Bluetooth 5.0, BLE
Wi-Fi	2.4 GHz and 5.0 GHz IEEE 802.11ac wireless
USB	2 USB 3.0 ports; 2 USB 2.0 ports
Ethernet	Gigabit Ethernet
HDMI	2 × micro-HDMI ports (up to 4kp60 supported)
Storage	MicroSD card slot
Power supply	5.1V 3A USB type C power
Dimensions	85.6 mm × 56.5 mm
Operating temperature	0 to 50 °C

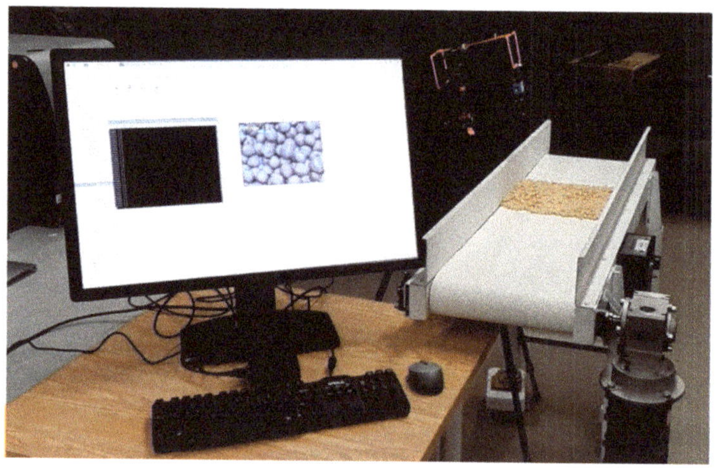

Figure 5. Experimental arrangement for model deployment in Raspberry Pi 4 using industrial camera.

3. Results and Discussion

3.1. Performance of the Models

The performance of the models is displayed in Figure 6. It is vital to monitor the progress of training a deep learning network. The training metrics for each iteration are shown in the training graphs (Figure 6). This visualisation not only displays the changes in network accuracy as the network is trained but it also displays any overfitting of the training data [9]. Figure 6 shows the classification accuracy and cross-entropy loss for the best overall classification accuracy of the pre-trained CNN networks. The holdout validation monitors the training progress and assists in model optimisation. The variation between their training and validation accuracy is negligible, which indicates the model has generalised well enough on the training dataset and performed well on the validation dataset. The figures revealed that NASNet Mobile (Figure 6a), MobileNetV3 (Figure 6b), and EfficientNetB0 (Figure 6c) appeared to have comparable outcomes. The models achieved validation accuracy values that were higher than their training accuracy (Table 7) and the figures indicated no overfitting, with the models converging within five epochs. Further, when the model was trained beyond five epochs, it resulted in model overfitting and poor accuracy.

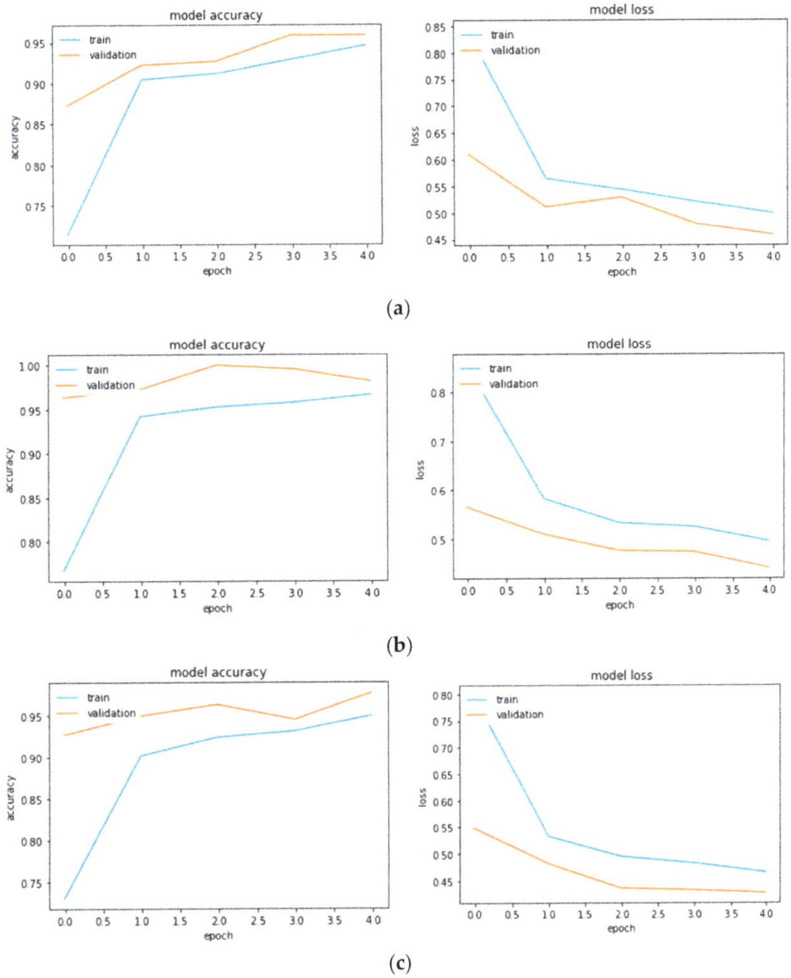

Figure 6. Accuracy and loss graph for each model: (**a**) NasNet-A (mobile); (**b**) MobileNet V3 (small); (**c**) EfficientNet B0.

The analysis of Table 7 indicated that the test accuracy obtained with NasNet-A, MobileNetV3, and EfficientNetB0 was 96.82%, 99.09%, and 98.18%, respectively. The high classification accuracy of NasNet can be ascribed to its architecture, where there are only two types of modules or cells. The normal cell extracts features and a reduction cell downsamples the input [27]. The ultimate architecture is created by stacking these cells in a specific pattern, resulting in faster convergence of the network with high accuracy. The good classification accuracy of the MobileNetV3 model can be attributed to their bottleneck residual block, which uses 1 × 1 convolutions to generate a bottleneck. When a bottleneck is used, the number of parameters and matrix multiplications is reduced, which makes the residual blocks as thin as possible in order to maximise depth while using fewer parameters [38]. The original EfficientNet's basic module is Mobile Inverted Bottleneck Convolution, which contains an expansion convolution to allow for a significantly larger number of channels for depth-wise convolution, resulting in higher accuracy [39]. Furthermore, these networks have a complex architecture and are classified as directed acyclic graph (DAG) networks, which have one layer that receives input from multiple layers and also outputs to multiple layers [9]. Since the chickpea varieties appeared to be very similar, it was critical for the network to learn the complexity among the varieties in order to perform well in classification.

Table 7. Training, validation, and testing results of the three models for chickpea varietal classification.

Model	Training		Validation		Testing		Time Per Image (ms)
	Loss	Accuracy	Loss	Accuracy	Loss	Accuracy	
NasNet-A (mobile)	0.499	0.946	0.459	0.959	0.455	0.968	38.36
MobileNetV3	0.465	0.966	0.427	0.981	0.425	0.990	26.82
EfficientNetB0	0.495	0.950	0.442	0.977	0.426	0.981	42.86

A useful tool for visualising the performance of the CNN networks is the confusion matrix. Tables 8–10 display the confusion matrix of the three models used to categorise the four varieties of chickpeas, making it simpler to identify the classes that led to the greatest amount of inaccuracy in the trained models. For a better understanding of the precision and sensitivity (recall) values acquired for each class and model, it is possible to determine the number of images that constitute true positives or false positives for each class. Table 8 of the confusion matrix illustrates this by showing that, of the 50 samples used to categorise CDC-Leader, 46 were correctly identified (TP = 46), whereas 4 were mistakenly classified as CDC Orion (FP = 4). As a result, the CDC Leader's accuracy was 92% and its false positive rate was 8%. From the confusion tables, it can be observed that the varieties CDC-Leader and CDC-Orion often mispredict each other's labels, which may be attributed to their close colour shades and subtle differences in seed texture.

Table 8. Confusion matrix of NasNet—A (mobile).

	Consul	Cory	Leader	Orion	Specificity	Sensitivity
Consul	57	0	0	0	1	1
Cory	0	54	0	0	1	1
Leader	0	0	46	4	0.982	0.921
Orion	0	0	3	56	0.975	0.949
Precision	1	1	0.938	0.933		

Table 9. MobileNetV3 (small) confusion matrix.

	Consul	Cory	Leader	Orion	Specificity	Sensitivity
Consul	61	0	0	0	1	1
Cory	0	58	0	0	1	1
Leader	0	0	50	0	0.988	1
Orion	0	0	2	49	1	0.961
Precision	1	1	0.961	1		

Table 10. EfficientNetB0 confusion matrix.

	Consul	Cory	Leader	Orion	Specificity	Sensitivity
Consul	49	0	0	0	1	1
Cory	0	50	0	0	1	1
Leader	0	0	54	2	0.988	0.964
Orion	0	0	2	63	0.987	0.969
Precision	1	1	0.964	0.969		

Further, the value of the area under the average receiver operating characteristic (ROC) curve, AUC, is provided in Figure 7a for all the three models. It was observed that the AUC values were the highest for the MobileNetV3 and EfficientNetB0 models followed by NasNet-Mobile. The high AUC values are an indicator of the excellent classification performance by the classifier models. Besides, the one-vs.-all classification approach was also conducted with ROC curve and AUC (Figure 7b). Since there are four chickpea varieties, four ROC curves were generated along with their AUC values. For each variety, it was taken as the positive class and the other varieties were jointly grouped as the negative class. The high AUC values in all the three models gives us an indication of the good performance of the models at classifying individual varieties.

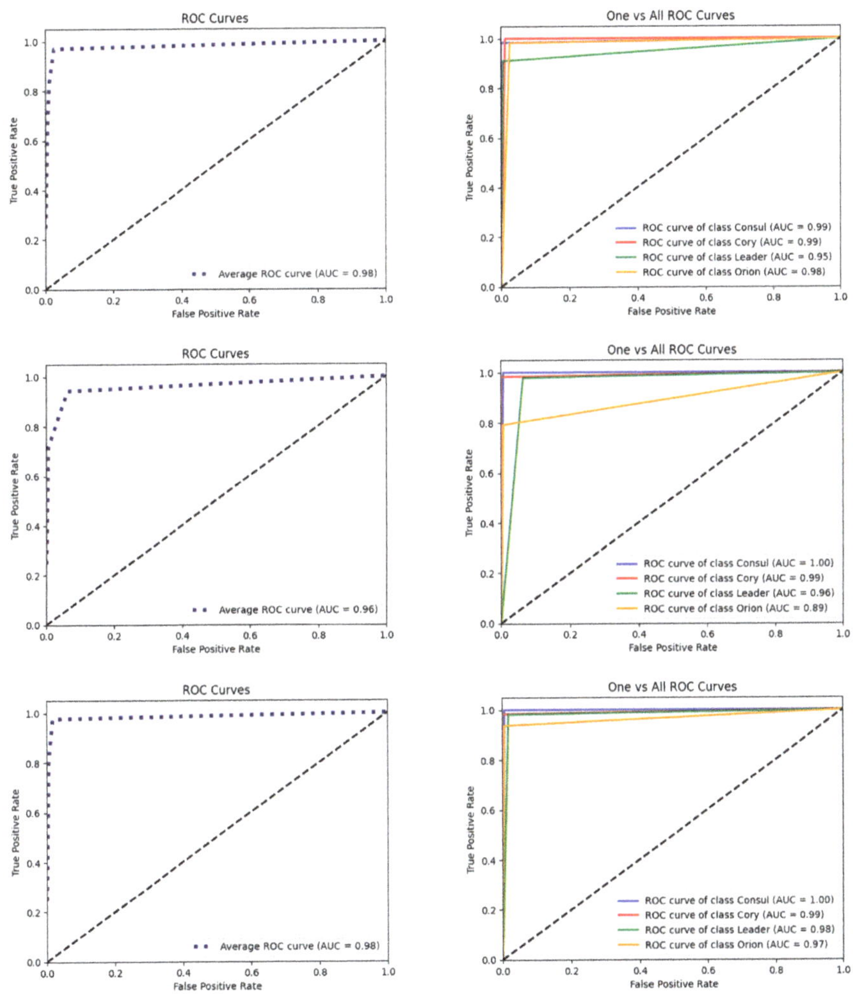

(**a**) (**b**)

Figure 7. (**a**) Average ROC curve and AUC value of MobileNetV3, NasNet-Mobile, and EfficientNetB0 models; (**b**) one-vs.-all ROC curves and AUC values of MobileNetV3, NasNet-Mobile, and EfficientNetB0 models.

It is crucial to determine the time taken by the various models to identify a test image because this reveals the model's real-time detection speed [40]. For a massive number of convolutional operations, deep CNN needs powerful computing power. The development of small models is urgently needed because smartphones and single-board computers frequently have resource limitations due to their small size. Smaller models can significantly increase detection speed while lowering resource costs. Table 7 shows the length of time required by various CNN models to categorise a single image. On mobile devices and single-board computers, MobileNetV3 outperformed NasNet-A and EfficientNetB0 in terms of speed. NasNet-A mobile (23 MB) and EfficientNetB0 (29 MB) had substantially larger models than MobileNetV3 (1.99 MB), although MobileNetV3′s detection speed (27 ms/image) was quicker than NasNet-A mobile (38 ms/image) and EfficientNetB0 (43 ms/image).

Hence, the classification of the images by MobileNetV3 took the least time among the compared models, making it the most effective model. The model, however, was assessed as a Keras model. Thereafter, the model was enhanced using post-training quantisation and transformed to a TensorFlow Lite model in order for it to be used in edge computing applications like Android mobile and Raspberry Pi 4.

3.2. *Performance of the Deployed MobileNetV3 Model*

The optimal machine-learning model chosen for deployment was MobileNetV3 based on its relative superior performance. To determine the deployment's performance, the smartphone's camera was held on top of a conveyor belt. The chickpea seeds were placed on a conveyor belt operating at a speed of 1 m/min. The smartphone identified each variety in real time, as can be seen in Figure 4. In order to record the results, the results of each frame were saved. As the testing experiment was conducted for 10 s at a speed of 40 fps, we got 400 labels. Thereafter, the confusion matrix was generated from the recorded results (Table 11). A similar approach was taken in the case of Raspberry Pi 4 deployment, where an industrial camera was used to capture the live images of chickpeas, classify them in real time, and display the results on the monitor as given in Figure 5. From Table 11, it is clear that the MobileNetV3 model can successfully classify the chickpea varieties during real-time deployment on both platforms.

Table 11. MobileNetV3 (small) confusion matrix on deployed platforms (Android mobile and Raspberry Pi 4).

	Consul	**Cory**	**Leader**	**Orion**	**Specificity**	**Sensitivity**
Consul	100	0	0	0	1	1
Cory	0	100	0	0	1	1
Leader	0	0	100	0	1	1
Orion	0	0	0	100	1	1
Precision	1	1	1	1		

4. Conclusions

The present study used a CNN-based model with the transfer-learning approach to distinguish the four different varieties of chickpea. The CNN models used for the study were NasNet-A, MobileNet-V3, and EfficientNet-B0. It was observed that the three models generalised well on the test dataset, with accuracy of 96.82%, 99.09%, and 98.18%, respectively. Further, the optimal MobileNetV3 model was deployed on two platforms, viz., Android mobile and Raspberry Pi 4, by converting trained models into the TensorFlow Lite version. The classification accuracy obtained was 100% on both deployment platforms. Compared to the other models, the MobileNetV3 is lightweight, inexpensive, and requires less time to train. In order to conduct deep learning tasks in rural areas without mobile networks, MobileNetV3-based models are ideal for integration into smartphone apps and IoT devices. However, this study is not applicable to mixtures of varieties and can only accurately identify the chickpea varieties before any mixing. Future studies may involve the application of this automated classification technique to beans and other legumes, and the information can act as a useful resource for bean and legume breeders.

Author Contributions: D.S.: Investigation, experimentation and data generation, software, validation, writing—original draft; M.P.M.: Software, image processing, validation, writing—review and editing; A.M.: Supervision, resources, writing—review and editing. All authors have read and agreed to the published version of the manuscript.

Funding: The funds received from CARE-AI, University of Guelph (Grant No: 300-05510) for conducting the study are gratefully acknowledged. This study is also partially supported by the Indian Council of Agricultural Research (ICAR), India (ICAR-IF 2018-19, F. No. 18(01)/2018-EQR/Edn). The authors are thankful for the funding from NSERC (Discovery Grant), Canada and Barrett Family Foundation, Canada.

Institutional Review Board Statement: Not Applicable.

Informed Consent Statement: Not Applicable.

Data Availability Statement: Data are available on request.

Acknowledgments: The authors are grateful to the Crop Development Centre (CDC), University of Saskatchewan, Canada for providing the chickpea varieties.

Conflicts of Interest: The authors declare no conflict of interest.

References

1. Food and Agriculture Organization (FAO). *FAOSTAT Statistical Database of the United Nation Food and Agriculture Organization (FAO) Statistical Division*; FAO: Rome, Italy, 2020.
2. Singh, N. Pulses: An overview. *J. Food Sci. Technol.* **2017**, *54*, 853–857. [CrossRef] [PubMed]
3. Havemeier, S.M.; Slavin, J. Pulses and Legumes: Nutritional Opportunities and Challenges. *Cereal Foods World* **2020**, *65*, 2.
4. Shevkani, K.; Singh, N.; Chen, Y.; Kaur, A.; Yu, L. Pulse proteins: Secondary structure, functionality and applications. *J. Food Sci. Technol.* **2019**, *56*, 2787–2798. [CrossRef] [PubMed]
5. Saha, D.; Senthilkumar, T.; Sharma, S.; Singh, C.B.; Manickavasagan, A. Application of near-infrared hyperspectral imaging coupled with chemometrics for rapid and non-destructive prediction of protein content in single chickpea seed. *J. Food Compos. Anal.* **2023**, *115*, 104938. [CrossRef]
6. Wood, J.A.; Grusak, M.A. Nutritional value of chickpea. In *Chickpea Breeding and Management*; CABI: Wallingford, UK, 2007; pp. 101–142.
7. Tiwari, U.; Bawa, A.S. Pulses: Quality standards and evaluation. In *Pulse Foods*; Academic Press: Cambridge, MA, USA, 2007; pp. 351–368.
8. Canadian Grain Commission. Pulse Crops Methods and Tests. Canadian Grain Commission: Winnipeg, MB, Canada, 2021. Available online: https://www.grainscanada.gc.ca/en/grain-research/export-quality/pulses/methods-tests.html (accessed on 20 May 2022).
9. Saha, D.; Manickavasagan, A. Chickpea varietal classification using deep convolutional neural networks with transfer learning. *J. Food Process Eng.* **2022**, *45*, e13975. [CrossRef]
10. Maes, W.H.; Steppe, K. Perspectives for remote sensing with unmanned aerial vehicles in precision agriculture. *Trends Plant Sci.* **2019**, *24*, 152–164. [CrossRef]
11. Xiao, Z.; Wang, J.; Han, L.; Guo, S.; Cui, Q. Application of machine vision system in food detection. *Front. Nutr.* **2022**, *9*, 888245. [CrossRef]
12. Saha, D.; Manickavasagan, A. Machine learning techniques for analysis of hyperspectral images to determine quality of food products: A review. *Curr. Res. Food Sci.* **2021**, *4*, 28–44. [CrossRef]
13. Rao, J.; Yang, F.; Mo, H.; Kollmannsberger, S.; Rank, E. Quantitative reconstruction of defects in multi-layered bonded composites using fully convolutional network-based ultrasonic inversion. *J. Sound Vib.* **2023**, *542*, 117418. [CrossRef]
14. Jiang, J.; Zhang, K.; Timofte, R. Towards flexible blind JPEG artifacts removal. In Proceedings of the IEEE/CVF International Conference on Computer Vision, Montreal, BC, Canada, 11–17 October 2021; pp. 4997–5006.
15. Saha, D.; Senthilkumar, T.; Singh, C.B.; Manickavasagan, A. Quantitative detection of metanil yellow adulteration in chickpea flour using line-scan near-infrared hyperspectral imaging with partial least square regression and one-dimensional convolutional neural network. *J. Food Compos. Anal.* **2023**, *120*, 105290. [CrossRef]
16. Osroosh, Y.; Khot, L.R.; Peters, R.T. Economical thermal-RGB imaging system for monitoring agricultural crops. *Comput. Electron. Agric.* **2018**, *147*, 34–43. [CrossRef]

17. Nasir, A.; Arshah, R.A.; Ab Hamid, M.R.; Fahmy, S. An analysis on the dimensions of information security culture concept: A review. *J. Inf. Secur. Appl.* **2019**, *44*, 12–22. [CrossRef]
18. Hsu, T.C.; Yang, H.; Chung, Y.C.; Hsu, C.H. A Creative IoT agriculture platform for cloud fog computing. *Sustain. Comput. Inform. Syst.* **2020**, *28*, 100285. [CrossRef]
19. Morais, R.; Silva, N.; Mendes, J.; Adão, T.; Pádua, L.; López-Riquelme, J.A.; Peres, E. mySense: A comprehensive data management environment to improve precision agriculture practices. *Comput. Electron. Agric.* **2019**, *162*, 882–894. [CrossRef]
20. AOAC. *Official Methods of Analysis, 18th ed*; Association of Official Analytical Chemists: Arlington, VA, USA, 2005.
21. Shorten, C.; Khoshgoftaar, T.M. A survey on image data augmentation for deep learning. *J. Big Data* **2019**, *6*, 60. [CrossRef]
22. Perez, L.; Wang, J. The effectiveness of data augmentation in image classification using deep learning. *arXiv* **2017**, arXiv:1712.04621.
23. Vallabhajosyula, S.; Sistla, V.; Kolli, V.K.K. Transfer learning-based deep ensemble neural network for plant leaf disease detection. *J. Plant Dis. Prot.* **2022**, *129*, 545–558. [CrossRef]
24. Tan, M.; Le, Q. Efficientnet: Rethinking model scaling for convolutional neural networks. In Proceedings of the International Conference on Machine Learning, PMLR, Long Beach, CA, USA, 9–15 June 2019; pp. 6105–6114.
25. Loni, M.; Sinaei, S.; Zoljodi, A.; Daneshtalab, M.; Sjödin, M. DeepMaker: A multi-objective optimization framework for deep neural networks in embedded systems. *Microprocess. Microsyst.* **2020**, *73*, 102989. [CrossRef]
26. Young, S.R.; Rose, D.C.; Karnowski, T.P.; Lim, S.H.; Patton, R.M. Optimizing deep learning hyper-parameters through an evolutionary algorithm. In Proceedings of the Workshop on Machine Learning in High-Performance Computing Environments, Austin, TX, USA, 15–20 November 2015; pp. 1–5.
27. Saxen, F.; Werner, P.; Handrich, S.; Othman, E.; Dinges, L.; Al-Hamadi, A. Face attribute detection with mobilenetv2 and nasnet-mobile. In Proceedings of the 2019 11th International Symposium on Image and Signal Processing and Analysis (ISPA), Dubrovnik, Croatia, 23–25 September 2019; IEEE: Piscataway, NJ, USA, 2019; pp. 176–180.
28. Koonce, B.; Koonce, B. *MobileNetV3. Convolutional Neural Networks with Swift for Tensorflow: Image Recognition and Dataset Categorization*; Apress: New York, NY, USA, 2021; pp. 125–144.
29. Montalbo, F.J.P.; Alon, A.S. Empirical analysis of a fine-tuned deep convolutional model in classifying and detecting malaria parasites from blood smears. *KSII Trans. Internet Inf. Syst.* **2021**, *15*, 147–165.
30. Lei, Y.; Hu, T.; Tang, K. Generalization performance of multi-pass stochastic gradient descent with convex loss functions. *J. Mach. Learn. Res.* **2021**, *22*, 1145–1185.
31. Chandel, N.S.; Chakraborty, S.K.; Rajwade, Y.A.; Dubey, K.; Tiwari, M.K.; Jat, D. Identifying crop water stress using deep learning models. *Neural Comput. Appl.* **2021**, *33*, 5353–5367. [CrossRef]
32. Zhou, L.; Zhang, C.; Liu, F.; Qiu, Z.; He, Y. Application of deep learning in food: A review. *Compr. Rev. Food Sci. Food Saf.* **2019**, *18*, 1793–1811. [CrossRef]
33. Gonzalez-Huitron, V.; León-Borges, J.A.; Rodriguez-Mata, A.E.; Amabilis-Sosa, L.E.; Ramírez-Pereda, B.; Rodriguez, H. Disease detection in tomato leaves via CNN with lightweight architectures implemented in Raspberry Pi 4. *Comput. Electron. Agric.* **2021**, *181*, 105951. [CrossRef]
34. Elwirehardja, G.N.; Prayoga, J.S. Oil palm fresh fruit bunch ripeness classification on mobile devices using deep learning approaches. *Comput. Electron. Agric.* **2021**, *188*, 106359.
35. Ye, J.; Li, X.; Zhang, X.; Zhang, Q.; Chen, W. Deep learning-based human activity real-time recognition for pedestrian navigation. *Sensors* **2020**, *20*, 2574. [CrossRef] [PubMed]
36. Zebin, T.; Scully, P.J.; Peek, N.; Casson, A.J.; Ozanyan, K.B. Design and implementation of a convolutional neural network on an edge computing smartphone for human activity recognition. *IEEE Access* **2019**, *7*, 133509–133520. [CrossRef]
37. Jain, S.; Vaibhav, A.; Goyal, L. Raspberry Pi based interactive home automation system through E-mail. In Proceedings of the 2014 International Conference on Reliability Optimization and Information Technology (ICROIT), Faridabad, India, 6–8 February 2014; IEEE: Piscataway, NJ, USA, 2014; pp. 277–280.
38. Abd Elaziz, M.; Dahou, A.; Alsaleh, N.A.; Elsheikh, A.H.; Saba, A.I.; Ahmadein, M. Boosting COVID-19 image classification using MobileNetV3 and aquila optimizer algorithm. *Entropy* **2021**, *23*, 1383. [CrossRef]
39. Mahbod, A.; Schaefer, G.; Wang, C.; Dorffner, G.; Ecker, R.; Ellinger, I. Transfer learning using a multi-scale and multi-network ensemble for skin lesion classification. *Comput. Methods Programs Biomed.* **2020**, *193*, 105475. [CrossRef]
40. Xie, W.; Wei, S.; Zheng, Z.; Yang, D. A CNN-based lightweight ensemble model for detecting defective carrots. *Biosyst. Eng.* **2021**, *208*, 287–299. [CrossRef]

applied sciences

MDPI

Article

UAV Hyperspectral Characterization of Vegetation Using Entropy-Based Active Sampling for Partial Least Square Regression Models

Donato Amitrano *, Luca Cicala, Marco De Mizio and Francesco Tufano

Italian Aerospace Research Centre, Via Maiorise snc, 81043 Capua, Italy
* Correspondence: d.amitrano@cira.it

Abstract: Optimization of agricultural practices is key for facing the challenges of modern agri-food systems, which are expected to satisfy a growing demand of food production in a landscape characterized by a reduction in cultivable lands and an increasing awareness of sustainability issues. In this work, an operational methodology for characterization of vegetation biomass and nitrogen content based on close-range hyperspectral remote sensing is introduced. It is based on an unsupervised active learning technique suitable for the calibration of a partial least square regression. The proposed technique relies on an innovative usage of Shannon's entropy and allows for the set-up of an incremental monitoring framework from scratch aiming at minimizing field sampling activities. Experimental results concerning the estimation of grassland biomass and nitrogen content returned RMSE values of 2.05 t/ha and 4.68 kg/ha, respectively. They are comparable with the literature, mostly relying on supervised frameworks and confirmed the suitability of the proposed methodology with operational environments.

Keywords: UAV; precision agriculture; hyperspectral imagery; active sampling; vegetation biomass; vegetation nitrogen content; partial least squares regression

Citation: Amitrano, D.; Cicala, L.; De Mizio, M.; Tufano, F. UAV Hyperspectral Characterization of Vegetation Using Entropy-Based Active Sampling for Partial Least Square Regression Models. *Appl. Sci.* **2023**, *13*, 4812. https://doi.org/10.3390/app13084812

Academic Editor: Yangquan Chen

Received: 1 March 2023
Revised: 27 March 2023
Accepted: 10 April 2023
Published: 11 April 2023

1. Introduction

Nowadays, adequate crop management is crucial to ensure high and sustainable food production. The increasing world population needs higher productivity, but at the same time, reduces the availability of agricultural lands [1]. Moreover, the increased awareness of the environmental impact of fertilizers and water supplies is modifying farming practices and regulations with the objective of minimizing waste of resources [2]. As a result, according to the second and the twelfth Sustainable Development Goals, i.e., zero hunger and responsible consumption and production, agri-food systems are asked to meet the challenge of increasing productivity in a sustainable manner [1].

In this context, automation and digitalization are key, because they allow for optimizing all the phases of the agricultural cycle, from sowing to harvesting. However, while automation, i.e., the exploitation of machines able to help and outperform humans in production tasks, has been widespread in agricultural practice for several years [1], digitalization, i.e., the creation and exploitation of digital twins of farmlands [3], is still limited.

Digital twins are virtual equivalents of physical objects [4] with whom they are connected in real time. Their intrinsic dynamic nature includes the representation of the current behavior of real objects, as well as the prediction of their future state [5]. In the case of agricultural systems, the parameters to be collected for building such a representation are related to both the environment (such as air temperature, humidity, precipitations, etc.) and the vegetation. These are traditionally retrieved through field measurements that, despite their precision, become unsustainable in presence of distributed targets. Therefore, the exploitation of remote [6] and close-range sensing technologies is essential [7–9]. Indeed, resolution issues and the flexibility of close-range remote sensing platforms, with particular reference to the possibility of carrying different sensors, make drones the preferred tools for precision agriculture. In particular, hyperspectral imagery is mostly exploited because of the spectral resolution and detail it can offer [10].

The approaches developed to retrieve vegetation biophysical parameters from remote sensing data can be classified as physical-based or empirical [11]. In the first case, a model based on radiative transfer theory is inverted. This requires significant knowledge of the scene, which has to be adequately parametrized both at canopy structure and atmospheric levels, since the solution of the problem is to find the best match between the simulated signal and the measured one. In turn, empirical methods rely on the direct calibration of a relationship between the measured signal and the biophysical variable of interest. These methods are constrained by how representative the calibration data is compared to the behavior of the object to be modeled [11]. However, if significant training samples are provided, empirical methods, such as the partial least squares regression (PLSR) [12], are able to deliver accurate predictions of many variables of agricultural interest. Despite some studies raised concerns about the ability of such methodologies to model nonlinear relation between remote sensing data and, as an example, the biomass [13], the literature widely exploited PLSR to estimate quantities such as grapevine yield and berry weight [8], grassland biomass [14], nitrogen [9] and phosphorus [15] content and carotenoids content in cotton crops [16].

Using empirical methods, the selection of significant samples is fundamental for model calibration. In the literature, this concept is usually expressed as active sample selection or active learning [17]. Its purpose is to extend the forecasting capability of an existing training dataset by adding a limited number of samples from newly available acquisitions [9,18] in order to allow new predictions with limited field work.

As stated in [18], active learning techniques suitable to solve regression problems can be categorized as based on uncertainty [19] or diversity criteria [20]. In the first case, available samples are ranked according to their uncertainty. The higher the uncertainty, the better their rank. In this family of methods, those based on variance-based pool of regressors (PAL) are probably the most interesting. The initial step in the PAL algorithm is the generation of n random subsets of the available training set. Each subset is used to train a regressor which delivers a prediction for the samples stored in the test set. This way, the samples in the test are coupled with n predictions, each one having its own variance with respect to the original training set. The higher the variance, the higher the uncertainty associated with a specific sample, which is therefore aggregated to the training set [18]. This methodology is thoroughly discussed in [21] and adopted, as an example, in [9].

Active learning techniques based on diversity criteria select samples based on the dissimilarities they introduce in the training datasets [18]. In this context, several metrics can be used to assess such dissimilarity like the Euclidean distance [22] or the cosine angle distance [23].

While the past literature mostly focused on the analysis of vegetation parameters, specific crops or growing stages, maybe not yet investigated using a given sensor [24–26], the purpose of this paper is the introduction of a crop monitoring framework, based on a PLSR model, iteratively updated as new measurements becomes available. The proposed approach is characterized by the use of a novel unsupervised and completely data-driven selection criterion for the choice of sampling areas for model calibration. It is assumed that any prior knowledge about the field or models to draw upon for active samples selection are available. In other words, monitoring activities are started from scratch and initialized by the first observation. Calibration data for PLSR are selected based on an innovative technique relying on the concept of entropy as defined by Shannon [27].

According to the active learning paradigm, the proposed methodology allows for (i) reducing field sampling, (ii) including newly acquired data within the model to improve its prediction capability and (iii) setting-up a continuous monitoring framework.

The work is organized as follows. Exploited data and the adopted methodology are introduced in Section 2. Section 3 is dedicated to the presentation of the obtained experimental results, which are discussed in Section 4. Conclusions are drawn at the end of the work.

2. Materials and Methods

2.1. Data

Data used in this study have been collected in two different campaigns using, as described by Franceschini et al. [9], the WageningenUR Hyperspectral Mapping System (HYMSY) [28]. It is a complex multisensory imaging system including, among other things, a hyperspectral sensor able to acquire data in 101 reflectance bands included in the range 450–950 nm with 5 nm interval. As declared in [9], data are provided in georeferenced (i.e., geometrically corrected) calibrated reflectance units.

The test site is a crop field cultivated with ryegrass (*Lolium perenne*) [29]. The study area was divided in 60 rectangular plots (see Figure 1) measuring 1.5 × 8 m. A total of 15 different fertilization treatments, with different nitrogen (N) amounts, were applied to the plots. In such way, each treatment was applied to 4 plots. In Figure 1, each treatment is associated with a different color.

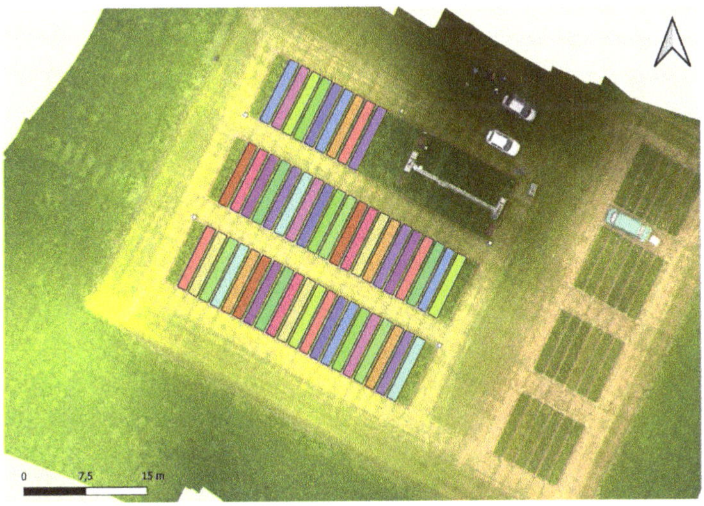

Figure 1. An orthomosaic of the study area. Colored rectangles indicate the plots. Each color refers to a fertilization treatment.

Ground measurements have been implemented concurrently with drone acquisitions with destructive methods on 15 May 2014 (average dry mass: 33.5 t/ha, average nitrogen content: 43.4 kg/ha), 14 October 2014 (average dry mass: 16.2 t/ha, average nitrogen content: 37.2 kg/ha), 9 May 2017 (average dry mass: 11.0 t/ha, average nitrogen content: 23.8 kg/ha), 29 August 2017 (average dry mass: 10.2 t/ha, average nitrogen content: 29.8 kg/ha) and 26 October 2017 (average dry mass: 9.5 t/ha, average nitrogen content 23.8 kg/ha) [9].

The objective of the study is the estimation of grassland biomass and nitrogen content. Ground data for both the quantities are available for each plot. Indeed, as reported in [9], only 49 plots were considered for the estimation, as some of them are partially obscured by a metallic structure altering their reflectance, as shown in Figure 1.

2.2. Methodology

The overall methodology is presented in Figure 2. Its rationale is to set-up a monitoring framework using no prior knowledge about the field and no models available to draw upon for active samples selection. A completely data-driven active sampling procedure allowing a prediction starting from scratch is proposed for the retrieval of calibration data necessary for PLSR.

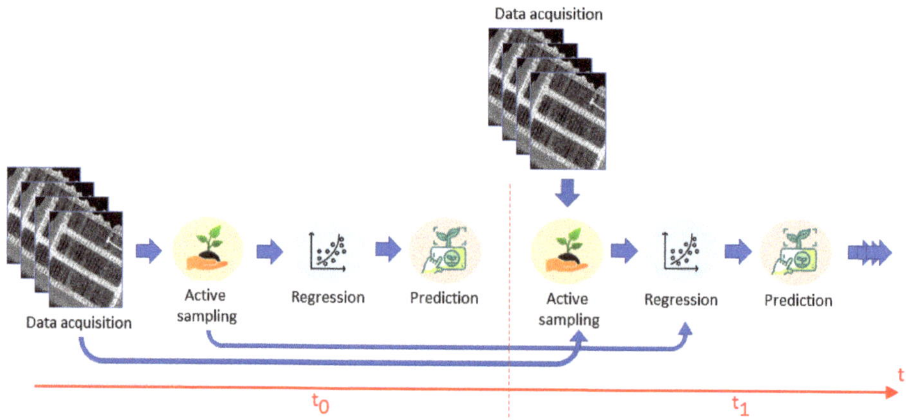

Figure 2. Proposed workflow. The active sampling procedure is fed by any observation available. Regression is implemented using all the data available at time t_i to generate the prediction of the biophysical variable of interest.

The prosed active sampling method is based on the concept of entropy H as defined by Shannon [27]

$$H = \sum_{i=1}^{N} -P_n \log_2 P_n, \tag{1}$$

where P_n is the normalized probability of the n-th histogram quantization level and N is total number of bins. Entropy is a measure of the quantity of information carried by a signal. The higher the entropy, the higher its information content.

In the literature, entropy is usually exploited as selection criterion for query by bagging (EQB). In this technique, the samples were selected based on the maximum disagreement between a committee of classifiers obtained by bagging. First, different training sets were defined by replacement within original training data. Then, each training set was used to train the selected classifier to predict labels for each unlabeled sample. Finally, the entropy of the distribution of the different labels associated to each sample was calculated with the purpose to evaluate the disagreement among the classifiers. The samples showing the maximum entropy, i.e., those with maximum disagreement among the classifiers, were added to the current training set [30]. For the purposes of this paper, it is particularly of interest as its calculation was very fast, even relative to large datasets. Moreover, being a histogram shape parameter, it is suitable to be used with vectors composed by heterogeneous data, such as spectral measurements and vegetation indices, without projection within an auxiliary homogeneous feature space.

According to the diversity paradigm of active learning techniques, the entropy is exploited as parameter to discriminate whether the sample under consideration is adding information to the set already collected. The hypothesis behind the method is that, within a given hypercube, reflectance differences between different areas of the image are due to variations of the biophysical parameter under investigation. These areas are represented by the plots, in which, according to [9], the spectral response was averaged in order to extract a single feature vector for each of them (see Figure 3). The histogram was calculated using all the available spectral features, starting from a plot randomly selected. At the time of the first acquisition, all the components of the feature vector were considered for the calculation of the histogram.

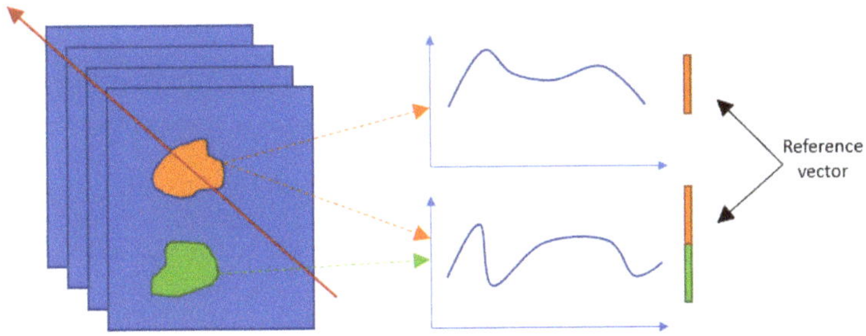

Figure 3. Schematic representation of the proposed active sampling methodology.

The histogram entropy was calculated according to Equation (1). A new plot was added by appending its spectral content to the vector constituted by the one previously considered. In other words, the vectors representing the spectral response of the plots were concatenated to form a longer one of $k \times n$ elements, where k is the number of available spectral features and n the number of plots considered. The entropy of the new histogram was calculated and, if higher than the previous value, the plot was marked as selected for field sampling. The procedure was repeated up to the end of available plots.

Marked plots are ideally sampled to retrieve ground data. They can be used to tune future sampling campaigns. In particular, the Pearson linear correlation coefficient [31] between the average plot response and ground data is computed in order to guide the successive active sampling procedure. When a new acquisition became available, the histogram was computed considering only the spectral features showing a significant correlation with the ground truth.

In synthesis, the proposed active sampling procedure can be summarized as follows.

1. At time zero, i.e., when just one image is available:
    a. Consider each plot of the field as cluster and make the spatial average for each spectral band;
    b. Calculate the histogram considering all the elements of the average spectral response starting from one randomly selected plot. Calculate the entropy H_1 of the histogram according to Equation (1);
    c. Add another plot to the dataset by appending its (average) spectral response to the vector constituted by the one of the first plot. Calculate the new histogram and its entropy H_2;
    d. If $H_2 > H_1$ mark the plot as informative and continue by adding new plots to the dataset without deleting those marked as not informative. The plots marked as informative will be sampled to retrieve model calibration data.
2. Use ground data collected with the above procedure to calculate the linear correlation with each spectral band.
3. When a new acquisition becomes available, repeat the procedure described at Point 1, this time using only the spectral features showing significant correlation with ground data (see point 2) for histograms calculation. A reasonable threshold to select features showing at least moderate correlation can be set to 0.4 [32]. All the available data should be used for samples selection. In other words, the new clusters should be appended to those already available from time zero in order to set-up an incremental sampling framework. This growing vector is the one used for entropy calculation at each iteration.
4. As the number of flights increases, the linear correlation analysis between the different spectral features and ground data continues to modulate the active sampling procedure. In this case, the computation of histograms will consider only the features showing, on average, a linear correlation coefficient higher than 0.4.

Following the diagram depicted in Figure 2, active sampling feeds the PLSR to estimate the value of specific biophysical parameter within the entire area of interest. The proposed model initially exploits a total of 141 variables constituted by the entire hyperspectral data-cube (101 reflectance bands) plus a selection of 40 vegetation indices, as suggested, among the others in [8,33]. For the ease of the reader, the list of the indices used in this study is reported in Table 1, which has a twofold purpose. The first is to collect all the formulas in one place, as it could be difficult to retrieve each of them from the literature. The second is to suggest a possible set of variables to exploit in similar studies.

Table 1. Hyperspectral vegetation indices used in this work as regression variables. In the formulas, the variable R_n refers to the central band wavelength expressed in nanometers. Indexing starts from 102 as band position indices from 1 to 101 are reserved to hyperspectral measurements.

Index	Full Name	Formula	Reference
102	Green normalized difference vegetation index	$GNDVI_1 = \frac{R_{850}-R_{540}}{R_{850}+R_{540}}$	[8]
103	Green normalized difference vegetation index	$GNDVI_2 = \frac{R_{780}-R_{550}}{R_{780}+R_{550}}$	[8]
104	Renormalized difference vegetation index	$RDVI = \frac{R_{800}-R_{670}}{(R_{800}+R_{670})^{0.5}}$	[34]
105	Red edge Position Index	$REP_{670,700} = 700 + 40\,\frac{\frac{R_{670}+R_{780}}{2}-R_{700}}{R_{740}-R_{700}}$	[35]
106	Red edge Position Index	$REP_{670,850} = 700 + 45\,\frac{\frac{R_{670}+R_{778}}{2}-R_{850}}{R_{735}-R_{695}}$	[36]
107	MERIS terrestrial chlorophyll index	$MTCI = \frac{R_{754}-R_{709}}{R_{709}+R_{681}}$	[37]
108	Modified chlorophyll absorption ratio index	$MCARI_{700,670} = [(R_{700}-R_{670}) - 0.2(R_{700}-R_{550})]\frac{R_{700}}{R_{670}}$	[38]
109	Modified Chlorophyll Absorption Reflectance Index/Optimized Soil Adjusted Vegetation Index	$\frac{MCARI_{700,670}}{OSAVI_{800,670}} = \frac{[(R_{700}-R_{670})-0.2*(R_{700}-R_{550})]\frac{R_{700}}{R_{670}}}{\frac{(1+0.6)(R_{800}-R_{670})}{R_{800}+R_{670}+0.16}}$	[33]
110	Transformed Chlorophyll Absorption Ratio Optimized Soil Adjusted Vegetation Index	$\frac{TCARI_{700,670}}{OSAVI_{800,670}} = \frac{3\left((R_{700}-R_{670})-0.2(R_{700}-R_{550})\frac{R_{700}}{R_{670}}\right)}{\frac{(1+0.6)(R_{800}-R_{670})}{R_{800}+R_{670}+0.16}}$	[33]
111	Transformed Chlorophyll Absorption Ratio	$TCARI_{700,670} = 3\left((R_{700}-R_{670}) - 0.2(R_{700}-R_{550})\frac{R_{700}}{R_{670}}\right)$	[38]
112	Modified Chlorophyll Absorption Ratio Index	$MCARI_{750,705} = \left(((R_{750}-R_{705}) - 0.2*(R_{750}-R_{550}))\frac{R_{750}}{R_{705}}\right)$	[39]
113	Modified Chlorophyll Absorption Reflectance Index/Optimized Soil Adjusted Vegetation Index	$\frac{MCARI_{750,705}}{OSAVI_{750,705}} = \frac{\left(((R_{750}-R_{705})-0.2*(R_{750}-R_{550}))\frac{R_{750}}{R_{705}}\right)}{\frac{(1+0.6)*(R_{750}-R_{705})}{R_{750}+R_{705}+0.16}}$	[39]
114	Transformed Chlorophyll Absorption Ratio	$TCARI_{750,705} = 3\left((R_{750}-R_{705}) - 0.2(R_{700}-R_{550})\frac{R_{750}}{R_{705}}\right)$	[40]
115	Transformed Chlorophyll Absorption Ratio/Optimized Soil Adjusted Vegetation Index	$\frac{TCARI_{750,705}}{OSAVI_{750,705}} = \frac{3\left((R_{750}-R_{705})-0.2(R_{700}-R_{550})\left(\frac{R_{750}}{R_{705}}\right)\right)}{\frac{(1+0.6)(R_{750}-R_{705})}{R_{750}+R_{705}+0.16}}$	[39]
116	Chlorophyll index red edge	$CI_{RE} = \left(\frac{R_{780}}{R_{710}}\right) - 1$	[33]
117	Chlorophyll index green	$CI_{G} = \left(\frac{R_{780}}{R_{550}}\right) - 1$	[33]
118	Normalized Difference Vegetation Index	$NDVI_{850,660} = \frac{R_{850}-R_{660}}{R_{850}+R_{660}}$	[7]
119	Normalized Difference Vegetation Index	$NDVI_{835,660} = \frac{R_{835}-R_{660}}{R_{835}+R_{660}}$	[7]
120	Normalized Difference Vegetation Index	$NDVI_{775,670} = \frac{R_{775}-R_{670}}{R_{775}+R_{670}}$	[7]
121	Modified Normalized Difference Vegetation Index	$mNDVI = \frac{R_{775}+R_{670}}{R_{775}+R_{670}}$	[41]
122	Soil Adjusted Vegetation Index	$SAVI = \frac{(1+0.5)\,(R_{802}-R_{660})}{(R_{802}+R_{660}+0.5)}$	[42]
123	Renormalized Difference Vegetation Index	$REDVI = \frac{R_{800}-R_{670}}{\sqrt{R_{800}+R_{670}}}$	[40]

Table 1. *Cont.*

Index	Full Name	Formula	Reference
124	Normalized Difference Red Edge Index	$NDRE_{720,795} = \frac{R_{795}-R_{720}}{R_{795}+R_{720}}$	[43]
125	Normalized Difference Red Edge Index	$NDRE_{750,770} = \frac{R_{770}-R_{750}}{R_{770}+R_{750}}$	[8]
126	NIR—red edge—red normalized difference vegetation index	$NRER = \frac{R_{850}-R_{695}}{R_{695}+R_{660}}$	[8]
127	Transformed Vegetation Index	$TVI = 0.5(120(R_{750} - R_{550}) - 200(R_{670} - R_{550}))$	[44]
128	Modified triangular vegetation index	$MTVI = 1.2[1.2(R_{800} - R_{450}) - 2.5(R_{600} - R_{540})]$	[8]
129	Enhanced Vegetation Index	$EVI_{850,660} = 2.5\left(\frac{(R_{850}-R_{660})}{((R_{850}+6* R_{660}-7.5* R_{505})+1)}\right)$	[43]
130	Enhanced Vegetation Index	$EVI_{800,670} = 2.5\left(\frac{(R_{800}-R_{670})}{((R_{800}+6* R_{670}-7.5* R_{508})+1)}\right)$	[8]
131	Leaf Chlorophyll Index	$LCI = \frac{R_{850}-R_{710}}{R_{850}+R_{680}}$	[45]
132	Modified Normalized Difference Vegetation Index	$MTCI_{var} = \frac{R_{850}-R_{680}}{R_{680}+R_{660}}$	[8]
133	Nitrogen Reflectance Index	$NRI = \frac{R_{555}-R_{550}}{R_{555}+R_{550}}$	[46]
134	Photochemical Reflectance Index	$PRI = \frac{R_{570}-R_{530}}{R_{570}+R_{530}}$	[47]
135	Spectral Polygon Vegetation Index	$SVPI = 0.4(3.7(R_{800} - R_{670}) - 1.2(R_{530} - R_{670}))$	[48]
136	Simple Ratio	$SR_{710} = \frac{R_{750}}{R_{710}}$	[49]
137	Simple Ratio	$SR_{680} = \frac{R_{800}}{R_{680}}$	[49]
138	Ratio Vegetation Index	$RVI = \frac{R_{810}}{R_{660}}$	[50]
139	Vogelmann Index	$VOG = \frac{R_{745}}{R_{720}}$	[51]
140	Gitelson and Merzlyak index	$GM = \frac{R_{750}}{R_{550}}$	[8]
141	Modified normalized difference	$MND = \frac{R_{750}-R_{705}}{R_{750}+R_{705}+2R_{508}}$	[8]

A tentative regression step was run using 10 latent variables (projected components). Its principal objective is to calculate the variable importance in projection (VIP) score. It was defined, for each variable j, as the sum, over latent variables f, of its PLS-weight value w_j weighted by the percentage of explained variance of the specific latent variable SSY_f according to relation [52]

$$VIP_j = \sqrt{\frac{J \times \left[\sum_{f=1}^{F} w_{fj}^2 \times SSY_f\right]}{\sum_{f=1}^{F} SSY_f}}, \tag{2}$$

where F is the total number of latent variables and J the total number of variables (in our case 141). The relation for the calculation of SSY_f is given by [53]

$$SSY_f = b_f^2 t_f^T t_f, \tag{3}$$

where b_f and t_f are the PLS inner relation coefficients and the score matrix relevant with the f-th latent variable, respectively.

The VIP score varies in a fixed range, being the sum of squared VIP equal to total number of variables. Therefore, it is common practice in the literature to assume as threshold for retaining informative variables a VIP score larger than one. This means that those variables have an above average influence on the building of the model explaining the observations [53].

The variables selected through VIP scoring are used for the final regression. In order to choose the optimal number of components [54], a leave-one-out cross-validation has been implemented [55]. This procedure consists in the building of a model leaving out from the calibration set one by one all the samples. In other words, if 20 samples are available, the model is built each time using 19 samples. The one left out changes at each iteration. The model built in such way is evaluated by using as test set the sample left out and the

RMSE with respect to the corresponding measure is calculated. The process is repeated using in PLSR a different number of components. For all the tries, the cumulative RMSE is calculated. Finally, the optimal number of components is selected as the one corresponding with the lowest cumulative RMSE.

3. Results

The proposed active sampling approach has been validated separately for dry mass and nitrogen content estimation purposes. Although the ground truth is available on all the plots, an operative scenario has been simulated with the objective to collect the minimum possible number of field samples. The proposed active sampling approach suggests how many and which area should be considered for field sampling. The purpose of the following experiments is to assess the regression performance of the iteratively calibrated model, showing the added value of the ground sampling at each calibration step.

The results obtained by applying the proposed methodology to the available data are reported in Table 2 and Figure 4 for the dry mass experiment and in Table 3 and Figure 5 for the nitrogen one. The variables exploited for each regression, determined via VIP scoring, are listed in Table 4.

Table 2. Dry mass estimation results using the proposed methodology, the greedy sampling and the random sampling approaches. The first five columns express, for each date, the number of samples exploited for PLSR model calibration, with the red-framed cell indicating the prediction date. Results are expressed in terms of the root mean squared error RMSE. The values $RMSE_p$ refer to predictions made using a model calibrated with data acquired up to D_{i-1}. The last column of the table refers to the average dry mass for each date. D_1: 15 May 2014, D_2: 14 October 2014, D_3: 9 May 2017, D_4: 29 August 2017, D_5: 26 October 2017.

						Proposed		Random Sampling		GSx	
D_1	D_2	D_3	D_4	D_5	Total	RMSE (t/ha)	$RMSE_p$ (t/ha)	$RMSE_{max}$ (t/ha)	$RMSE_{mean}$ (t/ha)	RMSE (t/ha)	M_{mean} (t/ha)
48					48	0.14	na	9.93	2.70	0.14	33.5
48	27				75	4.97	62.3	5.55	3.77	3.91	16.2
48	27	28			103	2.07	196	4.70	2.38	2.39	11.0
48	27	28	24		127	1.38	20.9	2.83	2.26	1.35	10.2
48	27	28	24	21	148	1.69	11.7	3.55	2.70	2.03	9.50

Table 3. Nitrogen content estimation results using the proposed methodology, the greedy sampling and the random sampling approaches. The first five columns express, for each date, the number of samples exploited for PLSR model calibration, with the red-framed cell indicating the prediction date. Results are expressed in terms of the root mean squared error RMSE. The values $RMSE_p$ refer to predictions made using a model calibrated with data acquired up to D_{i-1}. The column named as RMSE* reports values calculated on data averaged per-treatment in analogy with the study implemented in [9]. The last column of the table refers to the average nitrogen content for each date. D_1: 15 May 2014, D_2: 14 October 2014, D_3: 9 May 2017, D_4: 29 August 2017, D_5: 26 October 2017.

						Proposed			Random Sampling		GSx	
D_1	D_2	D_3	D_4	D_5	Total	RMSE (kg/ha)	RMSE* (kg/ha)	RMSE (t/ha)	$RMSE_{max}$ (kg/ha)	$RMSE_{mean}$ (kg/ha)	RMSE (kg/ha)	N_{mean} (kg/ha)
47					47	1.10	2.96	na	13.4	4.96	1.70	43.4
47	27				74	8.84	7.22	93.1	9.71	7.03	8.74	37.2
47	27	28			102	4.63	2.65	265	7.67	4.71	3.87	23.8
47	27	28	26		128	6.87	6.16	48.8	10.8	7.93	5.82	29.8
47	27	28	26	20	148	4.56	4.43	56.5	10.5	7.34	6.99	23.8

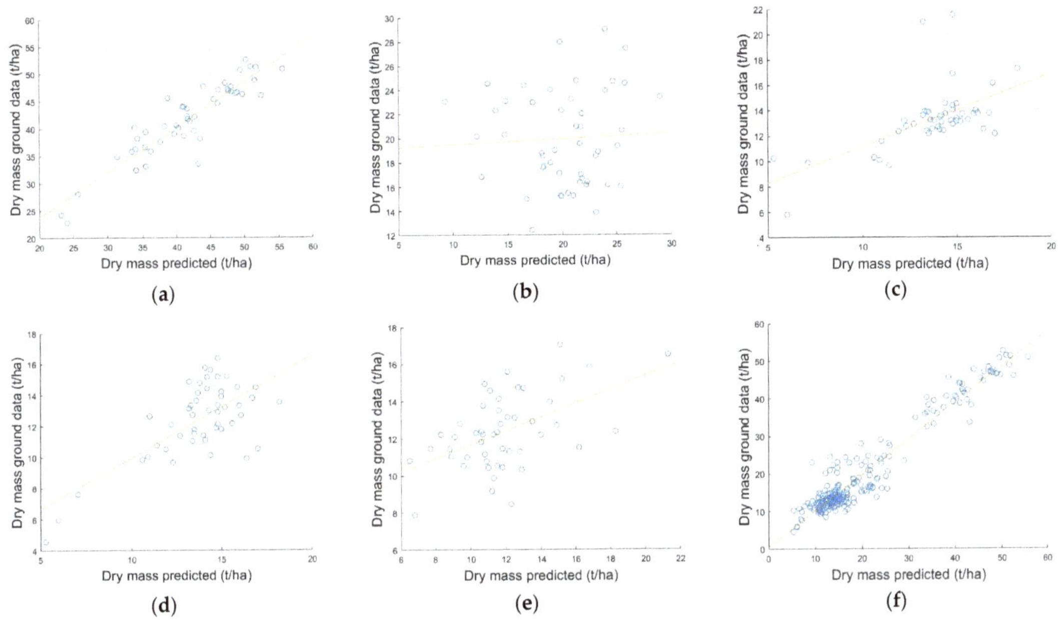

Figure 4. Scatter plots of the predicted dry mass against ground data. (**a**) D_1 (R^2 = 0.831), (**b**) D_2 (R^2 = 0.002), (**c**) D_3 (R^2 = 0.350), (**d**) D_4 (R^2 = 0.485), (**e**) D_5 (R^2 = 0.267), (**f**) all data (R^2 = 0.923).

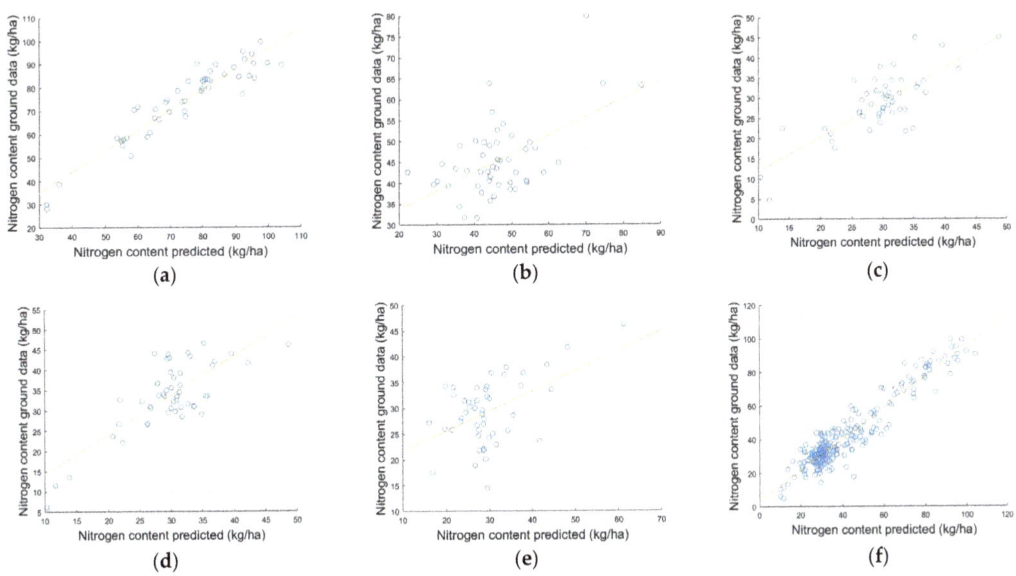

Figure 5. Scatter plots of the predicted nitrogen content against ground data. (**a**) D_1 (R^2 = 0.876), (**b**) D_2 (R^2 = 0.307), (**c**) D_3 (R^2 = 0.601), (**d**) D_4 (R^2 = 0.605), (**e**) D_5 (R^2 = 0.252), (**f**) all data (R^2 = 0.884).

Table 4. Variables used for regression determined via VIP scoring. NVIP stands for number of VIP variables.

	Dry Mass		Nitrogen	
	N_{VIP}	VIP	N_{VIP}	VIP
D_1	7	B112 (0.780), B113 (0.742), B114 (0.633), B115 (0.439), B127 (0.439), B128 (0.474), B135 (0.591)	8	B100 (0.732), B101 (0.754), B109 (−0.293), B112 (0.626), B113 (0.667), B114 (0.901), B115 (0.802), B127 (0.610)
D_2	5	B101 (0.158), B112 (0.065), B113 (0.115), B115 (0.266), B127 (0.333)	9	B65 (0.685), B99 (0.343), B100 (0.130), B101 (0.182), B112 (0.135), B113 (0.263), B114 (0.253), B115 (0.345), B127 (0.519)
D_3	9	B63 (0.774), B64 (0.699), B108 (0.028), B109 (−0.134), B112 (0.432), B113 (0.358), B114 (−0.543), B115 (−0.427), B127 (0.183)	11	B63 (0.780), B64 (0.767), B69 (0.719), B100 (0.255), B108 (−0.080), B109 (−0.242), B112 (0.526), B113 (0.441), B114 (−0.381), B115 (−0.243), B127 (0.232)
D_4	9	B63 (0.545), B64 (0.588), B108 (0.249), B109 (0.192), B112 (0.728). B113 (0.726), B114 (0.134), B115 (0.085), B127 (0.658)	10	B63 (0.435), B64 (0.477), B69 (0.527), B108 (0.136), B109 (0.080), B112 (0.669), B113 (0.661), B114 (0.233), B115 (0.187), B127 (0.579)
D_5	9	B63 (0.474), B64 (0.469), B108 (0.189), B109 (0.124), B112 (0.287). B113 (0.259), B114 (−0.168), B115 (−0.232), B127 (0.215)	11	B57 (0.278), B58 (0.420), B63 (0.547), B64 (0.538), B108 (0.253), B109 (0.190), B112 (0.340), B113 (0.317), B114 (−0.211), B115 (−0.280), B127 (0.280)

The figures show the scatter plots of the predicted dry mass and nitrogen content against the corresponding ground data. In both cases, the last one, i.e., the graph (f), refers to the whole dataset.

In Tables 2 and 3, the results were obtained using the proposed methodology and the literature approaches using greedy sampling (GSx) [56] and random sampling are reported. Acquisition dates are referred with the symbol D_i. The first five columns report, for each date, the number of samples exploited for PLSR model calibration. The red-framed cell indicates the prediction date. The sixth column expresses the total number of samples used for model calibration. Quantitative data are provided in terms of the root mean squared error (RMSE). The values reported in the last column of the tables ($RMSE_p$) refer to predictions for acquisition D_i made using a model calibrated using data collected up to the date D_{i-1}. Data concerning the random sampling approach refer to the maximum and average RMSEs obtained following 1000 experiments.

As for the dry mass experiment (see Table 2), the RMSE with respect to ground truth ranged between 0.14 t/ha obtained on 15 May 2014 and 4.97 t/ha registered on 14 October 2014. The average RMSE was of 2.05 t/ha. The number of samples selected for PLSR ranged from 48 for the acquisition made on 15 May 2014 to 21 referred to the flight made on 26 October 2017 with an average of about 30 samples per date.

The values of $RMSE_p$ were useful to assess the forecasting capability of the fitted model. They have been calculated by using the model fitted at time D_{i-1} with the data acquired at time D_i. From the Table, it arises that the model failed the prediction at time D_2 and D_3. Starting from time D_4, predictions show reasonable RMSE, although its value was much higher than that the one obtained using active learning.

Random sampling approach has been evaluated by computing the maximum RMSE and the average one. In the first case the range was between 2.83 t/ha and 9.93 t/ha. In the second, between 2.26 t/ha and 3.77 t/ha. The GSx approach returned an average RMSE of 1.96 t/ha.

As for the nitrogen content experiment, (see Table 3), the obtained RMSE ranged between 1.10 kg/ha, obtained on 15 May 2014, and 8.84 kg/ha, registered on 14 October 2014. The average value was 5.20 kg/ha. The number of samples selected for PLSR ranged from 47 for the acquisition made on 15 May 2014 to 20 referred to the flight made on 26'October 2017. For this experiment, further data about the RMSE calculated by averaging data based on the fertilization treatment (see Figure 1) have been reported in analogy with the results discussed in [9]. In this case, the obtained RMSE* ranged between 2.96 kg/ha, obtained on 9 May 2017, and 7.22 kg/ha, registered on 14 October 2014, with an average of 4.68 kg/ha. As for forecasting, the model calibrated with data collected up to time D_{i-1} fails in the prediction of the nitrogen content at D_2 and D_3. Starting from D_4, it was possible to make reasonable predictions, even with a significantly higher approximation with respect to active learning.

The random sampling test has been evaluated through calculation of the maximum RMSE and the average has been evaluated. In the first case, the range was between 7.67 kg/ha and 13.4 kg/ha. In the second case it was between 4.71 kg/ha and 7.93 kg/ha. The GSx approach returned an average RMSE of 5.42 kg/ha.

In Figure 6, the Pearson coefficient trends, with respect to available biomass (Figure 6a) and nitrogen (Figure 6b), data are reported. It is remarkable that the higher RMSE was registered for the acquisition, the one made on 14 October 2014 (see the orange line in the graphs) shows the lower correlation in the vegetation indices area.

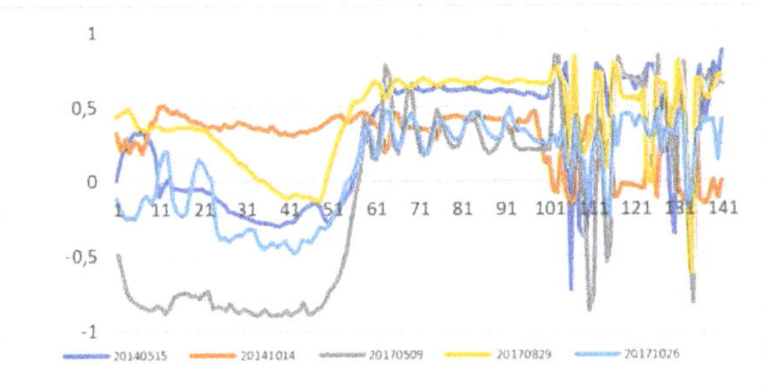

(a)

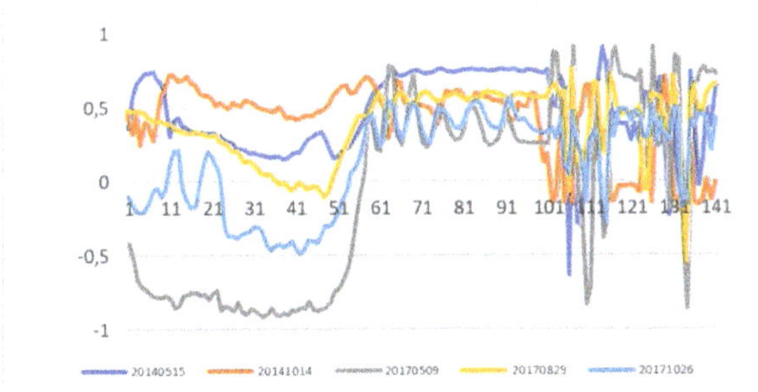

(b)

Figure 6. Linear correlation coefficient trends against the available ground data relevant with (**a**) dry mass and (**b**) nitrogen content. Indices on the x-axis from 1 to 101 refers to spectral bands. Those ranging from 102 to 141 correspond to the vegetation indices listed in Table 1.

In Tables 2 and 3 it is also reported the number of samples selected by the proposed active learning technique. Clearly, in an operational environment, the samples marked as informative for the estimation of all the biophysical parameters under investigation are collected simultaneously. With the help of Figure 7, it is possible to have a more precise idea about the amount of sampling necessary to implement the proposed framework.

In this graphics, each rectangle represents a plot. Each of them is divided in five parts, each one representative of an available acquisition. Sub-parts colored in green stand for selected samples. Figure 7a represents the sampling schema for biomass estimation, while Figure 7b depicts the one for nitrogen content investigation. By counting the green parts relevant to each acquisition, it arises that the number of necessary samples is 48, 29, 32, 30 and 24, for a total of 163, which corresponds approximately to the 66% of all the available samples.

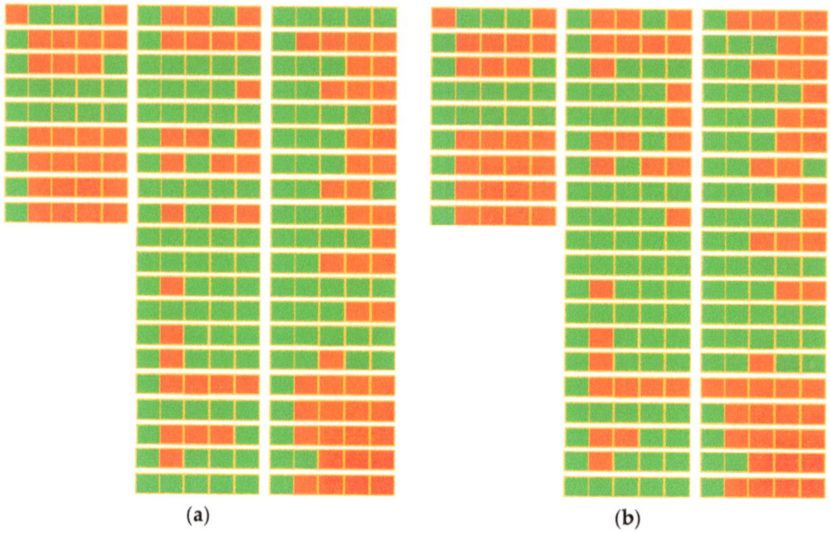

Figure 7. Schematic representation of the study area highlighting the samples selected by the proposed active learning technique. Each rectangle represents a plot. It is divided in 5 parts representing an acquisition date. If the part is colored in green, the sample is selected for regression. (a) Sampling schema for dry mass estimation. (b) Sampling schema for N-uptake estimation.

4. Discussion

Optimization of agricultural practice is key for facing the challenges of modern food production systems, which are expected to satisfy a growing demand, both in terms of quality and quantity, in a landscape characterized by a reduction of cultivable lands and an increasing awareness concerning sustainability issues.

Grassland can be accurately characterized by using UAV hyperspectral imagery. To this end, PLSR calibrated with an innovative active learning methodology has been exploited in this work.

The results presented here (using PLSR) revealed that predictions can be made with an average RMSE of 2.05 t/ha and 4.68 kg/ha for dry mass and N-uptake estimation, respectively. As shown in the previous section, they are mostly insensitive with respect to the validation sets used, as the results obtained using all the samples and only those not selected from the active sampling technique are quite similar.

As shown by data reported in Table 4, the variables used for regression mainly belong to NIR frequencies (as already observed in [9]) and to the family of chlorophyll absorption indices (CARI). By observation of the curves depicted in Figure 6 and of the linear correlation values in Table 4, it arises that these regressors are, on average, correlated with the biophysical variable under investigation. This contributes to reliable regressions in almost all the experiments, as shown in Figures 4 and 5. The only one which can be considered a failure is the D_2 dry mass regression. This experiment had a poor agreement with ground data, as testified by picture Figure 4b.

As a general comment, the acquisition D_2 is the one showing the lower correlation with ground data, especially in the vegetation indices area (see the orange line in Figure 6a). Moreover, among all the implemented regressions, the one implemented at D_2 has been implemented using the lowest number of VIP variables (five) which also showed poor correlation with ground data. This means that amount and the quality of information selected as valuable through VIP scoring was lower with respect to the other cases and this could have had an impact on the estimate. Moreover, looking at the average biomass values on the scene reported in Table 2, it is observed a significant decrease between D_1 and

D_2. This suggests the presence of nonlinearity in the data which could be not appropriately modeled by PLSR.

Overall, the obtained estimates for vegetation biomass and nitrogen content are comparable to those reported in the literature by studies using similar equipment. As an example, reference [57] claimed a RMSE of 3.75 t/ha in the estimation of maize biomass. Oliveira et al. [58] reported an RMSE of 16.99 kg/ha in the estimation of nitrogen content of grasslands. Reference [9] reported an RMSE of 3.25 t/ha and 6.50 kg/ha for biomass and N-uptake estimations of grassland, respectively, with the N-uptake result referring to an average made at fertilization treatment scale.

The results obtained using the proposed methodology have been compared against those returned by PLSR calibrated with GSx and random sampling, which is a common sampling strategy [59–61], especially when an extended ground truth is available [9].

From the comparison it arises that random sampling is, on average, less effective than the proposed methodology, as the average RMSE values tends to be higher than those reported in Section 3. This behavior is even more evident looking at the maximum RMSE values, which are constantly higher than the ones returned using the proposed methodology. These results suggest that active sampling is beneficial, especially when the dimensionality of the problem is increased by several acquisitions, as it is able to include in the calibration set the samples mostly representative of the variability of the data.

As for GSx, it can be argued that its performance is equivalent with that provided by the proposed entropy-based approach. However, GSx has one important parameter that should be set by the operator, i.e., the number of samples to be selected by active learning [56]. The experiments presented here have been implemented using the same number of samples output by the proposed approach, but it is in principle unknown. As the parameter affects the results of the calibration, this could be a serious drawback, especially in presence of extended targets. Moreover, GSx requires the projection of data in a homogeneous feature space for the computation of the distance between samples and this increases the computational complexity of the approach [56].

PLSR is a well-established linear statistical approach suitable for the analysis of multicollinear spectral dataset, thus making full use of redundant information [62]. This technique has been widely exploited to estimate crop biophysical variables from hyperspectral remote sensing data [24–26]. However, the literature also revealed that the large amount of data usually exploited in PLSR may contain irrelevant information, which could reduce the performance of the technique [62]. To cope with this, the VIP score calculated following a tentative regression has been used as discriminator and only the variable exhibiting a VIP value higher than one have been used for the final analysis [52].

Reference [9] allows for an analogy with the results here reported as its authors made available the data exploited for this study. However, significant differences in the methodology are present. In reference [9], the active sampling procedure relies on a technique developed in [21], opportunely reworked in order to improve the selection of the most representative samples. Two important aspects to be remarked about this framework are the necessity of supervision in active learning and the fact that it is based on predictive models already available, which are retrieved via bootstrapping [63]. In other words, the prediction model constituting the base for active sampling is built by splitting the calibration set in two parts. The first part is used to fit the prediction, while the second for validation. This process is repeated several times changing the composition of the two subsets. The final model is the one delivering the lowest RMSE.

The main innovation introduced in this paper is the active learning technique which, exploiting Shannon's entropy as diversity criterion, allows for the set-up of an unsupervised methodology suitable for the implementation of an incremental monitoring framework starting from scratch. The obtained results are fully comparable with the literature, which mainly relies on supervised techniques needing an already available model for active sampling implementation. This makes the proposed methodology very well suited for operational environments, in which predictive models are usually not available for specific crop fields and automation is highly requested.

In this context, a crucial aspect is represented by the amount of sampling necessary for the successful calibration of PLSR, as this could be a bottleneck for real-world implementation of monitoring activities. As reported in Section 3, the proposed methodology requires an initial massive sampling, as the entropy tend to increase when few calibration points are considered. However, as reported in Table 2, Table 3 and Figure 6, this behavior is strongly mitigated beginning with the second acquisition, in which the number of samples selected for calibration is almost halved and the trend is decreasing as soon new data are acquired. As reported in Section 3, if the whole dataset is considered, the amount of calibration data necessary to fit the predictions is about the 66% of the total.

In the literature, the amount of sampling requested for the calibration of machine learning techniques for predicting vegetation traits is not always explicitly declared. However, assuming that the area of interest has been preventively divided in sub-regions considered homogeneous against the parameter to be predicted (see as an example Figure 1), it is reasonable to collect ground data about the 70% of them [9,64] in order to implement an effective bootstrapping. This number is comparable with the one found by implementing the proposed methodology.

The last comment is about the prediction capability of the models, which is probably the most interesting aspect for end-users as it could cut out the need for field sampling. In the previous the prediction at date D_i has been determined using the data collected up to date D_{i-1}. The obtained results are different for the dry mass and the nitrogen content. In the first case, the RMSE values suggest that after few sampling campaigns it is possible to calibrate a model able to deliver reasonable predictions, although a longer time series should be studied in order to assess this consideration.

As for nitrogen content estimation, less stable RMSE values have been reported. This suggests that more calibration data are needed in order to build a model able to deliver reliable predictions. As a general comment, looking at the diagrams reported in Figures 4 and 5, it is possible to argue that, following two seasons of sparse sampling using the proposed active learning methodology, it is possible to retrieve a model with significant fitting with the ground truth. This confirms the findings of reference [9].

5. Conclusions

In this work, the characterization of vegetation biophysical parameters using unmanned aerial vehicles equipped with hyperspectral sensors has been discussed. The main innovation introduced is the active sampling technique for calibration of partial least squares regression models which, exploiting Shannon's entropy as diversity criterion, allows for setting-up an unsupervised incremental monitoring framework starting from scratch.

The proposed methodology has been tested by exploiting a dataset involving five flights made by a remotely piloted platform equipped with a hyperspectral sensor. The obtained results concerning the estimation of the ryegrass biomass (RMSE = 2.05 t/ha) and of content (RMSE = 4.68 kg/ha) revealed that the delivered prediction models are suitable for the purpose, although the relatively small dataset exploited for this study should be extended in order to draw conclusions about its generalization potential. Possible instabilities can be due to low correlation between measurements and ground data and or the presence of nonlinearity difficult to be modeled using partial least square regressions.

The amount of sampling needed is comparable with the numbers provided by the literature and therefore compatible with operational environments. In this regard, the obtained data also suggest that sampling activities can be drastically reduced as soon the monitoring is enriched with new acquisitions, thus making the proposed framework suitable for interoperability with agricultural digital twins.

Author Contributions: Conceptualization, D.A. and F.T.; methodology, D.A.; software, D.A.; validation, D.A.; formal analysis, D.A. and L.C.; investigation, D.A. and L.C.; resources, L.C., F.T. and M.D.M.; data curation, D.A.; writing—original draft preparation, D.A., L.C. and F.T.; writing—review and editing, L.C. and M.D.M.; supervision, M.D.M. and F.T.; project administration, F.T.; funding acquisition, M.D.M. All authors have read and agreed to the published version of the manuscript.

Funding: This work has been funded by the Italian Ministry of Economic Development under the aegis of the project "MONICAP—"Monitoraggio di colture agricole in persistenza"".

Institutional Review Board Statement: Not applicable.

Informed Consent Statement: Not applicable.

Data Availability Statement: Not applicable.

Conflicts of Interest: The authors declare no conflict of interest.

References

1. FAO. *The State of Food and Agriculture 2022*; FAO: Rome, Italy, 2022. [CrossRef]
2. Cisternas, I.; Velásquez, I.; Caro, A.; Rodríguez, A. Systematic literature review of implementations of precision agriculture. *Comput. Electron. Agric.* **2020**, *176*, 105626. [CrossRef]
3. Verdouw, C.; Tekinerdogan, B.; Beulens, A.; Wolfert, S. Digital twins in smart farming. *Agric. Syst.* **2021**, *189*, 103046. [CrossRef]
4. Fuller, A.; Fan, Z.; Day, C.; Barlow, C. Digital Twin: Enabling Technologies, Challenges and Open Research. *IEEE Access* **2020**, *8*, 108952–108971. [CrossRef]
5. Verdouw, C.; Beulens, A.; Reijers, H.; van der Vorst, J. A control model for object virtualization in supply chain management. *Comput. Ind.* **2015**, *68*, 116–131. [CrossRef]
6. Ranghetti, M.; Boschetti, M.; Ranghetti, L.; Tagliabue, G.; Panigada, C.; Gianinetto, M.; Verrelst, J.; Candiani, G. Assessment of maize nitrogen uptake from PRISMA hyperspectral data through hybrid modelling. *Eur. J. Remote. Sens.* **2022**. [CrossRef]
7. Sousa, J.J.; Toscano, P.; Matese, A.; Di Gennaro, S.F.; Berton, A.; Gatti, M.; Poni, S.; Pádua, L.; Hruška, J.; Morais, R.; et al. UAV-Based Hyperspectral Monitoring Using Push-Broom and Snapshot Sensors: A Multisite Assessment for Precision Viticulture Applications. *Sensors* **2022**, *22*, 6574. [CrossRef]
8. Matese, A.; Di Gennaro, S.F.; Orlandi, G.; Gatti, M.; Poni, S. Assessing Grapevine Biophysical Parameters From Unmanned Aerial Vehicles Hyperspectral Imagery. *Front. Plant Sci.* **2022**, *13*, 898722. [CrossRef]
9. Franceschini, M.H.D.; Becker, R.; Wichern, F.; Kooistra, L. Quantification of Grassland Biomass and Nitrogen Content through UAV Hyperspectral Imagery—Active Sample Selection for Model Transfer. *Drones* **2022**, *6*, 73. [CrossRef]
10. Adão, T.; Hruška, J.; Pádua, L.; Bessa, J.; Peres, E.; Morais, R.; Sousa, J.J. Hyperspectral imaging: A review on UAV-based sensors, data processing and applications for agriculture and forestry. *Remote Sens.* **2017**, *9*, 1110. [CrossRef]
11. Weiss, M.; Jacob, F.; Duveiller, G. Remote sensing for agricultural applications: A meta-review. *Remote Sens. Environ.* **2020**, *236*, 111402. [CrossRef]
12. De Jong, S. SIMPLS: An alternative approach to partial least squares regression. *Chemom. Intell. Lab. Syst.* **1993**, *18*, 251–263. [CrossRef]
13. Li, Y.; Li, M.; Li, C.; Liu, Z. Forest aboveground biomass estimation using Landsat 8 and Sentinel-1A data with machine learning algorithms. *Sci. Rep.* **2020**, *10*, 7155. [CrossRef]
14. Cho, M.A.; Skidmore, A.; Corsi, F.; van Wieren, S.E.; Sobhan, I. Estimation of green grass/herb biomass from airborne hyperspectral imagery using spectral indices and partial least squares regression. *Int. J. Appl. Earth Obs. Geoinf.* **2007**, *9*, 414–424. [CrossRef]
15. Ramoelo, A.; Skidmore, A.; Cho, M.; Mathieu, R.; Heitkönig, I.; Dudeni-Tlhone, N.; Schlerf, M.; Prins, H. Non-linear partial least square regression increases the estimation accuracy of grass nitrogen and phosphorus using in situ hyperspectral and environmental data. *ISPRS J. Photogramm. Remote Sens.* **2013**, *82*, 27–40. [CrossRef]
16. Yi, Q.; Jiapaer, G.; Chen, J.; Bao, A.; Wang, F. Different units of measurement of carotenoids estimation in cotton using hyperspectral indices and partial least square regression. *ISPRS J. Photogramm. Remote Sens.* **2014**, *91*, 72–84. [CrossRef]
17. Tuia, D.; Persello, C.; Bruzzone, L. Domain Adaptation for the Classification of Remote Sensing Data: An Overview of Recent Advances. *IEEE Geosci. Remote Sens. Mag.* **2016**, *4*, 41–57. [CrossRef]
18. Berger, K.; Caicedo, J.R.; Martino, L.; Wocher, M.; Hank, T.; Verrelst, J. A survey of active learning for quantifying vegetation traits from terrestrial earth observation data. *Remote Sens.* **2021**, *13*, 287. [CrossRef]
19. He, T.; Zhang, S.; Xin, J.; Zhao, P.; Wu, J.; Xian, X.; Li, C.; Cui, Z. An Active Learning Approach with Uncertainty, Representativeness, and Diversity. *Sci. World J.* **2014**, *2014*, 827586. [CrossRef] [PubMed]
20. Lu, X.; Zhang, J.; Li, T.; Zhang, Y. Incorporating Diversity into Self-Learning for Synergetic Classification of Hyperspectral and Panchromatic Images. *Remote Sens.* **2016**, *8*, 804. [CrossRef]
21. Douak, F.; Melgani, F.; Benoudjit, N. Kernel ridge regression with active learning for wind speed prediction. *Appl. Energy* **2013**, *103*, 328–340. [CrossRef]

22. Yuan, B.; Wu, Z.; Zhang, K.; Li, D.; Ma, Q. Application of Active Learning in Carbonate Lithologic Identification. In Proceedings of the 4th International Conference on Artificial Intelligence and Big Data, Chengdu, China, 28–31 May 2021; pp. 404–408. [CrossRef]
23. Demir, B.; Persello, C.; Bruzzone, L. Batch-Mode Active-Learning Methods for the Interactive Classification of Remote Sensing Images. *IEEE Trans. Geosci. Remote Sens.* **2011**, *49*, 1014–1031. [CrossRef]
24. Kira, O.; Linker, R.; Gitelson, A. Non-destructive estimation of foliar chlorophyll and carotenoid contents: Focus on informative spectral bands. *Int. J. Appl. Earth Obs. Geoinf.* **2015**, *38*, 251–260. [CrossRef]
25. Pan, W.-J.; Wang, X.; Deng, Y.-R.; Li, J.-H.; Chen, W.; Chiang, J.Y.; Yang, J.-B.; Zheng, L. Nondestructive and intuitive determination of circadian chlorophyll rhythms in soybean leaves using multispectral imaging. *Sci. Rep.* **2015**, *5*, 11108. [CrossRef]
26. Yu, K.-Q.; Zhao, Y.-R.; Zhu, F.-L.; Li, X.-L.; He, Y. Mapping of Chlorophyll and SPAD Distribution in Pepper Leaves During Leaf Senescence Using Visible and Near-Infrared Hyperspectral Imaging. *Trans. ASABE* **2016**, *59*, 13–24. [CrossRef]
27. Shannon, C.E. A Mathematical Theory of Communication. *Bell Syst. Tech. J.* **1948**, *27*, 623–656. [CrossRef]
28. Suomalainen, J.; Anders, N.; Iqbal, S.; Roerink, G.; Franke, J.; Wenting, P.; Hünniger, D.; Bartholomeus, H.; Becker, R.; Kooistra, L. A Lightweight Hyperspectral Mapping System and Photogrammetric Processing Chain for Unmanned Aerial Vehicles. *Remote Sens.* **2014**, *6*, 11013–11030. [CrossRef]
29. Capolupo, A.; Kooistra, L.; Berendonk, C.; Boccia, L.; Suomalainen, J. Estimating plant traits of grasslands from uav-acquired hyperspectral images: A comparison of statistical approaches. *ISPRS Int. J. Geo-Inf.* **2015**, *4*, 2792–2820. [CrossRef]
30. Patra, S.; Bruzzone, L. A Fast Cluster-Assumption Based Active-Learning Technique for Classification of Remote Sensing Images. *IEEE Trans. Geosci. Remote Sens.* **2011**, *49*, 1617–1626. [CrossRef]
31. Neter, J.; Wasserman, W.; Whitmore, G.A. *Applied Statistics*, 4th ed.; Allyn & Bacon: Boston, MA, USA, 2000.
32. Profillidis, V.; Botzoris, G. Statistical Methods for Transport Demand Modeling. In *Modeling of Transport Demand*; Elsevier: Amsterdam, The Netherlands, 2019; pp. 163–224. [CrossRef]
33. Clevers, J.G.P.W.; Kooistra, L. Using hyperspectral remote sensing data for retrieving canopy chlorophyll and nitrogen content. *IEEE J. Sel. Top. Appl. Earth Obs. Remote Sens.* **2012**, *5*, 574–583. [CrossRef]
34. Roujean, J.-L.; Breon, F.-M. Estimating PAR absorbed by vegetation from bidirectional reflectance measurements. *Remote Sens. Environ.* **1995**, *51*, 375–384. [CrossRef]
35. Guyot, G.; Baret, F. Utilisation de la Haute Resolution Spectrale pour Suivre L'etat des Couverts Vegetaux. In *Spectral Signatures of Objects in Remote Sensing*; European Space Agency: Paris, France, 1988; 279p.
36. Curran, P.J.; Windham, W.R.; Gholz, H.L. Exploring the relationship between reflectance red edge and chlorophyll concentration in slash pine leaves. *Tree Physiol.* **1995**, *15*, 203–206. [CrossRef]
37. Dash, J.; Curran, P. Evaluation of the MERIS terrestrial chlorophyll index (MTCI). *Adv. Space Res.* **2007**, *39*, 100–104. [CrossRef]
38. Daughtry, C.S.T.; Walthall, C.L.; Kim, M.S.; De Colstoun, E.B.; McMurtrey, J.E., III. Estimating Corn Leaf Chlorophyll Concentration from Leaf and Canopy Reflectance. *Remote Sens. Environ.* **2000**, *74*, 229–239. [CrossRef]
39. Wu, C.; Niu, Z.; Tang, Q.; Huang, W. Estimating chlorophyll content from hyperspectral vegetation indices: Modeling and validation. *Agric. For. Meteorol.* **2008**, *148*, 1230–1241. [CrossRef]
40. Haboudane, D.; Miller, J.R.; Pattey, E.; Zarco-Tejada, P.J.; Strachan, I.B. Hyperspectral vegetation indices and novel algorithms for predicting green LAI of crop canopies: Modeling and validation in the context of precision agriculture. *Remote Sens. Environ.* **2004**, *90*, 337–352. [CrossRef]
41. Jurgens, C. The modified normalized difference vegetation index (mNDVI) a new index to determine frost damages in agriculture based on Landsat TM data. *Int. J. Remote Sens.* **1997**, *18*, 3583–3594. [CrossRef]
42. Huete, A.R. A soil-adjusted vegetation index (SAVI). *Remote Sens. Environ.* **1988**, *25*, 295–309. [CrossRef]
43. Huete, A.; Didan, K.; Miura, T.; Rodriguez, E.P.; Gao, X.; Ferreira, L.G. Overview of the radiometric and biophysical performance of the MODIS vegetation indices. *Remote Sens. Environ.* **2002**, *83*, 195–213. [CrossRef]
44. Broge, N.H.; Leblanc, E. Comparing prediction power and stability of broadband and hyperspectral vegetation indices for estimation of green leaf area index and canopy chlorophyll density. *Remote Sens. Environ.* **2001**, *76*, 156. [CrossRef]
45. Datt, B. Remote Sensing of Water Content in Eucalyptus Leaves. *Aust. J. Bot.* **1999**, *47*, 909–923. [CrossRef]
46. Huang, W.; Yang, Q.; Pu, R.; Yang, S. Estimation of Nitrogen Vertical Distribution by Bi-Directional Canopy Reflectance in Winter Wheat. *Sensors* **2014**, *14*, 20347–20359. [CrossRef]
47. Gamon, J.A.; Peñuelas, J.; Field, C.B. A narrow-waveband spectral index that tracks diurnal changes in photosynthetic efficiency. *Remote Sens. Environ.* **1992**, *41*, 35–44. [CrossRef]
48. Vincini, M.; Frazzi, E.; Alessio, P. Angular Dependence of Maize and Sugar Beet VIs from Directional CHRIS/Proba Data. In Proceedings of the 4th ESA CHRIS PROBA Workshop, online, 19–21 September 2006; pp. 19–21.
49. Main, R.; Cho, M.A.; Mathieu, R.; O'Kennedy, M.M.; Ramoelo, A.; Koch, S. An investigation into robust spectral indices for leaf chlorophyll estimation. *ISPRS J. Photogramm. Remote Sens.* **2011**, *66*, 751–761. [CrossRef]
50. Gonenc, A.; Ozerdem, M.S.; Acar, E. Comparison of NDVI and RVI Vegetation Indices Using Satellite Images. In Proceedings of the 2019 8th International Conference on Agro-Geoinformatics (Agro-Geoinformatics), Istanbul, Turkey, 16–19 July 2019; pp. 1–4. [CrossRef]
51. Vogelmann, J.E.; Rock, B.N.; Moss, D.M. Red edge spectral measurements from sugar maple leaves. *Int. J. Remote Sens.* **1993**, *14*, 1563–1575. [CrossRef]

52. Farrés, M.; Platikanov, S.; Tsakovski, S.; Tauler, R. Comparison of the variable importance in projection (VIP) and of the selectivity ratio (SR) methods for variable selection and interpretation. *J. Chemom.* **2015**, *29*, 528–536. [CrossRef]
53. Cocchi, M.; Biancolillo, A.; Marini, F. Chemometric Methods for Classification and Feature Selection. *Compr. Anal. Chem.* **2018**, *82*, 265–299. [CrossRef]
54. Wold, S.; Sjostrom, M.; Eriksson, L. PLS-Regression: A Basic Tool of Chemometrics. *Chemom. Intell. Lab. Syst.* **2001**, *58*, 109–130. [CrossRef]
55. Molinaro, A.M.; Simon, R.; Pfeiffer, R.M. Prediction error estimation: A comparison of resampling methods. *Bioinformatics* **2005**, *21*, 3301–3307. [CrossRef] [PubMed]
56. Wu, D.; Lin, C.-T.; Huang, J. Active learning for regression using greedy sampling. *Inf. Sci.* **2019**, *474*, 90–105. [CrossRef]
57. Wang, C.; Nie, S.; Xi, X.; Luo, S.; Sun, X. Estimating the Biomass of Maize with Hyperspectral and LiDAR Data. *Remote Sens.* **2016**, *9*, 11. [CrossRef]
58. Oliveira, R.A.; Näsi, R.; Niemeläinen, O.; Nyholm, L.; Alhonoja, K.; Kaivosoja, J.; Jauhiainen, L.; Viljanen, N.; Nezami, S.; Markelin, L.; et al. Machine learning estimators for the quantity and quality of grass swards used for silage production using drone-based imaging spectrometry and photogrammetry. *Remote Sens. Environ.* **2020**, *246*, 111830. [CrossRef]
59. Tian, Q.; Gong, P.; Zhao, C.; Guo, X. A feasibility study on diagnosing wheat water status using spectral reflectance. *Chin. Sci. Bull.* **2001**, *46*, 666–669. [CrossRef]
60. Zhao, H.-S.; Zhu, X.-C.; Li, C.; Wei, Y.; Zhao, G.-X.; Jiang, Y.-M. Improving the Accuracy of the Hyperspectral Model for Apple Canopy Water Content Prediction using the Equidistant Sampling Method. *Sci. Rep.* **2017**, *7*, 11192. [CrossRef] [PubMed]
61. Verrelst, J.; Berger, K.; Rivera-Caicedo, J.P. Intelligent Sampling for Vegetation Nitrogen Mapping Based on Hybrid Machine Learning Algorithms. *IEEE Geosci. Remote Sens. Lett.* **2020**, *18*, 2038–2042. [CrossRef] [PubMed]
62. Jin, J.; Wang, Q. Evaluation of informative bands used in different pls regressions for estimating leaf biochemical contents from hyperspectral reflectance. *Remote Sens.* **2019**, *11*, 197. [CrossRef]
63. Afanador, N.; Tran, T.; Buydens, L. An assessment of the jackknife and bootstrap procedures on uncertainty estimation in the variable importance in the projection metric. *Chemom. Intell. Lab. Syst.* **2014**, *137*, 162–172. [CrossRef]
64. Schucknecht, A.; Seo, B.; Krämer, A.; Asam, S.; Atzberger, C.; Kiese, R. Estimating dry biomass and plant nitrogen concentration in pre-Alpine grasslands with low-cost UAS-borne multispectral data—A comparison of sensors, algorithms, and predictor sets. *Biogeosciences* **2022**, *19*, 2699–2727. [CrossRef]

applied sciences

MDPI

Article

An Improved Method to Obtain Fish Weight Using Machine Learning and NIR Camera with Haar Cascade Classifier

Samuel Lopez-Tejeida [1], Genaro Martin Soto-Zarazua [1,*], Manuel Toledano-Ayala [2], Luis Miguel Contreras-Medina [1], Edgar Alejandro Rivas-Araiza [2] and Priscila Sarai Flores-Aguilar [1]

1 Facultad de Ingeniería Campus Amazcala, Universidad Autónoma de Querétaro, Carretera a Chichimequillas S/N Km 1, Amazcala, El Marqués 76265, Querétaro, Mexico
2 Facultad de Ingeniería Centro Universitario, Universidad Autónoma de Querétaro, Cerro de las Campanas S/N, Santiago de Querétaro 76010, Querétaro, Mexico
* Correspondence: soto_zarazua@yahoo.com.mx or genaro.soto@uaq.mx; Tel.: +1-52-44-2332-9713

Abstract: The calculation of weight and mass in aquaculture systems is of great importance, since with this task, it is decided when to harvest; generally, the above is manipulating the body manually, which causes stress in the fish body. Said stress can be maintained in the fish body for several hours. To solve this problem an improved method was implemented using artificial intelligence, near-infrared spectroscopy camera, Haar classifiers, and a mathematical model. Hardware and software were designed to get a photograph of the fish in its environment in real conditions. This work aimed to obtain fish weight and fish length in real conditions to avoid the manipulation of fish with hands for the process mentioned, avoiding fish stress, and reducing the time for these tasks. With the implemented hardware and software adding an infrared light and pass band filter for the camera successfully, the fish was detected automatically, and the fish weight and length were calculated moreover the future weight was estimated.

Keywords: machine learning; Haar; NIR; aquaculture; fish growth

Citation: Lopez-Tejeida, S.; Soto-Zarazua, G.M.; Toledano-Ayala, M.; Contreras-Medina, L.M.; Rivas-Araiza, E.A.; Flores-Aguilar, P.S. An Improved Method to Obtain Fish Weight Using Machine Learning and NIR Camera with Haar Cascade Classifier. *Appl. Sci.* **2023**, *13*, 69. https://doi.org/10.3390/app13010069

Academic Editors: José Miguel Molina Martínez, Ginés García-Mateos and Dolores Parras-Burgos

Received: 29 November 2022
Revised: 9 December 2022
Accepted: 13 December 2022
Published: 21 December 2022

1. Introduction

Imaging object detection, nowadays, is widely used to identify objects. However, the most advantageous to the classification of moving objects [1], especially underwater object detection, has been a challenge because the water influences the recognition of the object [2]. The use of image analysis has produced applications that help in the agriculture to medicine fields [3]. On the other hand, in aquaculture, multiple tasks can be performed using imaging detection. However, there are also modern practices involving automation and intelligent technology [4]; technology such as the internet of thing (IoT) helps to obtain data from the tank culture with the use of sensors and avoid poor levels of oxygen, water contamination, parasites, or disease transmission [5]. Additionally, some examples are the regression methods used in aquaculture for the management and water quality predictions, time series methods, artificial neural network methods, and support machine methods [6]. Moreover, agriculture 4.0 contributes to the development of agriculture because crops can be cultivated much easier with less effort in less time, as well as every time the growing variables can be monitored by a sensor and displayed on a PC or cell phone screen: digital technologies such as cyber-physical (CPS), artificial intelligence (AI), wireless sensor networks (WSN), big data analytics (BDA), autonomous robots systems (ARS), and ubiquitous cloud computing (UCC) [7].

An essential aspect of aquaculture farms is fish growth; the fish's body length and width are related to fish weight. This information calculates the feeding ratio, fish size classification, and harvest [8]. Regularly, these data are collected manually, transforming into an expensive and tiring, as well as a waste of time, experience for intensive systems [9]; moreover, handling the fish with hands causes stress to it, and stress causes cortisol. It is accepted that long-term elevation in cortisol results in reduced food intake and growth in fish [10]. In addition to avoiding stress fish, there are different non-invasive methods, such as mathematical models or the length–weight relationship (LWR). They are based on numerical observations that can estimate fish biomass [11]; moreover, there is substantial demand for automatic fish monitoring systems, because these could improve efficiency and reduce the labor requirements in the aquaculture industry [12]. A machine vision system (MVS) consists of an image acquisition system, image processing, and statistical analysis [13], and MVS is a non-invasive technique for estimating fish mass and size, which has been attracting the interest of researchers because it avoids stress and injury [14]. Another vital aspect of MVS is that these fish sampling systems involve two tasks: fish detection, which discriminates fish from non-fish objects in underwater videos, and fish species classification, which identifies the species of the detected fish [15]. The quality of the images will depend on the light conditions at the time of the shooting; otherwise, if this factor is ideal, more accurate processing results will be achieved [16].

Some aquaculture installations have poor lighting, and near-infrared (NIR) computer vision technology is not affected by visible light intensity. It can obtain good imaging results in relatively dim light environments [17]. The NIR had other advantages: autofluorescence, deeper tissue penetration, and minimum photodamage to biological samples [18]. Additionally, the NIR has the rapid acquisition of sample data, the potential for simultaneous determination of several parameters, and the ability to replace destructive, expensive, and time-consuming conventional reference methods [19]. Another tool is Haar cascading by machine learning method; this uses a classifier from many positive and negative photos [20]; the advantage of using the machine learning method is short detection time, high detection rate, and strong adaptability to light changes [21]. In the case of machine vision system, ref. [1] used convolutional neural network and machine vision to provide an automatic method for grading fish feeding intensity; additionally, ref. [2] used a machine vision system including a multi-column convolution neural network (MCNN) and deeper dilated convolution neural network (DCNN) for counting fishes; finally, ref. [3] used machine vision, acoustics. and sensors to analyze fish behavior in pro of production and management decisions.

This article aims to show a non-invasive method to obtain fish weight using image NIR plus Haar cascading classifiers. These tools are used to segment the fish and obtain fish length, then obtain accurate weight measurements with a mathematical model. Generally, the mathematical models used for aquaculture are logistics, exponential, Michaelis-Menten, Gompertz, Von Bertalanffy, and Janoschechk. Finally, this work is different from others; in the first instance, it used an infrared camera adding a pass-band filter lens to reduce the wavelength, obtain more focus in capturing the fish, and avoid the noise that can interfere; it also used Haar classifier to identify the fish in the culture system, and compared with other works, they used more complex analysis such as the convolution neural network, multi-column convolution neural network, artificial neural network, or wavelet; lastly, the system used a mathematical model to estimate fish future weight and length.

2. Materials and Methods

2.1. System Design

An essential aspect of the experiment was how to get the pictures; for this task, a way to place all the used elements was implemented, and it is shown in Figure 1.

Each number specifies an element:

1. PC was used with the NIR camera and MATLAB 9 software to program the Haar cascade classifier.
2. NIR camera iDS brand, made in Germany, Dimbacher Str. 10, UI-3240CP-NIR-GL model, this camera is infrared sensitive in the near-infrared range. The equipment was obtained in the web page of the brand.
3. Kowa lens sensor size 2/3", focal length 1.4–12 mm, C mound, and this lens has a passband filter. The equipment was obtained in Midwest Optical System.

4. The species used in this case for the experiment was tilapia used.
5. Glass fishbowl 30 cm long × 20 cm wide × 22 cm high.
6. Infrared lamp INF-L-DBLIR082/16H.
7. Lighting controller, CCS Inc. (Tokyo, Japan) PD3-5024-4-EI(A); lastly, to subject the camera, two tripods were used, one DOLICA ST650 brand with clamp and another of the Manfrotto brand obtained the webpage

The operation of the system is as follows. First, the image is captured by the NIR camera this camera is connected to the PC, the information captured by the camera is used by an algorithm done with Haar cascade classifier and mathematical models. Lastly, the information related to fish actual length, future length, the actual weight, and future weight is shown in the user interface.

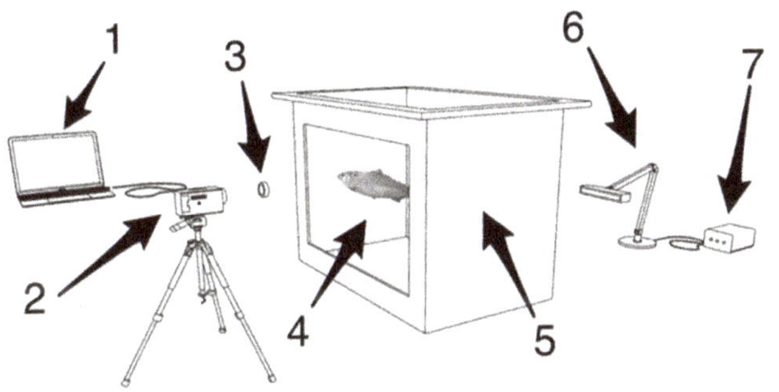

Figure 1. Arrangement of materials.

2.2. Image Acquisition System

An essential aspect of obtaining images is the light from the environment, which is the reason for the use of a NIR camera. To obtain the collection fish data, the fish species used was tilapia *Oreochromis niloticus*; a total of 1200 pictures were obtained with the NIR camera; the resolution of 1.3 MP, 1280 × 1024, and the image format used was PNG, later the software was trained whit these pictures. The camera was collocated in two places to get the pictures; first, the camera was placed in front of the fishbowl; an infrared lamp stood behind; this one was placed behind the fishbowl to increase the intensity of light images. The fishbowl was filled with water, and some fish got into it; this camera configuration is shown in Figure 2. The procedure to take the pictures was necessary to connect the NIR camera to the PC through the USB port. In the pc was installed, as well as the NIR infrared camera software, and using this software, the images were taken, and we saved them in the PC.

Figure 2. First system configuration.

The second configuration considered the camera above the fishbowl and the infrared lamp above the fishbowl; because fish culture tanks are made of black geomembrane, it is impossible to set the NIR camera in front of the fish tank and the infrared lamp opposite to the fish tank Figure 3.

Figure 3. Second system configuration.

2.3. Fish Growth Models

An essential aspect of the experiment was to choose a growth model; fish growth is any change in size or amount of body. These changes could be positive or negative, and the fish length can be obtained the weight with the equation:

$$W = aL^n \tag{1}$$

where W is the fish weight in grams; L is the fish length in cm; a is the intersection in the y-axis, and n is the exponent to estimate the fish's well-being.

A relationship to know the fish's fatness or well-being and the condition factor [22]. This relationship is shown in Equation (1). Additionally, can be transformed to:

$$\log(W) = \log(a) + b\log(L) \tag{2}$$

The formula expresses the following, in the case of b, it might not vary from the ideal value of 3.0, depicting an isometric growth; otherwise, if b has a value less than 3.0, the fish is turning slimmer with increasing length, and growth is considered negative allometric. On the other hand, if b is greater than 3.0, the fish is getting heavier, which is proof of positive allometric growth, which shows an optimum condition. However, depending on the kind of fish feed, this relates to the quantity of protein in the fish feed [23].

There is another model that uses some variables, precisely the temperature [24], taking into account that fish is a poikilothermic species; Equation (3) is shown a model growth based on temperature:

$$\Delta L = -1.6707 + 0.09682\ T \tag{3}$$

where L is the length in cm and T is the temperature in °C.

The result obtained in Equation (3) can be transformed to weight by Equation (4)

$$W = 1.861 \times 10^{-8} \cdot L^3 \tag{4}$$

where W is weight in grams and L is the length in cm.

Mainly for the experiment was interested in the models that predict the growth of the fish, depending on the age and then thrown length in cm; the logistic, Michaelis-Menten, Gompertz, Von Bertalanffy, and Janoscheck models were the basis. Taking into account that the artificial vision device coupled with the processing software would show the length of the fish, the models used were the Michaelis-Menten model; the equation used in the algorithm was:

$$Length = \frac{0.009 * 235.27^{2.983} + 630.14 * day^{2.983}}{235.27^{2.983} + day^{2.983}} \tag{5}$$

The other mathematical model used was the Logistic, shown in Equation (6), unlike the Michaelis-Menten model, which predicts the weight in grams based on age in days.

$$Weight = \frac{427.64}{1 + 85.9133 * e^{(-0.023612*day)}} \tag{6}$$

2.4. Algorithm Training

The software used for the experiment was MATLAB; this software has a tool called Haar Clasiffier, and this software can obtain the classification in real-time. The training algorithm was performed with 1200 images taken by the camera. The classifier has pre-trained classifier objects, but generally, they refer to body parts, so it is impossible to use the pre-trained classifiers for this work. For the experimental object detector, 1200 fish images were loaded into the algorithm; these were considered positive samples, and the algorithm also needed negative samples. The negative samples were pictures taken with the NIR camera, but these images were not fish and were considered negative. With this training, the algorithm could detect regions of interest and differentiate between fish and non-fish. Additionally, the algorithm can automatically generate negative samples to increase the training level, which improves the algorithm to detect fishes and objects that are not fishes. Figure 4 shows a positive image, and Figure 5 shows a negative image. After the training, the system could automatically identify the fish in the fishbowl using the Haar classifier. Then, the system will be collocated in an aquaculture system to be proven. Lastly, the image labeler was used, which provides an easy way to identify the positive samples by interactively specifying rectangular regions of interest. Additionally, the positive images previously saved can be made more positive by adding rotation noise or by varying brightness or contrast. In the case of negative images, as more stages are added, the detector's overall false positive rate decreases, causing the generation of negative samples to be more difficult. An aspect to take into consideration is that Haar classifier can throw false positives. In this training, the false positive test was performed with the camera taking objects that can be in the fish environment, but are not fish, these were classified as negative images. In the first instances, there could be many false positives, but as the training continues, this false positive rate starts to reduce. The impact on the test result is that, in the beginning, the software could throw false positives, but with the use of the software, this will decrease.

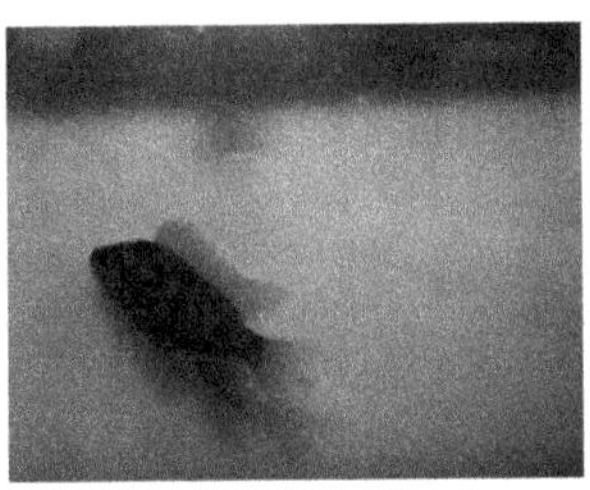

Figure 4. Positive image.

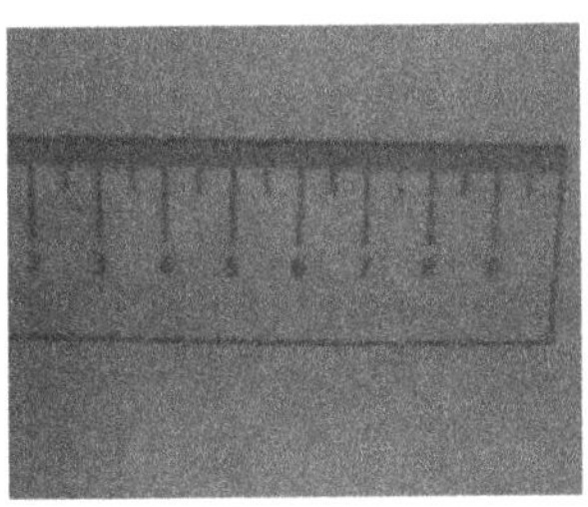

Figure 5. Negative image.

For this reason, it is helpful to supply as many negative images as possible. To improve training accuracy, we supplied negative images with backgrounds typically associated with the objects of interest. Additionally, we included negative images that contained non-objects similar in appearance to the objects of interest.

The strong classifier used in this work was the Adaboost for searching a small number of features related to fish and not having a significant variation.

2.5. Image Processing

After algorithm training, the system was put to work in real-time; for this, the NIR camera was used to set communication with the computer through USB protocol and show live video in the software interface implemented in the previously designed PC. First, the system was tested with the first configuration; the way to do this was after the NIR camera and the infrared lamp were collocated in their respective positions. The software interface has the option to capture video, and when this is performed automatically in an interface implemented, it appears on the length and weight of the fish. Figure 6 shows how the algorithm performs the image processing.

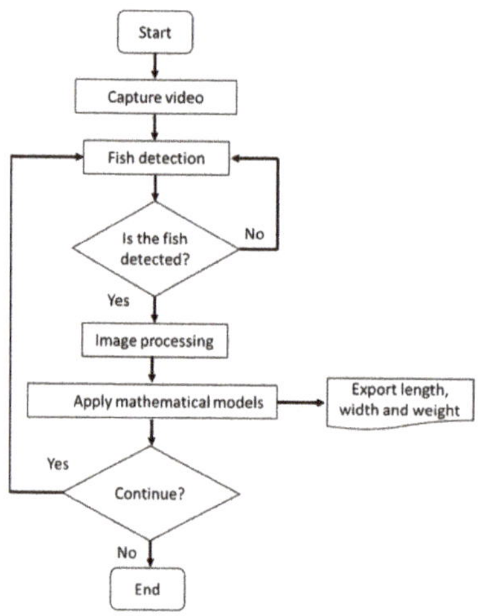

Figure 6. Flow chart of the image processing.

As shown in Figure 6, the main algorithm comprises two main blocks: the pre-processing and the estimation. It starts when the NIR camera plays the captured video, and the algorithm instantly detects if the objects that appear in the camera are fish or not, and it is possible to perform by using Haar classifiers in the algorithm using only a region of interest for the investigation. The fish detector generates a box containing the fish detected correctly; that is, when the fish is entirely in a lateral position, this is shown in Figure 7. This performed through a process that is transparent for the user. Once the region of interest is delimited, the image is extracted and processed to extract its contour, obtaining the width and length measurements of the fish in pixel dimensions.

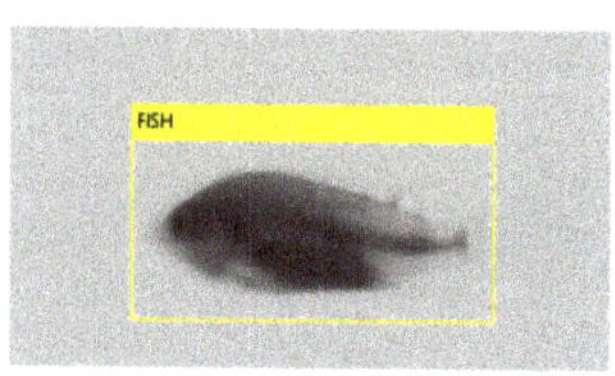

Figure 7. Fish detection by the algorithm.

This kind of processing is considered a machine learning approach because later it could automatically identify the objects of interest by marking them with a box chart, as shown in Figure 7; the MATLAB software was used to train the algorithm. The way that the algorithm identifies the fish apart from the positive and negative images requires a way to identify the object rapidly; in this regard, the Haar classifiers are used. The image is detected, then the algorithm applies the Haar classifiers, and instantly the parts are discarded from the image that do have not the object of interest; this is made with statistics, and the Haar classifiers implemented in the algorithm use rectangles (Figure 8) to identify the objects of interest.

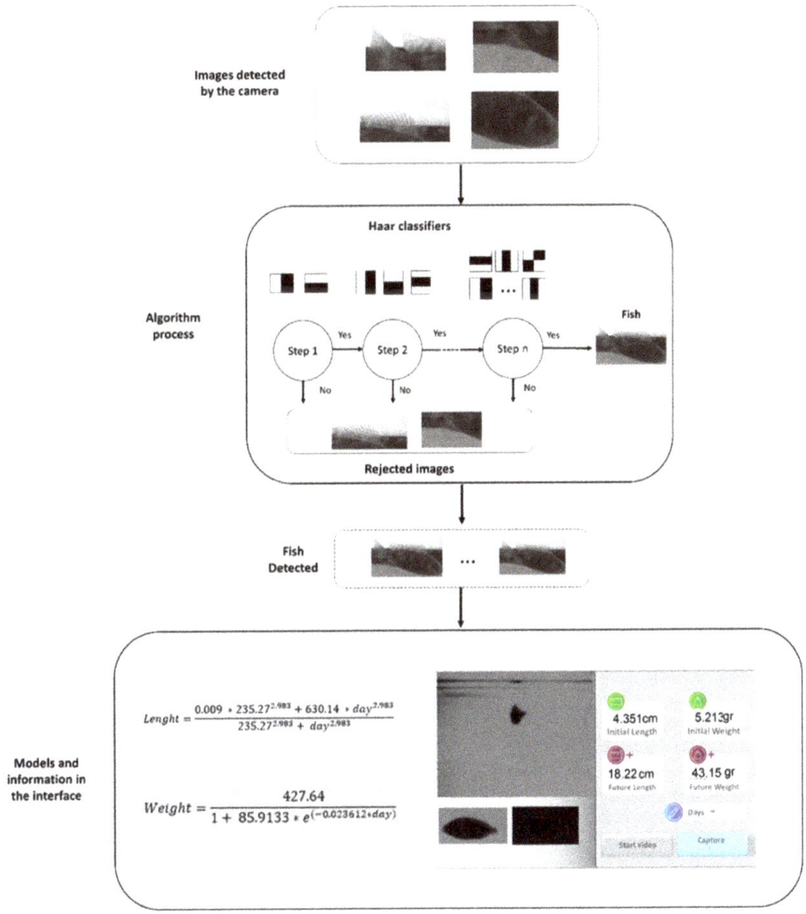

Figure 8. The process to identify a fish.

The below algorithm performs using the integral image at locations x, y that contains the sum of pixels above and left x, y, adding.

$$ii(x, y) = \sum_{x' \leq x, y' \leq y} i(x', y'), \tag{7}$$

where $ii(x, y)$ is the integral image and $i(x, y)$ is the original image, and the following occurrences are used:

$$s(x, y) = s(x, y - 1) + i(x, y) \tag{8}$$

$$ii(x,y) = ii(x-1,\ y) + s(x,y) \tag{9}$$

where (x,y) is the cumulative row sum, $s(x,y) = 0$, and $ii(-1,\ y) = 0$ the integral image can be compounded in one pass over the original image [4]. All the basis to this processing was initiated by [25]

2.6. User Interface

A user interface was made to make the system easy to use; with a NIR camera, an image was obtained, then the initial weight, initial length, future weight, and future length appeared in the interface. Additionally, an option was needed for the number of days to estimate the future weight and length; for this, mathematical models were used. Figure 9 shows the user interface.

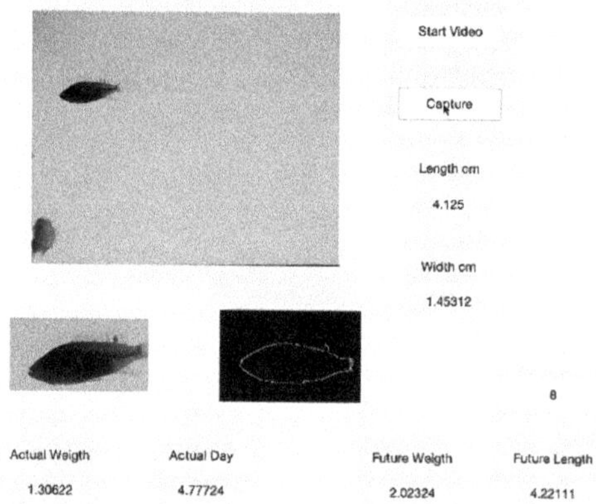

Figure 9. User interface.

3. Results

3.1. NIR Camera

The distance between the camera and the center of the fish tank was adjusted to 30 cm to obtain an image that could capture both juvenile and adult fish. During the development of the prototype, tests were carried out at the laboratory under controlled conditions to prove the concept and verify the correct functioning of the algorithm and the integral system; they were also made with different lighting levels, and lamp positions are shown in Figure 10. The best position for the laboratory setting was opposite the fishbowl; in the case of the 500 L fish tank the best position was on the top.

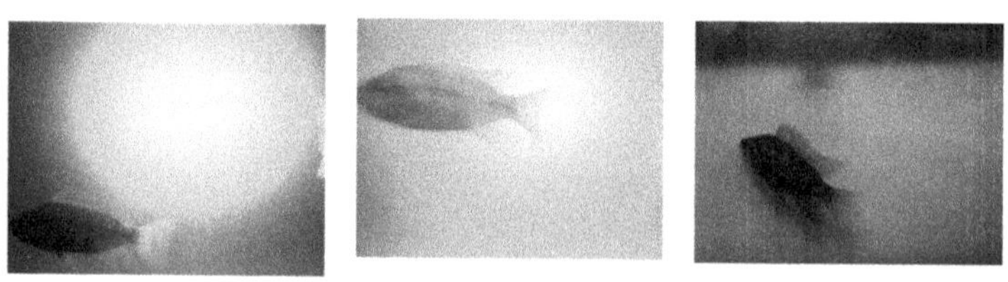

Figure 10. Captures from NIR camera.

To verify if the algorithm made the correct conversion from pixel to centimeters, Figure 11 shows the capture obtained with the NIR camera using a tape measure.

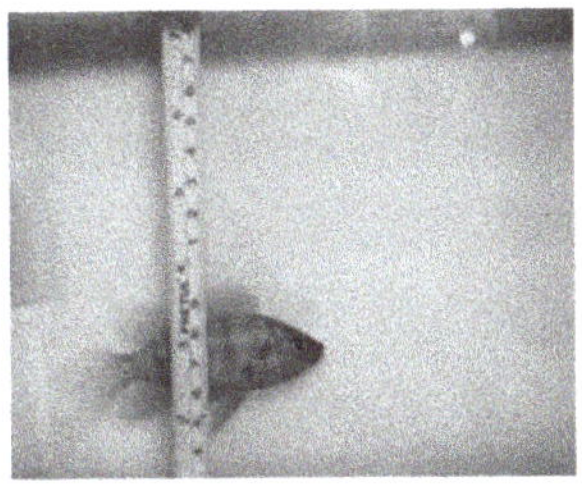

Figure 11. Pixels to centimeters conversion.

Infrared lighting and the infrared camera allow a high contrast between the object of interest and the rest of the habitat, thus allowing for elements that can be considered noise to be ruled out, with said noise being the color of the water in the fish tank or suspended particles. With the band-pass filter, it was possible to block everything below or above 850 nm wavelength, so all components within the visible spectrum of light, such as colors, were removed.

3.2. Haar Cascade Fish Detection

The work considered a non-intensive density, 20 kg/m^3, and in an intensive system, the density was 80 kg/m^3; the system was tested with both densities. The algorithm was capable of detecting fish within their typical development habitat. Processing time was around 1.5 s from image capture to fish measurement display.

The algorithm contributes to fish detection because the fish recognition is automatic, due to the camera or video [5], the algorithm can analyze a large amount of data in a short time, and the data previously was saved in the algorithm [6], also the algorithm allows segmenting the object in more difficult environments [7]; moreover, the algorithm can be combined with other methods to improve it [8].

The testing dataset was performed after the training, and this was performed using the first system configuration shown in Figure 12; several repetitions were made, which consisted of detecting the fish in the small fishbowl. The validation stage was performed using a 500 L fish tank, and in these tanks, a certain number of fish were placed, and the validation was performed on different days by taking photographs from different tanks. Additionally performed were weight and length measurements to corroborate the validation of the algorithm. The recognition performance obtained from the validation phase resulted in an average accuracy of 92%, a true positive rate equal to 95%, and finally, a false positive rate equal to 12%.

Figure 12. Weight and length measurement.

Several measurements were performed to validate the prototype and graphical interface, using the prototype and recording data in different tanks with fish of the same size and age. These measurements were performed at different times; Figure 13 shows the tanks and the camera collocated nearby by the tanks, Figure 14 shows the data interface with the algorithm.

Figure 13. The system used to test the algorithm.

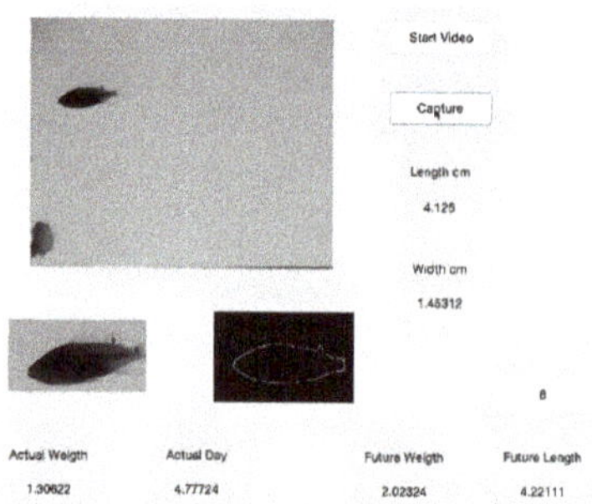

Figure 14. Interface and the algorithm.

3.3. Mathematical Model

The two models used in the experiment to obtain an equation based on the length in centimeters that can estimate the weight in grams were combined in a regression, taking the logistic model that estimates the weight based on age in days as the dependent variable. As an independent variable, the Michaelis-Menten model estimated the length based on age in days. Figure 15 shows the regression analysis.

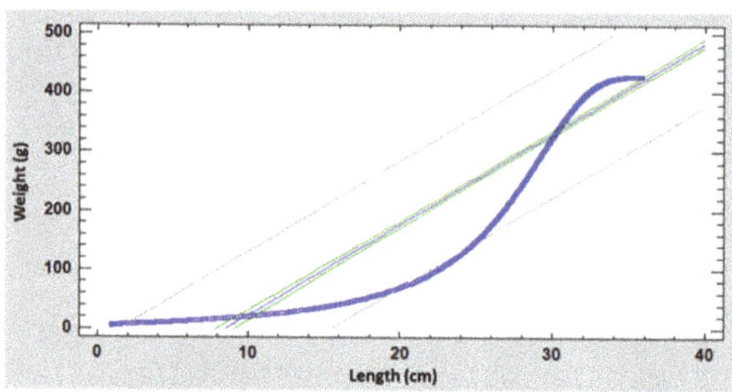

Figure 15. Regression analysis between two models (Logistic and Michaelis-Menten).

As previously mentioned, the algorithm and the interface were used to have a complete system; the algorithm evaluated the images obtained with the camera, and the interface showed the length and weight that appeared in Figure 14.

With the previous regression obtained, the equation related the length in centimeters and the weight in grams

$$Weight = (-0.339061 + 0.482898 * length)^2 \quad (10)$$

Subsequently, the length in the previous equation was used to estimate the fish culture day. Additionally, a regression analysis was performed, taking the day as the dependent variable and length as rgw independent variable; the equation obtained was:

$$Time = \left(4.86558 + 0.013364 * length^2\right)^2 \tag{11}$$

With the day entered by the user, the interface returned the fish's future weight and future length. Additionally, it took into account the resulting length and weight that the interface returns by the artificial vision system for these Equations (5) and (6) were used. Figure 16 shows the graph in which the future weight adjustment is made, taking into account the weight thrown by the interface based on the length of the artificial vision system.

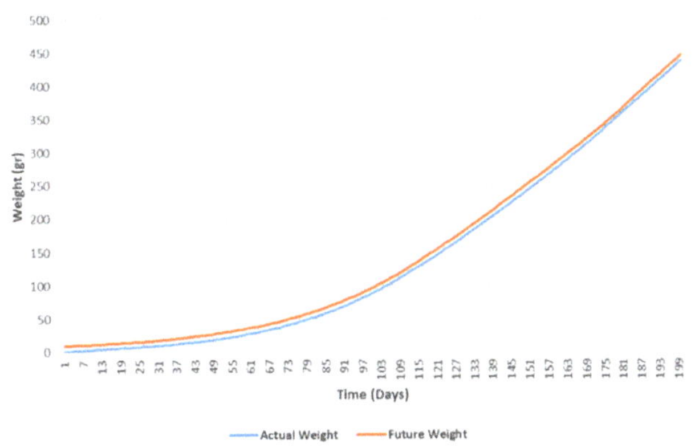

Figure 16. Fish weight adjustment.

An adjustment was made, considering the length thrown by the interface. With these adjustments, the weight and length prediction were more accurate (Figure 17).

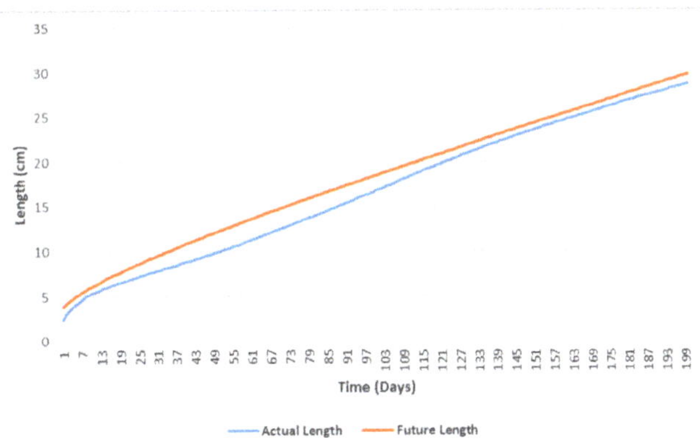

Figure 17. Fish length adjustment.

In this part, the weight and length error analysis were performed with statistical software that estimated the r-square.

4. Discussion

Selecting an appropriate distance and illumination for using the NIR camera is essential and has to be fixed throughout the experiment. As used by [14], the distance from level water to the camera was 140 cm, whereas [17] used a distance of 150 cm. For the experiment, 150 cm was the distance, and the tank used had a depth of 50 cm; in this case, the system can only capture fish on the water's surface. Another limitation is density, but if the system has the required flow water speed, this is not a limitation. The NIR camera has a harmless wavelength between 700 to 1400 nm and can penetrate biological tissues. Additionally, this is used in different fields, for example, to monitor leaf area [26], for iris recognition [27], automobile tasks [28], fire detection [29], detect chemical compounds [30], and finger veins recognition [31].

As previously mentioned, the Haar classifier is considered in machine learning tools; unlike other tools, such as convolutional works, big data analytics, or the internet of things, the Haar classifier just needs a one pc. For the other tools, an internet connection or cloud space is required. In the experiment, a pass band filter was used to reduce the wavelength, get focus in a specific wavelength, and help to have less range of experimentation.

The Haar classifiers are mainly used for face detection, but some applications are used for different challenges, such as automotive applications [32,33], agriculture [34], and cataract detection [35]. Generally, the Haar classifier can be adapted for every application, such as in this work. However, the essential part is the training part, and if this training is complex, the Haar classifier can be used for any application. Regarding this work, the training part can be more complicated, but this implies the waste of resources because, for aquaculture applications, the environment has fish and is less likely to have negative images considered as no fish.

The mathematical models were an essential part of this experiment. They helped to obtain a relation between length and weight. Models in aquaculture are used for feeding control [36], production in general [37], harvest [38], energy calculation [39], and monitoring culture variables [40], and the use of these models helps the aquaculture system to have a precision culture. However, the most crucial aspect of aquaculture is growth, such as the model used in this experiment. The mathematical model used in the experiment had good results, as shown in Figures 16 and 17.

5. Conclusions

The method first allowed for the recognition of the fish and then the processing to obtain fish length and fish weight to avoid the fish stress because of the handling. The measurement task by hand takes consideration time; this method consumes less time for this task. If the environment of the fish changes or the species is different, new training in the software will be required.

As time goes, applications related to artificial intelligence will be more common; this technology allows to have a better control in the majority of systems of production.

Author Contributions: S.L.-T.: writing—original draft preparation. G.M.S.-Z.: review. M.T.-A.: software. L.M.C.-M.: validation. E.A.R.-A.: data curation. P.S.F.-A.: resources. All authors have read and agreed to the published version of the manuscript.

Funding: Consejo Nacional de Ciencia y Tecnologia (CONACY), and scholarship provided for the same institution.

Institutional Review Board Statement: Not applicable.

Informed Consent Statement: Not applicable.

Conflicts of Interest: The authors declare no conflict of interest.

References

1. Kiaee, N.; Hashemizadeh, E.; Zarrinpanjeh, N. Using GLCM features in Haar wavelet transformed space for moving object classification. *IET Intell. Transp. Syst.* **2019**, *13*, 1148–1153. [CrossRef]
2. Arvind, C.S.; Prajwal, R.; Bhat, P.N.; Sreedevi, A.; Prabhudeva, K.N. Fish Detection and Tracking in Pisciculture Environment using Deep Instance Segmentation. In Proceedings of the 2019 IEEE Region 10 Conference/TENCON, Kochi, India, 17–20 October 2019. [CrossRef]
3. Saleh, A.; Sheaves, M.; Rahimi Azghadi, M. Computer vision and deep learning for fish classification in underwater habitats: A survey. *Fish Fish.* **2022**, *23*, 977–999. [CrossRef]
4. Zhao, S.; Zhang, S.; Liu, J.; Wang, H.; Zhu, J.; Li, D.; Zhao, R. Application of machine learning in intelligent fish aquaculture: A review. *Aquaculture* **2021**, *540*, 736724. [CrossRef]
5. Chiu, M.-C.; Yan, W.-M.; Bhat, S.A.; Huang, N.-F. Development of smart aquaculture farm management system using IoT and AI-based surrogate models. *J. Agric. Food Res.* **2022**, *9*, 100357. [CrossRef]
6. Yu, H.; Yang, L.; Li, D.; Chen, Y. A hybrid intelligent soft computing method for ammonia nitrogen prediction in aquaculture. *Inf. Process. Agric.* **2021**, *8*, 64–74. [CrossRef]
7. Abbasi, R.; Martinez, P.; Ahmad, R. An ontology model to represent aquaponics 4.0 system's knowledge. *Inf. Process. Agric.* **2022**, *9*, 514–532. [CrossRef]
8. Yu, C.; Fan, X.; Hu, Z.; Xia, X.; Zhao, Y.; Li, R.; Bai, Y. Segmentation and measurement scheme for fish morphological features based on Mask R-CNN. *Inf. Process. Agric.* **2020**, *7*, 523–534. [CrossRef]
9. Rahman, A.; Xi, M.; Dabrowski, J.J.; McCulloch, J.; Arnold, S.; Rana, M.; George, A.; Adcock, M. An integrated framework of sensing, machine learning, and augmented reality for aquaculture prawn farm management. *Aquac. Eng.* **2021**, *95*, 102192. [CrossRef]
10. Mccormick, S.D. General and Comparative Endocrinology Effects of long-term cortisol treatment on growth and osmoregulation of Atlantic salmon and brook trout. *Gen. Comp. Endocrinol.* **2021**, *308*, 113769. [CrossRef]
11. Dash, G.; Sen, S.; Pradhan, R.K.; Ghosh, S.; Josileen, J.; Jayasankar, J. Modeling framework for establishing the power law between length and weight of fishes and a meta-analysis for validation of LWRs for six commercially important marine fishes from the northwestern Bay of Bengal. *Fish. Res.* **2023**, *257*, 106496. [CrossRef]
12. Terayama, K.; Shin, K.; Mizuno, K.; Tsuda, K. Aquacultural Engineering Integration of sonar and optical camera images using deep neural network for fish monitoring. *Aquac. Eng.* **2019**, *86*, 102000. [CrossRef]
13. Azarmdel, H.; Mohtasebi, S.S.; Jafari, A.; Muñoz, A.R. Developing an orientation and cutting point determination algorithm for a trout fish processing system using machine vision. *Comput. Electron. Agric.* **2019**, *162*, 613–629. [CrossRef]
14. Saberioon, M.; Císa, P. Automated within tank fish mass estimation using infrared reflection system. *Comput. Electron. Agric.* **2018**, *150*, 484–492. [CrossRef]
15. Salman, A.; Maqbool, S.; Hannan, A.; Jalal, A.; Shafait, F. Ecological Informatics Real-time fish detection in complex backgrounds using probabilistic background modelling. *Ecol. Inform.* **2019**, *51*, 44–51. [CrossRef]
16. Rohani, A.; Taki, M.; Bahrami, G. Artificial Intelligence in Agriculture Application of arti fi cial intelligence for separation of live and dead rainbow trout fish eggs. *Artif. Intell. Agric.* **2019**, *1*, 27–34. [CrossRef]
17. Zhou, C.; Zhang, B.; Lin, K.; Xu, D.; Chen, C.; Yang, X.; Sun, C. Near-infrared imaging to quantify the feeding behavior of fish in aquaculture. *Comput. Electron. Agric.* **2017**, *135*, 233–241. [CrossRef]
18. Guo, R.; Ma, Y.; Tang, Y.; Xie, P.; Wang, Q.; Lin, W. Talanta A novel mitochondria-targeted near-infrared (NIR) probe for detection of viscosity changes in living cell, zebra fishes and living mice. *Talanta* **2019**, *204*, 868–874. [CrossRef]
19. Zhou, J.; Wu, X.; Chen, Z.; You, J.; Xiong, S. Evaluation of freshness in freshwater fish based on near infrared reflectance spectroscopy and chemometrics. *LWT-Food Sci. Technol.* **2019**, *106*, 145–150. [CrossRef]
20. Shetty, A.B.; Rebeiro, J. Facial recognition using Haar cascade and LBP classifiers. Glob. *Transit. Proc.* **2021**, *2*, 330–335. [CrossRef]
21. Wang, G.; Zhang, L.; Sun, H.; Zhu, C. Longitudinal tear detection of conveyor belt under uneven light based on Haar-AdaBoost and Cascade algorithm. Meas. *J. Int. Meas. Confed.* **2021**, *168*, 108341. [CrossRef]
22. Rodriguez, C.; Galli, O.; Olsson, D.; Tellechea, J.; Norbis, W. Length-weight relationships and condition factor of eight fish species inhabiting the Rocha Lagoon, Uruguay. *Braz. J. Oceanogr.* **2017**, *65*, 97–100. [CrossRef]
23. Jisr, N.; Younes, G.; Sukhn, C.; El-Dakdouki, M.H. Length-weight relationships and relative condition factor of fish inhabiting the marine area of the Eastern Mediterranean city, Tripoli-Lebanon. *Egypt. J. Aquat. Res.* **2018**, *44*, 299–305. [CrossRef]
24. Taylor, P.; Soderberg, R.W. A Linear Growth Model for Nile Tilapia in Intensive Aquaculture. *North Am. J. Aquac.* **2011**, *68*, 37–41. [CrossRef]
25. Viola, P.; Jones, M. Rapid Object Detection using a Boosted Cascade of Simple Features. In Proceedings of the Accepted Conference on Computer Vision and Pattern Recognition, Kauai, HI, USA, 8–14 December 2001.
26. Fan, X.; Kawamura, K.; Guo, W.; Xuan, T.D.; Lim, J.; Yuba, N.; Kurokawa, Y.; Obitsu, T.; Lv, R.; Tsumiyama, Y.; et al. A simple visible and near-infrared (V-NIR) camera system for monitoring the leaf area index and growth stage of Italian ryegrass. *Comput. Electron. Agric.* **2018**, *144*, 314–323. [CrossRef]
27. Nguyen, D.T.; Pham, T.; Lee, Y.; Park, K.R. Deep learning-based enhanced presentation attack detection for iris recognition by combining features from local and global regions based on NIR camera sensor. *Sensors* **2018**, *18*, 2601. [CrossRef]

28. Naqvi, R.A.; Arsalan, M.; Batchuluun, G.; Yoon, H.; Park, K.R. Deep learning-based gaze detection system for automobile drivers using a NIR camera sensor. *Sensors* **2018**, *18*, 456. [CrossRef]
29. Burnett, J.D.; Wing, M.G. A low-cost near-infrared digital camera for fire detection and monitoring. *Int. J. Remote Sens.* **2018**, *39*, 741–753. [CrossRef]
30. Zhou, X.; Geng, W.; Li, J.; Wang, Y.; Ding, J.; Wang, Y. An Ultraviolet–Visible and Near-Infrared-Responded Broadband NIR Phosphor and Its NIR Spectroscopy Application. *Adv. Opt. Mater.* **2020**, *8*, 1–8. [CrossRef]
31. Kim, W.; Song, J.; Park, K.R. Multimodal biometric recognition based on convolutional neural network by the fusion of finger-vein and finger shape using near-infrared (NIR) camera sensor. *Sensors* **2018**, *18*, 2296. [CrossRef]
32. Anggadhita, M.P.; Widiastiwi, Y. Breaches Detection in Zebra Cross Traffic Light Using Haar Cascade Classifier. In Proceedings of the 2020 International Conference on Informatics, Multimedia, Cyber and Information System (ICIMCIS), Jakarta, Indonesia, 19–20 November 2020; pp. 272–277. [CrossRef]
33. Hakim, I.M.; Christover, D.; Marindra, A.M.J. Implementation of an image processing based smart parking system using haar-cascade method. ISCAIE 2019. In Proceedings of the 2019 IEEE 9th Symposium on Computer Applications & Industrial Electronics (ISCAIE), Kota Kinabalu, Sabah, Malaysia, 27–28 April 2019; pp. 222–227. [CrossRef]
34. Marzan, C.S.; Marcos, N. Towards tobacco leaf detection using Haar cascade classifier and image processing techniques. *ACM Int. Conf. Proceed. Ser.* **2018**, *173*, 63–68. [CrossRef]
35. Jacob, I.J. Data Intelligence and Cognitive Informatics. 2020. Available online: https://link.springer.com/10.1007/978-981-16-6460-1 (accessed on 20 November 2022).
36. Zhou, C.; Xu, D.; Lin, K.; Sun, C.; Yang, X. Intelligent feeding control methods in aquaculture with an emphasis on fish: A review. *Rev. Aquac.* **2018**, *10*, 975–993. [CrossRef]
37. Føre, M.; Frank, K.; Norton, T.; Svendsen, E.; Alfredsen, J.A.; Dempster, T.; Eguiraun, H.; Watson, W.; Stahl, A.; Sunde, L.M.; et al. Precision fish farming: A new framework to improve production in aquaculture. *Biosyst. Eng.* **2018**, *173*, 176–193. [CrossRef]
38. Konovalov, D.A.; Saleh, A.; Efremova, D.B.; Domingos, J.; Jerry, D.R. Automatic Weight Estimation of Harvested Fish from Images. In Proceedings of the 2019 Digital Image Computing: Techniques and applications (DICTA), Perth, Australia, 2–4 December 2019. [CrossRef]
39. Tirkolaee, E.B.; Goli, A.; Weber, G.W. Fuzzy Mathematical Programming and Self-Adaptive Artificial Fish Swarm Algorithm for Just-in-Time Energy-Aware Flow Shop Scheduling Problem with Outsourcing Option. *IEEE Trans. Fuzzy Syst.* **2020**, *28*, 2772–2783. [CrossRef]
40. Parra, L.; Rocher, J.; Escrivá, J.; Lloret, J. Design and development of low cost smart turbidity sensor for water quality monitoring in fish farms. *Aquac. Eng.* **2018**, *81*, 10–18. [CrossRef]

applied sciences

MDPI

Article

Digital Transformation of Beekeeping through the Use of a Decision Making Architecture

Jean-Charles Huet [1,*], Lamine Bougueroua [1], Yassine Kriouile [1,2], Katarzyna Wegrzyn-Wolska [1,2] and Corinne Ancourt [2]

1 EFREI Paris, Efrei Research Lab, 30/32 Avenue de la République, 94800 Villejuif, France
2 Mines ParisTech, PSL University, Centre de Recherche en Informatique, 35 rue Saint Honoré, 77300 Fontainebleau, France
* Correspondence: jean-charles.huet@efrei.fr

Abstract: The use of information and communication technologies (ICT) in agriculture is far from their potential. In this article, we consider how to facilitate and systematize the process of transforming traditional agriculture into digital agriculture; Agriculture 4.0. Among the different technologies, we focus on the IoT aspects. In the article, we propose a new approach for the design of intelligent agricultural management and supervision systems. The proposed approach is illustrated as an example of application in the beekeeping sector. Indeed, this sector is affected by a crisis due to the disappearance of bees and the different actors need support to make their decisions. As an example of decisions that can be made, we can cite: treatment planning or policy planning. An architecture based on sensors and open data is proposed to help them make decisions. An implementation of it is shown; it is based on a device with sensors, as well as an interface to collect the data on beehives and show notifications and alerts to beekeepers. The proposed architecture is flexible, and it can be used in the context of different levels of technology maturity. The final objective is to develop a reusable architecture for Agriculture 4.0.

Keywords: intelligent system; agriculture 4.0; smart beehives; maturity model

Citation: Huet, J.-C.; Bougueroua, L.; Kriouile, Y.; Wegrzyn-Wolska, K.; Ancourt, C. Digital Transformation of Beekeeping through the Use of a Decision Making Architecture. *Appl. Sci.* **2022**, *12*, 11179. https://doi.org/10.3390/app122111179

Academic Editors: José Miguel Molina Martínez, Ginés García-Mateos and Dolores Parras-Burgos

Received: 29 September 2022
Accepted: 28 October 2022
Published: 4 November 2022

1. Introduction

Recently, the beekeeping industry has been threatened by a decline in the number of bee populations and a drop in production. Beekeepers grapple with many issues related to the health of bee colonies. Scientific work is focused on a deep understanding of the causes of these phenomena in order to help beekeepers make the right decisions [1]. In France, a study on the mortality rate of bee colonies during the winter of 2020–2021 was carried out with more than 13,000 participating beekeepers (Platform ESA. National survey on winter mortality of bee colonies, 2021. Accessed on 28 September 2022. https://siteweb.esa.inrae.fr/en/node/536). In general, the death rate is estimated at 24.8%.

According to a FranceAgremer (FranceAgrimer: The National Establishment of Agricultural and Sea Products) study [2], sales of honey and derivatives (pollen, pure royal jelly, honey and royal jelly, etc.) in supermarkets are estimated in 2020 at 16,200 tons, an increase of nearly 11% compared to 2019. The increasing consumer interest is explained by the positive image of the product (healthy, natural, etc.). For more than 10 years, honey production in France has been increasing, but despite this, it is not enough to cover an increasingly strong demand. Consequently, France imports honey to fill the gap between consumption and production. The volumes of imported honey thus increased by 36% between 2010 and 2020 (6% in 2020).

It is now evident that environmental pollution and climate change are a great danger to our planet. Experts have confirmed that they pose a very dangerous threat to many species of terrestrial and marine fauna. However, the death of the bee population necessary for pollination is considered the most important warning for humans. According to FranceAgrimer

(FranceAgrimer. Observatory of the production of honey and royal jelly, July 2022. Accessed on 28 September 2022. https://www.franceagrimer.fr/fam/content/download/69152/document/SYN-API-Observatoire_Miel_et_Gel%C3%A9e_Royale_2021.pdf?version=3), the total number of beekeepers has increased over the past 6 years and for the first year, the Observatory has highlighted a decline in the number of beekeepers in France. Thus, in 2021, there were 70,847 declared beekeepers, compared to 71,273 in 2020. Nevertheless, the number of beekeepers with more than 50 hives continues to grow (+288 beekeepers, or more than a 5% improvement). France has 5708 beekeepers with more than 50 hives.

A solution to tackle these issues is to connect and digitalize the beehives. Our goal is to help the beekeepers and decision-makers (politician, responsible for cooperative, etc.) to make their decisions and to supervise one or several beehives. These decisions could be from a day to a year. More specifically, beekeepers keep a record or journal of their hives. In this record, the beekeepers note the information about all treatments, handling, feeding, visits, harvesting of the honey, etc.

In addition, many applications in manufacturing systems require a combination of new technologies, resulting in the emergence of Industry 4.0. The most well-known technologies used by Industry 4.0 are IoT, Cloud Computing, Big data, and cyber-physical systems (CPS). CPS are mechanisms that allow monitoring through communication, data storage, and computational capabilities directly incorporated into objects [3]. More precisely, CPS has embedded software that uses sensors and actuators. To allow humans and software to communicate between them, standard interfaces have been developed. The use of data coming from sensors or from the network is supported by this software. They have storage and processing abilities.

Agriculture systems have the same challenges as those encountered by manufacturing systems. Theses challenges are related to their sizing, to the understanding of their mechanisms, to the improvement of their productivity, and to their performance evaluation. Theses challenges produce difficulties to make decisions for all the actors in the agriculture systems.

Digitalization processes are currently being increasingly introduced into various sectors of the national economy, including the agricultural sector [4]. Digital Transformation (DT) refers to the combination of digital technology and areas of the business to finally realize the transformation of the management model and business model [5]. Indeed, DT is often referred to either as evolution, or the creation of entire new business models in companies or business sectors [6]. Digitalization can be considered as part of DT [7]. It describes socio-technical processes and their impacts on human activities that result from the use of interconnected digital technologies [8]. DT is a convergence of hard (technology) and soft (people and businesses) forces and movements from which additional value emerges [9]. According to the authors, one of the disruptive technologies adopted in DT implementations is "IoT connected devices". In the same way, the following categories were used for coding the DT [10]: *Technologies*, *Management/Processes*, and *People*. The authors also list the elements within each category. Among these elements, in this paper, we meanly use IoT (*Technology*) in order to create new services and products (*Management/Processes*) for workforce, stakeholders, and partners (*People*). Thus, our process respects the properties of the DT, with an implementation in beekeeping. As new services, we focus on the development of decision support systems. This new service allows us to improve the business models of the different actors and to help save the bees.

In our context, we depict a new approach to apply the DT in the beekeeping sectors and help the actors of this area to make their decisions. To this end, different proposals are made:

- A proposal of Maturity Model (MM) Levels for smart beehives;
- A proposal for analysing all the decisions that can be made in the beekeeping sectors using spatio-temporal matrix;
- A proposal of generic architecture that corresponds to the decisions analysed in the matrix.

The paper is structured as follows: Section 2 presents a synthetic review of work about smart agriculture and more specifically on beekeeping with a focus on IoT. Section 3 proposes a Maturity Model to categorize the digital maturity of smart beehive systems and presents classification of the literature on smart beehives following this Maturity Model. Section 4 depicts a spatio-temporal matrix to analyse the possible decision-making on smart beehives. Section 5 explains the global view and the generic architecture for the proposed smart beehives. An implementation is shown in the Section 6. Finally, Section 7 presents the conclusions and perspectives.

2. Context and Related Work

2.1. From Industry 4.0 to Agriculture 4.0

One of the main objectives of our approach is to facilitate the use of IT communication and information tools in agriculture. As an analogy to Industry 4.0, digital agriculture, smart agriculture, intelligent agriculture, or Agriculture 4.0 can be defined as a set of modern technologies that today meet the needs of communication, storage, automatism, computing, and security.

Industries, as well as researchers and decision-makers around the world, have increasingly called for a fourth industrial revolution to enter a new era of digital and connectivity [11].

According to Danjou et al. [12], 10 technology groups can be considered to implement Industry 4.0, such as cloud computing, IoT, CPS, and big data. The use of these technologies allows the transmission of information throughout the entire system and enables better control and monitoring. Practically, they are adapted to the operations in real time according to varying requirements and demand [11].

From these concepts, we proposed to reuse them in the domain of Agriculture 4.0 (Figure 1) similar to Ref. [13], who proposed a framework to manage traceability in Agriculture 4.0. Agriculture becomes a very important sector to apply the concept of CPS to [14]. In fact, the agriculture sector follows the industrial revolution and moves from traditional systems to the implementation of a full range of modern systems. These systems enable data management of the geographic information, production experiments, climate information, and other data. The beekeeping sector is obviously concerned by this revolution. Figure 1 shows the reuse of 10 technologies from Industry 4.0. The orange circles represent the technologies used in our project.

In the context of our project and paper, we are only interested in five technologies:

- Internet of Things, because decisions need data collection done through connected objects like smart hives;
- Cyber-Physical System, a set of primordial principles that allow defining the hierarchy of decision makers and decision cycle;
- Big Data, because data is the most important thing on which decisions are based;
- Cloud computing, which refers to the delivery of resources and services on demand over the Internet, i.e., the access to data from different resources and high computing power via the internet. It is an important computing step that facilitates the deployment of intelligent systems;
- Artificial Intelligence: in fact, nowadays, artificial intelligence is a powerful tool that can be used to help users in the process of making decisions.

Among the concepts of the agriculture 4.0 presented in this section, the main concept used in this paper and our system is IoT. The next section presents this concept.

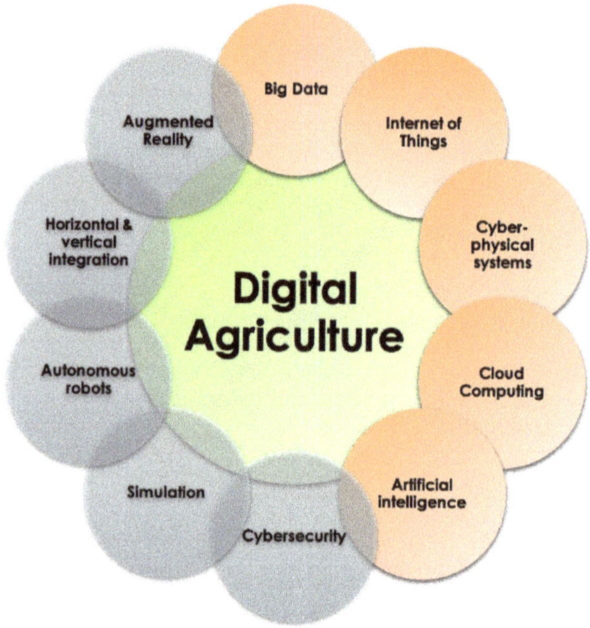

Figure 1. Reuse of the 10 technologies of Industry 4.0.

2.2. IoT for Digital & Green Agriculture

In an era of the Information Technology (IT) revolution, IoT is being utilized in various forms such as smart pills, personal devices, smart cities, robotics, smart monitoring devices, and real time monitoring systems. IoT can be explained by the ever growing interconnecting that connects hardware, computing devices, sensors, interfacing software, people of different networks, to exchange data and communication [15]. It facilitates and eases the process of communication and interaction between the elements connected through the network. The technology of IoT was derived from Radio Frequency Identification (RFID) and Wireless Sensor Networks (WSN) [16]. The data collected from RFID and WSN can be communicated easily to the different nodes of the interconnected sensors. The technology of IoT is being widely used in various devices such as home appliances, phones, gadgets, vehicles and other networking objects. Simply put, it means that all networking devices can be interconnected with each other globally for a real-time data exchange without the interference of human beings. IoT usage is growing rapidly in the IT industry, driven by the use of connected device applications. However, the COVID-19 pandemic will have a major impact on global IoT spending in 2022. IDC forecasts that global IoT spending will grow at a compound annual growth rate (CAGR) of 11.3% over the 2020–2024 forecast period (Worldwide Spending on the Internet of Things Will Slow in 2020 Then Return to Double-Digit Growth, According to a New IDC Spending Guide. https://www.idc.com/getdoc.jsp?containerId=US49578922, August 2022. Accessed on 28 September 2022).

According to IDC, there will be 41.6 billion connected objects generating 79.4 zetta bytes of data in 2025 (Steve Ranger. What Is the IoT? Everything you need to know about the internet of things right now. https://www.zdnet.com/article/what-is-the-internet-of-things-everything-you-need-to-know-about-the-iot-right-now/, February 2020. Accessed on 28 September 2022). There are many different types of connected sensors that can be used in IoT applications to make the process simpler and efficient [17]. Appliances such air conditioning, security cameras, etc., can be controlled and audited from anywhere through connected devices. Smart cities are also being developed through the help of IoT by the transportation data analysis. The technology of IoT is used by supply chain systems also when they use trackers to track deliveries in real time. Similarly, applications such as ApplePay, PayPal, etc., are also forms of connected devices that are used by banking systems to enable transactions using smartphones. Smart devices are used to collect data with the help of sensors and collected data can be transferred to the network layer through cloud systems. It is used for collecting various types of data that is useful to monitor and track various activities. The data collected through the different networks is converted into useful data, which become very important elements in the decision process.

What is the landscape of the sensors and IoT in agriculture? Agriculture benefits from interconnected objects, as the use of smart devices in the agriculture sector enables the monitoring of different important values and activities of farmers. The applications of connected devices are being utilized to track and record different kinds of data such as temperature, humidity, pollution, etc. Smart and IoT applications entered agriculture with a delay in relation to other fields such as medicine or industry. However, these solutions have started to increase over the years for better monitoring. This paves way for better practices in the form of better quality services and cost reduction. IoT-based approaches in the agriculture and agro-industry have proved to be very useful and have become necessary to propose better adapted services. More precisely, most initiatives in the agriculture and agro-industry focus not only on increasing the production but also on improving services such as sustainability, biodiversity vigilance, etc.

IoT is another very important paradigm used in the agro-industrial and environmental field [18]. The authors in Ref. [19] explain that one of the main requirements for devices used in IoT projects is that they must be energy efficient. The main point is to be able to evaluate the energy impact of the proposed architecture. The authors also explain the interest of Edge Computing in the future of agriculture. The IoT driven by AI-based recommendation model is one of the greatest promises of the development [18].

Several research activities have been carried out to study the use of IoT concept and agriculture. The authors of Ref. [20] presented an interesting survey of IoT solutions and demonstrated how IoT can be integrated into the smart agriculture sector. Ref. [21] presented the design of a smart system based on low-cost IoT sensors and popular data storage services and data analytics services on the cloud for crop production. The authors of Ref. [22] presented an overview on the recent trends on sensors and IoT systems for irrigation in Precision Agriculture. On the same scope, Ref. [23] proposed a data acquisition platform that is easily reusable. Other authors [24] have focused on the communication between the sensors and a NoSQL database through the use of an Interoperable Platform. The goal is to support the actors to help them manage their crops. In Ref. [25], the authors present a comprehensive review of emerging technologies for IoT-based smart agriculture. They provide a classification of IoT applications for smart agriculture into several categories. They propose to list the connected smart agriculture sensors that enable the IoT.

3. Maturity Model Levels and Smart Beehives

Many authors have already dealt with smart beehives with different levels of technological maturity. We therefore propose to focus on technological maturity for smart beehives and then analyse the articles that cover this field.

3.1. Maturity Model Levels for Smart Beehives

Beekeepers, depending on their financial means, will have different levels of technological maturity. The more connected beekeepers are, the more precise and relevant the notifications they receive will be.

To design our maturity levels, we were inspired by the literature about Maturity Model (MM) in Industry 4.0. We have therefore redefined MM so that it is compatible with a system similar to ours. Indeed, the literature proposes different Maturity Model levels to evaluate the digitalization of manufacturing systems [26–28].

Among the different existing maturity levels, we were interested in those corresponding to IoT and CPS. Westermann et al. [29] offered a maturity model for all CPS systems. We then adapted this model to our system. Table 1 proposes to summarize this adaptation.

The system is partially interlinked with other systems in horizontal and vertical dimensions [29]. Horizontal means communication with systems along the value chain, e.g., up- or down-stream systems such as other machines or other smart beehives in the studied domain. Vertical means interlinking with superordinate systems such as central decision centres or ERP in industry.

The system at the hive level will be between levels 1 and 5. Depending on the CPS set up by the beekeeper, the level of alerts and notifications will be different (Table 1).

Table 1. Maturity model levels adapted for smart beehives.

Maturity Model Levels	1	2	3	4	5
Data Input	Manual or vocal input (semi-automatic) and global connection to environmental data	Automatic input by sensors and global connection to environmental data	Automatic input by sensors and global connection to environmental data	Automatic input by sensors and local/global connection to environmental data	Automatic input by sensors and local/global connection to environmental data
Intelligence	Centralized	Centralized	Centralized and locally limited	Centralized + Local (including history of decisions made) + Local analysis of the data and automatic alerts sending	Centralized + Local cooperation and negotiation between beehives
Communication	Vertical with synchronization for several days or weeks	Vertical with synchronization for several days or weeks	Vertical with Real Time Synchronization	Vertical with Real Time Synchronization	Horizontal and Vertical with Real Time Synchronization

Regarding the agriculture, Büyük et al. [30] dealt with the Digital Maturity Assessment Model for Smart Agriculture. It consists of a model for companies that evaluates their competencies in Smart Agriculture. It is possible to calculate the digital maturity levels by using the proposed evaluation model. Actors in the agricultural sector can also analyse their degree of success in relation to the requirements of Industry 4.0. With the proposed criteria, the human and technological aspects are taken into account. The whole cycle (from the farm to the industry) could be analysed. Only the technological and farm aspects of this evaluation model correspond to our project.

3.2. Smart Beehives Literature Classified by Maturity Model Levels

Even though using technology in the beekeeping domain is still in its infancy, some works attempt to exploit certain technology aspects in order to improve the beekeeper practices and performance. Table 2 presents some of these works, each of them corresponding to a system that has input data and output data, and is based on special hardware, software, and algorithms. The last column of the table indicates the maturity of the system, as described in Section 3.

Table 2. Literature about smart beehives analysed with the Maturity Model presented in the Table 1.

Authors	Date	Input Data	Output Data	Hardware	Software	MM
Aksoy et al. [31]	2018	company age, company province, level of education, status of membership of an association, other activities, hives number, race of bees, frequency of queen change	honey production		Data Mining algorithms: CART, CHAID, MARS	1
Dineva and Atanasova [32]	2018	beehive internal data (temperature, humidity, weight, noise), external data (temperature, humidity, CO2, polluting air)	relation between data, prediction of future events	sensors	OSEMN	2
Braga et al. [33]	2020	temperature, weight, weather data, dew point, wind direction, wind speed, rainfall, luminosity	colony health status: good health, bad health, or collapse of colony health	sensors	K nearest neighbors method, Random decision forests, Neural Networks	2
beeinformed.org [34]	2021	weight, number of varroas, virus, nosema	statistics	sensors	web application, open data, diagrams	2
Edwards-Murphy et al. [35]	2016	CO2, O2, polluting gases, temperature, humidity, acceleration	colony health state, beehive internal weather data and external weather data	sensors (temperature, humidity, acceleration, air), network devices (3G, ZigBee)	Machine learning: decision tree	3
Markovic et al. [36]	2016	temperature	critical events	temperature sensors	Complex event processing (CEP)	3
Balta et al. [37]	2017	hive data (temperature, humidity, weight, bees traffic), external data (weather, location information)	bees counting, management of devices, resources and sensors, management of beekeeping data (beehive, apiary and region), detecting dangerous situation (robbing, swarming and colony losses)	Raspberry Pi, camera, sensors	sqlite, mongoDB, Node.js, Express.js, Angular.js, java on android, MQTT protocol, RESTful protocol, Cloud computing (Microsoft's Azure and IBM's Bluemix), Rule-based algorithm, XML, JSON, CSV	3
Zogović et al. [38]	2017	weight, temperature, O2, CO2, vibration, sound, humidity, foragers traffic, thermal images of colonies, atmospheric pressure	decision making aid, automatic actions performed by robots	scale, sensors, robot, actor, camera	OODA cycle: observe, orient, decide, act	3
Cazier (beeculture.com) [39]	2018	weight, temperature, humidity, sound, images, knowledge of good beekeeping practices	alerts, suggestions of actions, management practices, optimisation (colony health, production, pollination performance)		use of the beeXML standard, continuous learning and integration of good practices	3
Mrozek et al. [40]	2021	stream video of bees at hive entrance	bee infestation status	RaspberryPi, Camera, GSM modem	Neural Networks, Google Cloud, AWS Cloud, AWS Lambda, AWS DynamoDB	3
Debauche et al. [41]	2018	external data: temperature, humidity, atmospheric pressure, luminosity. Beehive internal data: humidity, temperature, acceleration, air contaminants	monitoring of bees behaviour, information for comprehension and analysis of the extinction of bees, colony health state	Network protocols (LPWAN, 3GPP), sensors, PyCom microcontroller	Lambda Architecture, REDIS, PostgreSQL, Apache kafka, Apache Samza, HDFS, Docker	3

Table 2. *Cont.*

Authors	Date	Input Data	Output Data	Hardware	Software	MM
Zetterman [42]	2018	temperature, weight, sound, humidity, number of bees, acceleration, CO2	data collection, monitoring, graphics	sensors	web application	3
Latioui et al. [43]	2019	expert rules, social network data (images, texts), beekeeper voice, GPS	alerts, recommendations, voice detection	sensors	Chatbot, deep learning, text mining	3

None of the articles presented in Table 1 provide a description of the digital maturity level. Most of those articles have proposed architectures to manage the data exchange between components. However, these architectures are very technical and have a lack of global view and genericity. Thus, they are hardly reusable. We propose in this paper, using the digital maturity level and the analysis made in the next section, to approach the subject in a more global and generic way.

4. Analysis with a Spatio-Temporal Matrix

To design an architecture for a decision-making system, a relevant methodology must be used. Among methodologies that can be used in the case of beekeeping, there is the spatio-temporal "3 × 3 matrix". It consists of filling a matrix composed of two axes: the temporal horizon and the modelling level. There are several works that try to use a two-dimensional grid to design a decision-making system, for instance, Chabrol et al. [44] proposed to couple the modelling approaches to the time horizons to design decision aid-tools for the hospital domain. Comelli et al. [45] propose a methodology based on studying horizon levels and system flows to evaluate the supply chain. The suggested framework in Ref. [46] has the advantages of handling decisions on two dimensions. It allows the designer to have a global view and to achieve a more complete architecture, taking into account different temporal and spatial levels. As explained in that paper, Figure 2 shows the matrix axis and its matching levels for the case of beekeeping.

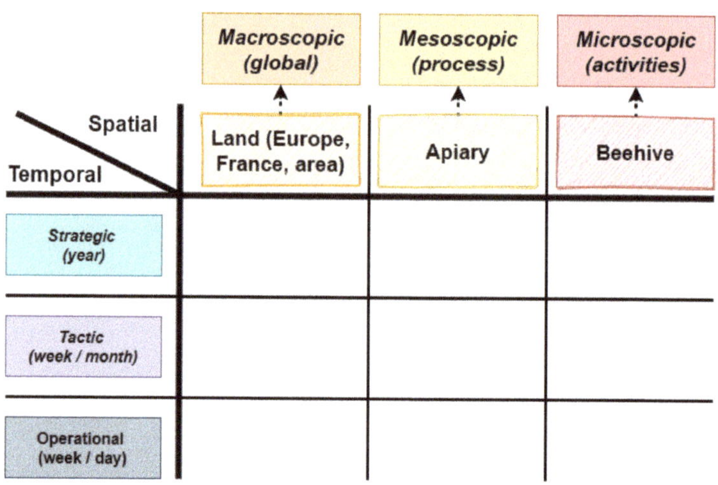

Figure 2. Spatio-temporal matrix.

The authors of Ref. [46] define the spatio-temporal matrix, show its matching levels for the case of beekeeping, and illustrate its application by an example of web application for managing apiary data. This tool allows its user to take decisions at the beehive or microscopic level. These decisions are either tactic, such as planning activities for the next week or month, or operational, such as evaluating results and adapting the activities. Figure 3 shows the application of the 3×3 matrix for that use case.

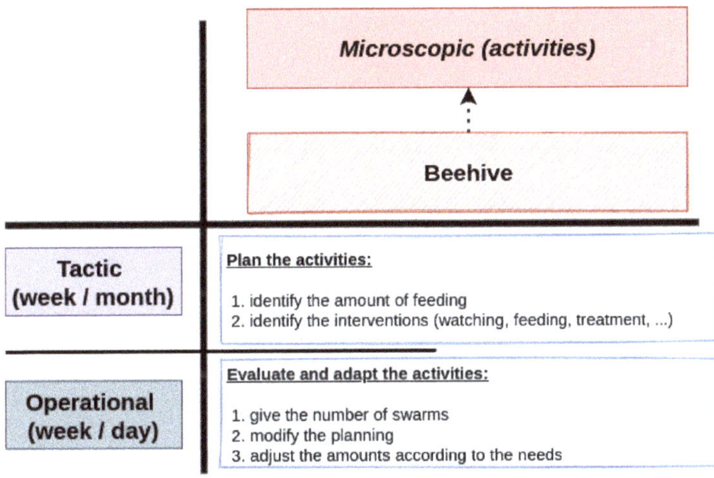

Figure 3. Application of the spatio-temporal matrix to the beekeeping case.

Our suggested architecture implements some levels of the matrix.

5. Decision Support Systems for Smart Beehives

According to Ref. [47], a Decision Support System (DSS) is a computer-based system intended for use by a particular manager or usually a group of managers at any organizational level in making a decision in the process of solving a semi structured decision. The DSS produces an output in the form of periodic or special report or the results of mathematical simulations. For the famous Cambridge Dictionary (Cambridge Dictionary. https://dictionary.cambridge.org/fr/dictionnaire/anglais/decision-support-system. Accessed on 13 October 2022), DSS is "a computer program that can arrange and sort large amounts of data, and that is used to help people in companies and organizations make important decisions based on the data: A successful decision support system is one that assists rather than replaces the human decision-maker".

By following these definitions, Figure 4 shows our global view of how to implement DSS to help beekeepers manage their beehives. The three levels of decision-making, described in Figure 4, correspond to the spacial dimension of the matrix depicted in Section 4. The data used to make the decision come from two sources:

- Sensors in the embedded system of the beehive;
- Open data about the weather (temperature, humidity, etc.) and pollen. The pollen is to know the state of the flora in a sectioned area.

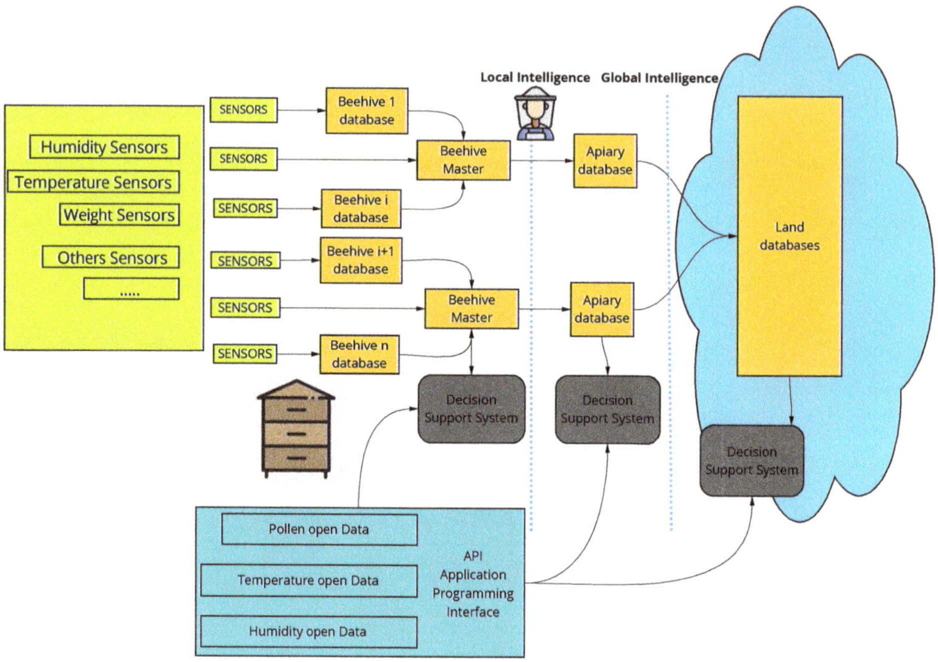

Figure 4. Global view of the decision support systems for smart beehives.

5.1. Generic Architecture for Smart Beehives

Our solution is inspired by a standard API (Application Programming Interface) for managing contextual information (NGSI-LD API) [48]. This standard enables near real-time access to information from many different sources (not only IoT data source). NGSI-LD open specification released by ETSI defines the context information model and the API to produce, consume, and subscribe to context information. In our solution, different sensors are used to measure and quantify the indicators of the beehives. The most used are weight and temperature. The weight can be used to see the increase or decrease in the weight of the hives. It makes it possible to monitor the production of honey and to trigger alerts if the variation in weight is significant. Weight can also indicate the best feeding and processing strategy applied to a hive, particularly comparing weight over time and between different hives/apiary. Outdoor weather conditions change over time, and the temperature, humidity, and wind speed must be taken into consideration. All these different types of data are used to optimize beekeeping decisions. One of the major problems in beekeeping is the varroa destructor mite. It is the most destructive enemy of the Western honeybee (Apis mellifera). The varroa is a serious threat to bee health. A bee colony infested with mites will typically die off in these regions within three years. Until now, it has been difficult to deal with this problem. Indeed, counting varroa mites is interesting when quantifying how many bees are infected per hive. However this task remains very complicated because it is often carried out manually. Our platform offers an automatic counting service based on photos of frame bees. This service is based on deep learning methods to process images and detect objects. In fact, two neural networks for object detection and image segmentation are used. Faster RCNN is used for detecting bee locations; for each bee in the image it predicts a bounding box surrounding the object. Figure 5 illustrates the network outputs. To locate varroas, the U-NET takes as input an image containing one bee and segments it into two regions: varroa and background. Figure 6 illustrates the U-NET outputs. Estimating the infestation level is performed by combining these two detections; first the frame image is given to the system, then the bees are identified, and after that the presence of varroa is predicted for each bee. The infestation percentage is calculated by dividing the number of infected bees by the number of bees. This system needs data for learning; in fact, bee images were retrieved from project partners, from camera photos, and from the Internet. Thanks to annotation tools, the images were annotated by bounding boxes for bees and ellipses for varroas.

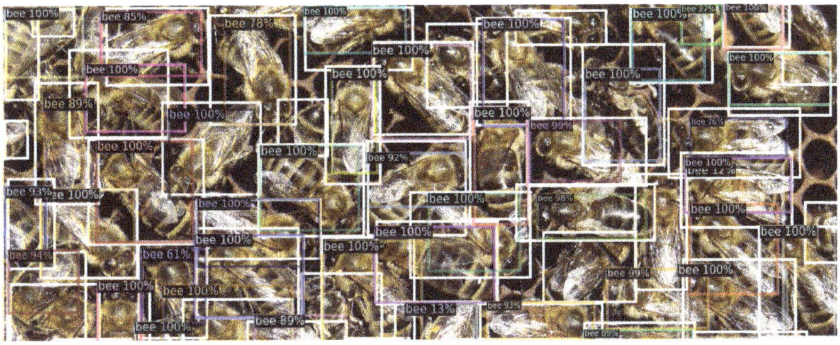

Figure 5. Detected bees by Faster RCNN. Coloured bounding boxes are predicted objects and white bounding boxes are ground-truth objects.

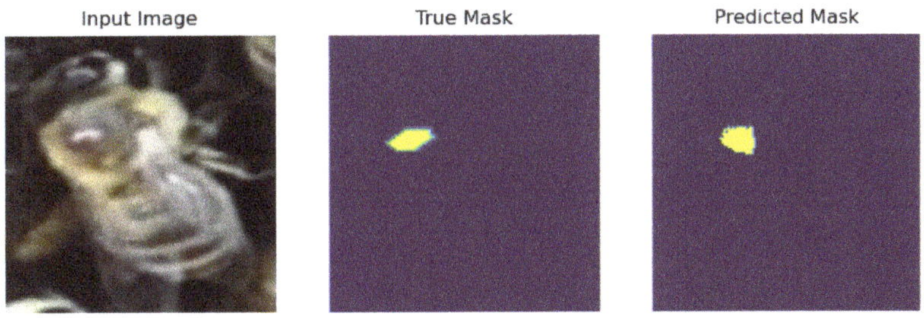

Figure 6. Varroa segmentation by U-NET. The yellow region represents varroa. True Mask is ground-truth segmentation and Predicted Mask is the predicted segmentation.

5.2. Databases

Our system is able to send valuable information to subscribed beekeepers thanks to the collection and extraction of data from different data sources such as sensors. The collected data will help train machine learning models capable of generating advice or alerts to users.

A survey was carried out by our partner (ITSAP—Bee Institute). The survey is based on multiple questionnaires on the several activities and needs of beekeepers (Survey "Beekeepers and digital technology". https://itsap.asso.fr/pages_thematiques/numerique/enquete-apiculteurs-numerique/, 31 January 2020. accessed on 28 September 2022).

The number of beekeepers participating in the survey is 415 (43% have more than 70 hives).The most important productions are honey and breeding. According to the survey, data collection appears to be a relatively common practice among beekeepers. While certain information must necessarily be collected to meet administrative requirements (especially for the breeding register), others show a desire to collect objective information on the colonies. However, a handwritten input remains the majority in most cases, despite the advantage of having a computer version concerning the ease of consultation of the data and their use. Only 60% of beekeepers who responded to the survey monitor varroa infestation.

The most expected features consist of fairly simple data processing, namely graphic productions and indicator tables. In addition to data visualization, beekeepers are looking for an intelligent alert service capable of reporting relevant information (Table 3).

Table 3. Expectations of beekeepers regarding the functionalities to be integrated into a management platform.

Functionality	Interested Beekeepers (%)
Data visualization	81
Alert service	78
Voice input	77
Data storage	72
Data sharing between beekeepers	71

The proposed data model is based on the results of this survey concerning the useful and necessary information that should be managed on an apiary.

The implementation of this architecture is done in collaboration with beekeepers. This will lead to the development of a comprehensive and flexible system that would allow the beekeeping community to share relevant information and knowledge, and make appropriate decisions. All information is modelled from data models that allow the unification of data structures using standards and also ensure further extensibility, if necessary.

Figure 7 presents an excerpt from our data model. We have developed a new model that would meet the needs of our beekeepers. Conceptually, the model presented is based on two levels:

- Spatial dimensions: sensors/hive/apiary/zone;
- Temporal dimensions: real-time /operational(day /week)/tactic (week/month) /strategic (year).

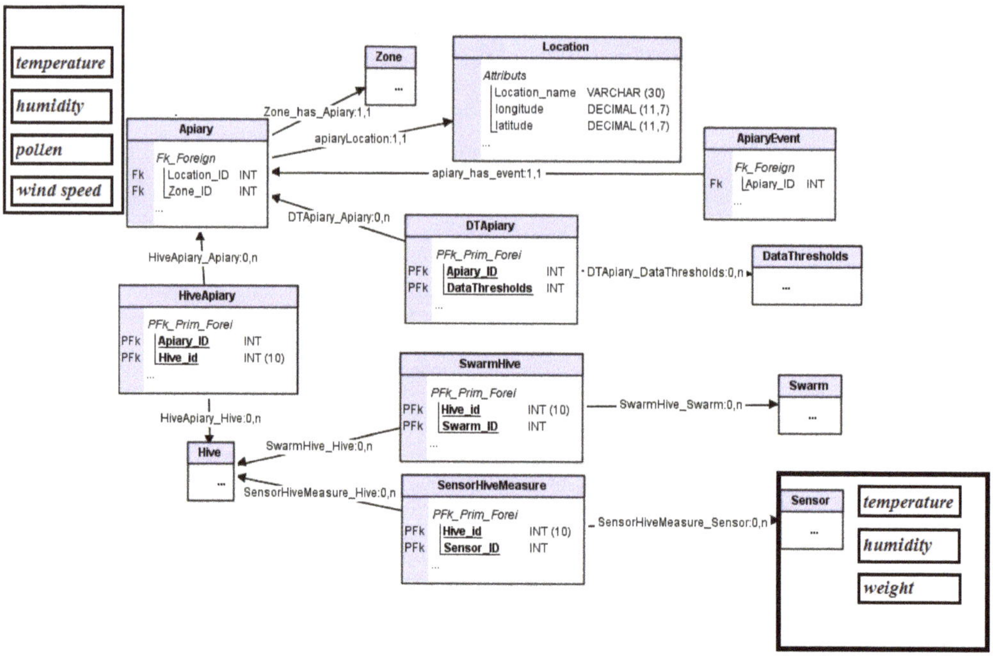

Figure 7. Excerpt from the data model.

Figure 8 presents an excerpt from our central data mode. The figure shows an example of a service offered, collecting data by integrating the time dimension.

As an example, for an apiary:

- Real-time: quantity of honey;
- Operational: varroa counting;
- Tactic: varroa treatment;
- Strategic: define the type and the amount of treatment and write a health risk management.

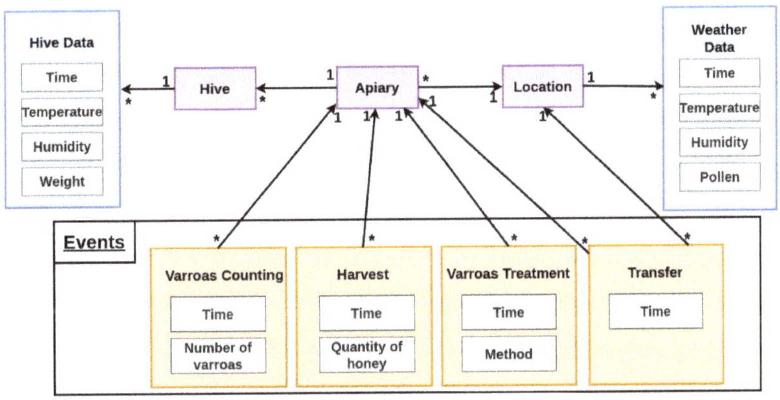

Figure 8. Excerpt from the central data model.

The selection of materials used to build the embedded system is a compromise between the cost and the accuracy of the sensors. Indeed, our goal is to encourage the beekeepers to make their digitalization and the cost of the system is a very important criterion.

6. Case Study

The case study developed in this project corresponds to level 3 of the MM presented in Section 3 and in Table 1. Table 4 presents our case study with the MM.

Table 4. Maturity Model levels of the case study.

Maturity Model Levels	1	2	3	4	5
Beehive connected	-	-	yes	-	-

We have chosen the agile approach because it is flexible based on principles of collaboration, adaptability, and continuous improvement. This approach is ideal for our project, which is service-oriented and allows for the rapid adaptation based on stakeholder feedback.

The information is transferred by using smartphone or web applications to the data model for later decisions, as shown in Figure 9. Along with the best practices and modern design techniques, REST architectural style allows building an API that is both extendable and flexible. Indeed, the use of API HTTP REST provides various possible interactions with third-party platforms in both directions. REST APIs should accept the JSON format for the request payload and send responses to JSON. JSON is a standard for transferring data. Almost every networked technology can use it.

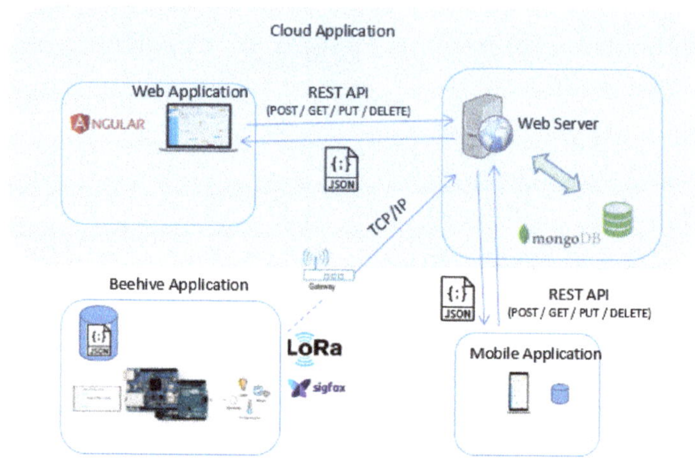

Figure 9. An example of the proposed architecture.

The criteria for this project are:

- Low-power board computer, because the system has to be embedded on the apiary during several weeks;
- The materials have to be low-cost because we need to convince the beekeepers to use it. It means we have to make a compromise between accuracy and price.

From these constraints, we selected the materials below:

- Arduino: the best embedded low-power single board computer;
- Sensors for the humidity and temperature: **AZDelivery DHT11**;
- Sensors for the Weight: **HX711 Module**;
- Sensors for communication: **Z-Delivery ESP8266 ESP-01 Serial Wireless WLAN WiFi Transceiver Module Transceiver**.

We made our weight sensor by combining four single strain gauges (load sensor). We made boxes to encapsulate the sensors using a 3D printer. Figure 10 shows: single strain gauge (a) arranged in a Wheatstone bridge configuration (c). A load sensor combinator from sparkfun (b), which is a module that directly integrates a Wheatstone bridge configuration to which we only have to connect the four sensors according to the diagram (c). The combinatory board is hooked up to an amplifier (HX711 module). Load cell measurements can be off due to a range of things (temperature, vibration). In our tests, we obtained calculation variations between +2% and −3.5%.

At the top of Figure 11, we can see the day's indicators (specified at the top left), such as the maximum and minimum temperatures of the hive, its maximum and minimum humidity levels, as well as its weight. These indicators turn red if they are too high, and blue if they are too low. Each indicator is represented as a curve below, showing the evolution of the data over a chosen period (which year, month, or number of days can be changed in the filters at the top right). We also see on each of the curves the forecast of the potential evolution of the data for the remainder of the month. Finally, the dotted lines represent the warning threshold that a curve must not exceed. In Figures 11–13, the measurements were made until 19 February. After this date, it is a question of prediction.

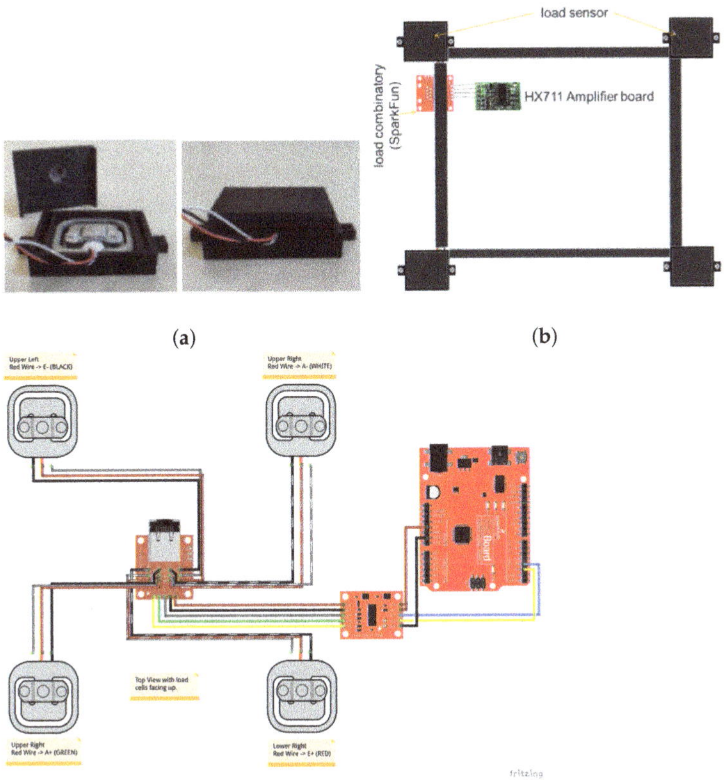

(a) (b)

(c)

Figure 10. Weight sensor: (**a**) single strain gauge, (**b**) a load sensor combinator from sparkfun, (**c**) Wheatstone bridge configuration © https://learn.sparkfun.com/tutorials/load-cell-amplifier-hx711-breakout-hookup-guide/all. Accessed on 28 September 2022.

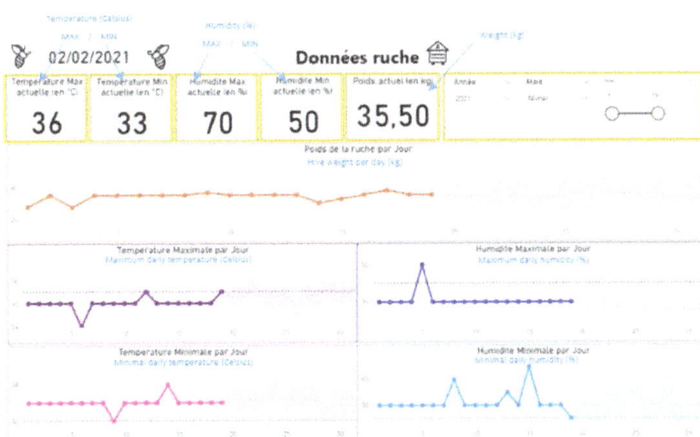

Figure 11. Dashboard: data.

The figures also show a prediction of the weight as well as the weather. We use the linear regression algorithm.

Very similar to Figure 11, Figure 12 shows the indicators for the external data. The ambient temperature as well as the outdoor humidity levels have been taken into account. Unlike the previous part, the indication of the wind force as well as the weather forecast for the day are shown.

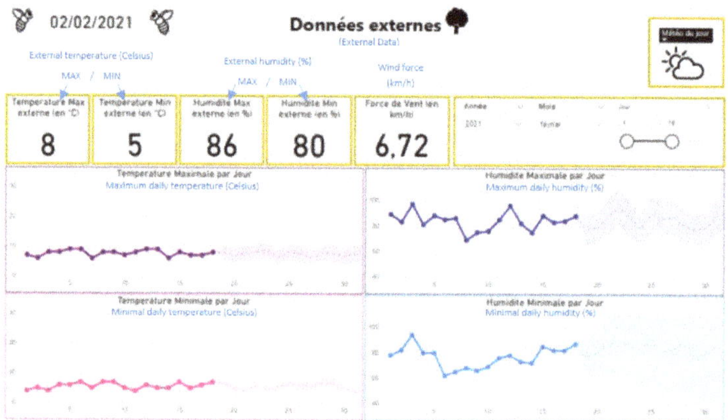

Figure 12. Dashboard: environmental data.

In Figure 13, on the left, the total pollination rate by region (selectable in the filter below for a global view in France, or for a focus on a particular region) over a chosen period is shown. For more details, on the right, we can see the most common types of pollen (according to their pollination rate in number of grains/m^3 of air) in France, in a particular region. It is possible via the filters below to choose a region, or a particular type of pollen.

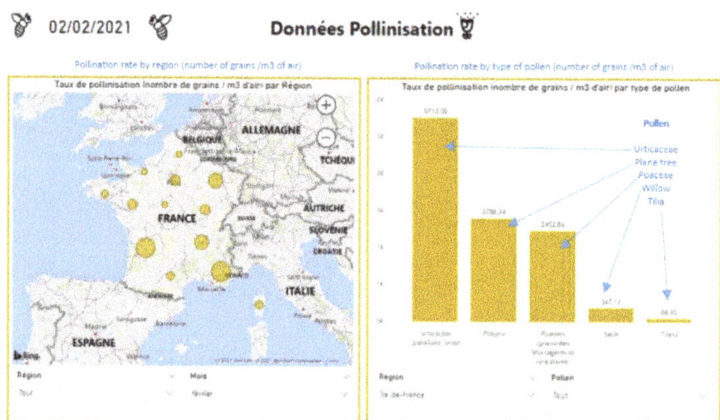

Figure 13. Dashboard: pollination data.

The Alert Manager is a component of our system. Its mission is to address beekeepers with possible alerts or valuable information based on data provided by all other parts of the system (e.g., sensors and predictions).

Figure 14 shows the nature of the data to be managed and collected by the system, as well as the information output from the system. In the context of our case study, a system of alerts and notifications has been developed. A screenshot with an example of use is shown in Figure 15. The **Today** column presents the alerts or notifications from the current day data. The **Forecast** column shows the alerts and notifications for future days from forecast analysis. The severity rate is represented by different colours (orange for warnings and red for the alerts). The parameter settings for each hive are changed.

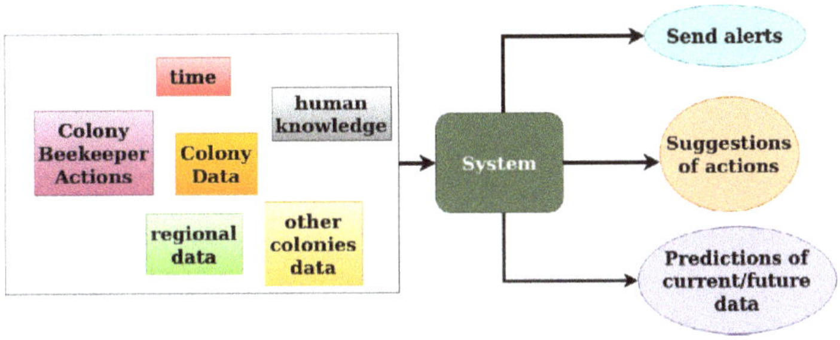

Figure 14. System data and information.

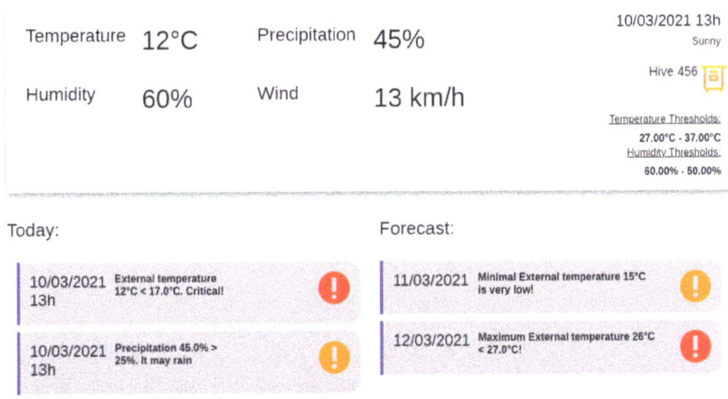

Figure 15. Example of alerts and notifications from environmental data.

7. Conclusions

This paper presents a new approach for modelling, monitoring, and deploying Agriculture 4.0 systems. As for Industry 4.0, agriculture is going through its digital revolution. In fact, Agriculture 4.0 is based on solutions and technologies developed for Industry 4.0. In this context, smart beehives need some technological solutions to build smart hive motoring systems. The goal is to help different stakeholders to make the best decisions. We studied through a spatio-temporal matrix with the time and the geographical horizons all kinds of decisions that can be made. Indeed, beekeeping is affected by a crisis due to the disappearance of bees, and the different stakeholders need support to make their decisions. Our approach used a part of these technologies: Big Data, the Internet of Things, Cyber-Physical Systems, Cloud Computing, and Artificial Intelligence. We have implemented a Digital Transformation (DT) process to help the actors of the beekeeping sector. The idea was to develop new services for beekeepers and more generally for the actors of the apiary using these technologies. Among the elements of the DT, the "people" become essential, and even more important than anything else [10]. It is why beekeepers and their cooperatives were fully involved. We used their vocabulary and we worked together to understand their culture. Concretely, a reusable and flexible architecture based on sensors, data from beekeepers, and Open Data has been introduced. To illustrate our new approach, a specialized software application has been developed. This application notifies and alerts beekeepers to help them to take care of their hives. We also proposed a Maturity Model

(MM) to diagnose the level of digitalization in the area of agriculture. We analysed the literature on smart beehives regarding this MM.

A long term perspective is to move toward the development of a methodology to address issues related to Agriculture 4.0 and support agriculture in their DT in general. Another important perspective is to use this approach to build a global architecture. It allows to register all the data at different hierarchical levels (land, apiary, and beehive databases). Such an architecture would be based on Industry 4.0 and Fog computing (Edge Computing). We will analyse architectures proposed by these concepts [49]. Moreover, our use case will evolve toward more local decisions, and will correspond to level 4 of our MM (Table 1). It means that the decisions will be faster, more relevant, efficient, and accurate. Concretely, this evolution will be achieved thanks to the development of a data analysis module inside the beehives. This will allow us to cross-check and analyse the history of the data acquired. Another future line of research is the development of cooperation between the embedded systems of smart beehives of the same apiary. It will generate better accuracy for the notification and alert system. To achieve all these evolutions, we need to implement machine and deep learning algorithms.

Author Contributions: Conceptualization, J.-C.H., L.B. and Y.K.; Methodology, L.B. and Y.K.; Project administration, J.-C.H.; Software, L.B. and Y.K.; Supervision, C.A.; Validation, K.W.-W. and C.A.; Writing—original draft, J.-C.H., L.B., Y.K. and K.W.-W.; Writing—review & editing, J.-C.H., L.B., Y.K., K.W.-W. and C.A. All authors have read and agreed to the published version of the manuscript.

Funding: This research was funded partially by French Chambers of Agriculture development programs (CASDAR) and the APC was funded by EFREI Paris (engineering school of digital technologies).

Institutional Review Board Statement: Not applicable.

Informed Consent Statement: Not applicable.

Data Availability Statement: Not applicable.

Acknowledgments: The authors gratefully acknowledge the Ministry of Agriculture and Food, who are financing PNAPI through CASDAR (the special appropriation account "Agriculture and Rural Development") under the project number 18 ART 1831 as well as the support and help of Alexandre Dangléant, ITSAP (Technical and Scientific Institute of Beekeeping and Pollination).

Conflicts of Interest: The authors declare no conflict of interest.

Abbreviations

The following abbreviations are used in this manuscript:

API	Application Programming Interface
CAGR	Compound Annual Growth Rate
CPS	Cyber-Physical System
DSS	Decision Support Systems
ERP	Enterprise Resource Planning
IoT	Internet of Thing
IT	Information Technology
JSON	JavaScript Object Notation
MM	Maturity Model
REST	REpresentational State Transfer
RFID	Radio Frequency Identification
WSN	Wireless Sensor Networks

References

1. Latioui, Z.E.; Bougueroua, L.; Moretto, A. Social media chatbot system-beekeeping case study. In Proceedings of the International Conference on Hybrid Intelligent Systems, Porto, Portugal, 13–15 December 2018; pp. 302–310.
2. FranceAgreMer. *Bilan de Campagne Miel 2020*; Technical Report; FranceAgreMer, 2021. Available online: https://www.franceagrimer.fr/content/download/67599/document/BIL_API_2021_Bilan_campagne_2020.pdf (accessed on 28 September 2022).

3. Cardin, O. Classification of cyber-physical production systems applications: Proposition of an analysis framework. *Comput. Ind.* **2019**, *104*, 11–21. [CrossRef]
4. Nezamova, O.; Olentsova, J. The main trends of digitalization in agriculture. *IOP Conf. Ser.: Earth Environ. Sci.* **2022**, *981*, 032018. [CrossRef]
5. Schallmo, D.R.; Williams, C.A. *Digital Transformation Now! Guiding the Successful Digitalization of Your Business Model*; Springer: Berlin/Heidelberg, Germany, 2018.
6. Berman, S.J. Digital transformation: Opportunities to create new business models. *Strategy Leadersh.* **2012**, *40*, 16–24. [CrossRef]
7. Gabriel, A.; Gandorfer, M. Adoption of digital technologies in agriculture—An inventory in a european small-scale farming region. In *Precision Agriculture*; Springer: Berlin/Heidelberg, Germany, 2022; pp. 1–24. [CrossRef]
8. Nambisan, S.; Lyytinen, K.; Majchrzak, A.; Song, M. Digital Innovation Management: Reinventing Innovation Management Research in a Digital World. *MIS Q.* **2017**, *41*, 223–238. [CrossRef]
9. Ebert, C.; Duarte, C.C. Digital Transformation. *IEEE Softw.* **2018**, *35*, 16–21. [CrossRef]
10. Verina, N.; Titko, J. Digital transformation: Conceptual framework. In Proceedings of the International Scientific Conference "Contemporary Issues in Business, Management and Economics Engineering", Vilnius, Lithuania, 9–10 May 2019.
11. Moeuf, A.; Pellerin, R.; Lamouri, S.; Tamayo-Giraldo, S.; Barbaray, R. The industrial management of SMEs in the era of Industry 4.0. *Int. J. Prod. Res.* **2018**, *56*, 1118–1136. [CrossRef]
12. Danjou, C.; Rivest, L.; Pellerin, R. Douze positionnements stratégiques pour l'Industrie 4.0: Entre processus, produit et service, de la surveillance à l'autonomie (Twelve strategic positioning for Industry 4.0: Between process, product and service, from monitoring to autonomy). In Proceedings of the 12e Congrès International de Génie Industriel CIGI, Compiègne, France, 3–5 May 2017.
13. Corallo, A.; Latino, M.E.; Menegoli, M. From industry 4.0 to agriculture 4.0: A framework to manage product data in agri-food supply chain for voluntary traceability. *Int. J. Nutr. Food Eng.* **2018**, *12*, 146–150.
14. Ahmed, S.H.; Kim, G.; Kim, D. Cyber Physical System: Architecture, applications and research challenges. In Proceedings of the 2013 IFIP Wireless Days (WD), Valencia, Spain, 13–15 November 2013; pp. 1–5. [CrossRef]
15. Gubbi, J.; Buyya, R.; Marusic, S.; Palaniswami, M. Internet of Things (IoT): A vision, architectural elements, and future directions. Including Special sections: Cyber-enabled Distributed Computing for Ubiquitous Cloud and Network Services & Cloud Computing and Scientific Applications—Big Data, Scalable Analytics, and Beyond. *Future Gener. Comput. Syst.* **2013**, *29*, 1645–1660. [CrossRef]
16. Atzori, L.; Iera, A.; Morabito, G. The Internet of Things: A survey. *Comput. Netw.* **2010**, *54*, 2787–2805. [CrossRef]
17. Firouzi, F.; Rahmani, A.; Mankodiya, K.; Badaroglu, M.; Merrett, G.; Wong, P.; Farahani, B. Internet-of-Things and big data for smarter healthcare: From device to architecture, applications and analytics. *Future Gener. Comput. Syst.* **2018**, *78*, 583–586. [CrossRef]
18. Vincent, D.R.; Deepa, N.; Elavarasan, D.; Srinivasan, K.; Chauhdary, S.H.; Iwendi, C. Sensors driven AI-based agriculture recommendation model for assessing land suitability. *Sensors* **2019**, *19*, 3667. [CrossRef] [PubMed]
19. Talavera, J.M.; Tobón, L.E.; Gómez, J.A.; Culman, M.A.; Aranda, J.M.; Parra, D.T.; Quiroz, L.A.; Hoyos, A.; Garreta, L.E. Review of IoT applications in agro-industrial and environmental fields. *Comput. Electron. Agric.* **2017**, *142*, 283–297. [CrossRef]
20. Quy, V.K.; Hau, N.V.; Anh, D.V.; Quy, N.M.; Ban, N.T.; Lanza, S.; Randazzo, G.; Muzirafuti, A. IoT-Enabled Smart Agriculture: Architecture, Applications, and Challenges. *Appl. Sci.* **2022**, *12*. [CrossRef]
21. Colombo-Mendoza, L.O.; Paredes-Valverde, M.A.; Salas-Zárate, M.d.P.; Valencia-García, R. Internet of Things-Driven Data Mining for Smart Crop Production Prediction in the Peasant Farming Domain. *Appl. Sci.* **2022**, *12*, 1940. [CrossRef]
22. García, L.; Parra, L.; Jimenez, J.M.; Lloret, J.; Lorenz, P. IoT-Based Smart Irrigation Systems: An Overview on the Recent Trends on Sensors and IoT Systems for Irrigation in Precision Agriculture. *Sensors* **2020**, *20*, 1042. [CrossRef]
23. Lee, S. A Novel Design and Implementation of Data Acquisition and Preprocessing System for Precision Agriculture. *Agric. Environ. Perspect. Intell. Syst.* **2019**, *24*, 52.
24. López-Morales, J.A.; Martínez, J.A.; Skarmeta, A.F. Digital Transformation of Agriculture through the Use of an Interoperable Platform. *Sensors* **2020**, *20*, 1153. [CrossRef]
25. Friha, O.; Ferrag, M.A.; Shu, L.; Maglaras, L.; Wang, X. Internet of Things for the Future of Smart Agriculture: A Comprehensive Survey of Emerging Technologies. *IEEE/CAA J. Autom. Sin.* **2021**, *8*, 718–752. [CrossRef]
26. Felch, V.; Asdecker, B.; Sucky, E. Maturity Models in the Age of Industry 4.0–Do the Available Models Correspond to the Needs of Business Practice? In Proceedings of the 52nd Hawaii International Conference on System Sciences, Maui, HI, USA, 8–11 January 2019.
27. Menon, K.; Kärkkäinen, H.; Lasrado, L.A. Towards a Maturity Modeling Approach for the Implementation of Industrial Internet. In Proceedings of the PACIS 2016; Association for Information Systems, AIS Electronic Library (AISeL), Chiayi, Taiwan, 27 June–1 July 2016; p. 38.

28. Jæger, B.; Halse, L.L. The IoT Technological Maturity Assessment Scorecard: A Case Study of Norwegian Manufacturing Companies. In Proceedings of the Advances in Production Management Systems. The Path to Intelligent, Collaborative and Sustainable Manufacturing, Hambourg, Germany, 3–7 September 2017; Lödding, H., Riedel, R., Thoben, K.D., von Cieminski, G., Kiritsis, D., Eds.; Springer International Publishing: Cham, Switzerland; 2017; pp. 143–150.
29. Westermann, T.; Anacker, H.; Dumitrescu, R.; Czaja, A. Reference architecture and maturity levels for cyber-physical systems in the mechanical engineering industry. In Proceedings of the 2016 IEEE International Symposium on Systems Engineering (ISSE), Edinburgh, UK, 3–5 October 2016; pp. 1–6.
30. Büyük, A.M.; Ateş, G.; Burghli, S.; Yılmaz, D.; Temur, G.T.; Sivri, Ç. Digital Maturity Assessment Model for Smart Agriculture. In Proceedings of the The International Symposium for Production Research, Antalya, Turkey, 24–26 September 2020; Springer: Berlin/Heidelberg, Germany, 2020; pp. 289–301.
31. Aksoy, A.; Ertürk, Y.E.; Erdoğan, S.; Eyduran, E.; Tariq, M.M. Estimation of Honey Production in Beekeeping Enterprises from Eastern Part of Turkey through Some Data Mining Algorithms. *Pak. J. Zool.* **2018**, *50*, 2207. [CrossRef]
32. Dineva, K.; Atanasova, T. OSEMN process for working over data acquired by IoT devices mounted in beehives. *Curr. Trends Nat. Sci.* **2018**, *7*, 47–53.
33. Rafael Braga, A.; Gomes, D.G.; Rogers, R.; Hassler, E.E.; Freitas, M.B.; Cazier, J.A. A method for mining combined data from in-hive sensors, weather and apiary inspections to forecast the health status of honey bee colonies. *Comput. Electron. Agric.* **2020**, *169*, 105161. [CrossRef]
34. Bee Informed Partnership—Using Beekeepers' Real World Experience to Solve Beekeepers' Real World Problems. Available online: https://beeinformed.org/ (accessed on 28 September 2022).
35. Edwards-Murphy, F.; Magno, M.; Whelan, P.M.; O'Halloran, J.; Popovici, E.M. b+WSN: Smart beehive with preliminary decision tree analysis for agriculture and honey bee health monitoring. *Comput. Electron. Agric.* **2016**, *124*, 211–219. [CrossRef]
36. Markovic, D.; Pesovic, U.; Djurasevic, S.; Randjic, S. Decision support system for temperature monitoring in beehives. *Acta Agric. Serbica* **2016**, *21*, 135–144. [CrossRef]
37. Balta, A.; Dogan, S.; Ozmen Koca, G.; Akbal, E. Software Modeling of Remote Controlled Beehive Design. In Proceedings of the International Conference on Advances and Innovations in Engineering ICAIE, Elazig, Turkey, 10–12 May 2017; pp. 1133–1138.
38. Zogović, N.; Mladenović, M.; Rašić, S. From primitive to cyber-physical beekeeping. In Proceedings of the 7th International Conference on Information Society and Technology, Kopaonik, Serbia, 13–16 March 2017.
39. Cazier, J. The Genius Hive Will Be Able to Tell You What It Needs to Do Better. Available online: https://www.beeculture.com/peering-into-the-future-a-path-to-the-genius-hive/ (accessed on 28 September 2022).
40. Mrozek, D.; Gòrny, R.; Wachowicz, A.; Małysiak-Mrozek, B. Edge-Based Detection of Varroosis in Beehives with IoT Devices with Embedded and TPU-Accelerated Machine Learning. *Appl. Sci.* **2021**, *11*, 11078. [CrossRef]
41. Debauche, O.; Moulat, M.E.; Mahmoudi, S.; Boukraa, S.; Manneback, P.; Lebeau, F. Web Monitoring of Bee Health for Researchers and Beekeepers Based on the Internet of Things. *Procedia Comput. Sci.* **2018**, *130*, 991–998. [CrossRef]
42. Zetterman, B.E.A. Beekeepers Usage of IoT: Data Collection, Sharing and Visualization in the Domain of Beekeeping. Master's Thesis, Linnaeus University, Vaxjo, Sweden, 2018.
43. Latioui, Z.E.; Bougueroua, L.; Moretto, A. Social Media Chatbot System—Beekeeping Case Study. In *Hybrid Intelligent Systems;* Advances in Intelligent Systems and Computing; Madureira, A.M., Abraham, A., Gandhi, N., Varela, M.L., Eds.; Springer International Publishing: Cham, Switzerland, 2020; Volume 923; pp. 302–310.[CrossRef]
44. Chabrol, M.; Gourgand, M.; Rodier, S. A modeling methodology and its application to the design of decision-making aid tools dedicated to the hospital systems. In Proceedings of the RCIS'08: International Conference on Research Challenges in Information, Marrakech, Morocco, 3–6 June 2008.
45. Comelli, M.; Féniès, P.; Tchernev, N. A combined financial and physical flows evaluation for logistic process and tactical production planning: Application in a company supply chain. Special Section on Recent Developments in the Design, Control, Planning and Scheduling of Productive Systems. *Int. J. Prod. Econ.* **2008**, *112*, 77–95. [CrossRef]
46. Huet, J.C.; Bougueroua, L.; Kriouile, Y.; Moretto, A. Toward an intelligent system architecture for smart agriculture: Application to smart beehives. In Proceedings of the 20th International Conference on Intelligent Systems Design and Applications—ISDA 2020, Online, 12–15 December 2020.
47. McLeod, R. *Information Systems*, 1st ed.; Prentice Hall PTR: Hoboken, NJ, USA, 1989.
48. ETSI. Context Information Management (CIM); NGSI-LD API. Technical Report; ETSI. 2020. Available online: https://www.etsi.org/deliver/etsi_gs/CIM/001_099/009/01.03.01_60/gs_CIM009v010301p.pdf (accessed on 28 September 2022).
49. Rafique, W.; Qi, L.; Yaqoob, I.; Imran, M.; Rasool, R.U.; Dou, W. Complementing IoT Services Through Software Defined Networking and Edge Computing: A Comprehensive Survey. *IEEE Commun. Surv. Tutor.* **2020**, *22*, 1761–1804. [CrossRef]

applied sciences

MDPI

Article

A Visual Feedback for Water-Flow Monitoring in Recirculating Aquaculture Systems

Krzysztof Okarma [1,*], Piotr Lech [1], Darius Andriukaitis [2], Dangirutis Navikas [2], Agata Korzelecka-Orkisz [3], Adam Tański [3] and Krzysztof Formicki [3]

1 Department of Signal Processing and Multimedia Engineering, West Pomeranian University of Technology in Szczecin, Sikorskiego 37, 70-313 Szczecin, Poland
2 Department of Electronics Engineering, Kaunas University of Technology, Studentu St. 50, LT-51368 Kaunas, Lithuania
3 Department of Hydrobiology, Ichthyology and Biotechnology of Reproduction, Faculty of Food Sciences and Fisheries, West Pomeranian University of Technology in Szczecin, Kazimierza Królewicza 4, 71-550 Szczecin, Poland
* Correspondence: okarma@zut.edu.pl

Abstract: The optimal water flow in fish breeding tanks is one of the crucial elements necessary for the well-being and proper growth of fish, such as salmon or trout. Considering the round tanks and the uneven distribution of water-flow velocity, ensuring a nearly optimal flow is an important task that may be performed using various sensors installed to monitor the water flow. Nevertheless, observing the rapid development of video analysis methods and considering the increasing availability of relatively cheap cameras, the use of video feedback has become an interesting alternative that limits the number of sensors inside the water tanks in accordance with the requirements of fish breeders. In this paper, an analysis of the use of optical flow algorithms for this purpose is performed and an estimation method based on their features is proposed. The results of the flow estimation using the proposed method are verified experimentally and compared with the measurement results obtained using the professional water-flow meter, demonstrating a high correlation, exceeding 0.9, confirming the proposed solution as a good alternative in comparison to the use of expensive sensors and meters.

Keywords: video analysis; visual feedback; water monitoring; optical flow; aquaculture; recirculating aquaculture systems

Citation: Okarma, K.; Lech, P.; Andriukaitis, D.; Navikas, D.; Korzelecka-Orkisz, A.; Tański, A.; Formicki, K. A Visual Feedback for Water-Flow Monitoring in Recirculating Aquaculture Systems. *Appl. Sci.* **2022**, *12*, 10598. https://doi.org/10.3390/app122010598

Academic Editor: Ginés García-Mateos

Received: 14 September 2022
Accepted: 17 October 2022
Published: 20 October 2022

1. Introduction

The application of video analysis has recently become one of the leading interdisciplinary trends in modern aquaculture. Some of the evident reasons for this situation are the increasingly available and affordable video cameras as well as the growth of the computational power of the popular devices that, in some applications, may replace typical personal computers, such as, e.g., Raspberry Pi, FPGA boards, nVidia Jetson Nano and similar more sophisticated hardware platforms. An analogous situation also exists in the automotive industry and some other areas of technology, which is typical for the rapid development of Industry 4.0 solutions, which is becoming Agriculture 4.0 as well [1].

The most typical applications of video analysis in fish farming seem to be motion tracking [2], as well as the recognition and classification of fish [3], which has recently also utilized popular convolutional neural networks (CNNs) [4]. The application of such video systems may also be helpful for the determination of fish behavior [5–8] as well as in the development of intelligent feeding systems [9–11]. Another interesting area of research related to these topics is water-quality monitoring [12]. Some other recent works focus on the applications of artificial-intelligence methods, including mobile remote-control systems and IoT solutions applied in fish farms [13,14].

Many of these research areas are motivated by the willingness to limit the number of sensors and additional equipment mounted inside the ponds or tanks, which is also essential for economic reasons. Such video-based approaches do not affect fish directly, reducing the possibility of injuring individual fish, additionally simplifying the hardware architecture of systems containing sensors and their necessary connections. Therefore, a similar approach may also be beneficiary in monitoring the water flow in recirculating aquaculture systems (RAS).

Proper water flow in fish-farming ponds and tanks is one of the most crucial requirements in the fish-farming centers in which species such as salmon and trout are bred. These fish behave in a specific way, positioning themselves parallel to the water current. Therefore, the correct flow velocity of water determines the proper development of the fish as well as the welfare of the whole group of them present in the pond. Much work has been carried out in recent years to determine the optimal flow speeds for maintaining overall fish health, but the water velocity distribution inside circular tanks is often very heterogeneous, particularly near their walls (edges). Usually, these works are focused on analyzing the effect of design parameters on the distribution of water velocities inside circular aquaculture tanks [15]. An example of such a model can be found in the paper [16], where the proposed model estimates the velocity distribution by determining the angular momentum per unitary mass close to the tank wall and around the central axis. The model depends on the water inlet and outlet flow rates, water inlet velocity, reservoir radius, water depth, and three reservoir-specific parameters that must be determined experimentally. It also takes into account the influence of the wall roughness, the characteristics of the water inlet devices, and the presence of individual elements at the bottom of the tank causing friction losses. However, it is not always necessary to use such advanced models, as in most cases simplified models of flow distribution are sufficient.

The optical flow methods analyzed in this paper may also be used for the estimation of the water-flow velocity in rivers [17–19], replacing previously used particle image velocimetry (PIV) methods [20] as well as typical motion-tracking methods based on point detectors [21]. Nevertheless, in such types of video recordings, there is no ground truth data; hence, the evaluation of the obtained results may be conducted with the use of trajectory reconstruction. In recent years, the accuracy and efficiency of such techniques have improved significantly [22], making it also possible to apply some of the methods in real-time applications [23]. Some of the proposed approaches utilize feature matching applied for orthorectification and velocimetry, such as airborne feature matching velocimetry (AFMV) [18], whereas the applicability of the optical-flow methods has also been verified in experimental irrigation stations. It led to promising results, with average relative errors less than 6.5%, applying the time-averaged surface velocity of water flows based on the Farneback optical flow method [24].

With an appropriate measuring environment, it is possible to prepare spatial distributions of the flow in the tank [25]. Unfortunately, breeders are usually opposed to the excessive installation of sensors that can injure fish, so the most commonly available information is the amount of water leaving the water pump supplying the fish-farming tank. The spatial distribution is constant for the static configuration of the tank, that is: the number and settings of nozzles, pump performance and the shape of the tank. It changes as the amount of fish and floating biomass increases, which is problematic if it is not possible to place the sensors in the water together with the fish.

The video measurement method developed and presented in the article is non-invasive, with the quality of results comparable to those obtained by sensors immersed in the tank and a flow meter measuring the performance of the pump feeding the tank.

2. Materials and Methods

2.1. Experimental Site

One of the essential parts of a system that uses vision feedback to determine the behavior of fish in a breeding tank is the estimation of the water-flow velocity. Such studies can be carried out by various image analysis methods, which, however, require verification with a calibrated water-flow meter used in experimental studies. To verify the suitability of individual methods of video sequence analysis for this purpose, experimental tests were carried out by recording video material for the determined values of water flows in a round tank (SDK RT 29-68 type). The experiments were initially carried out without the fish in the tank in order to prevent possible disturbance caused by their movements.

Due to the fact that the water in the tank is never perfectly clean and transparent, visible small objects may appear on its surface, making it easier to estimate the flow velocity thanks to their tracking. Additionally, it may be useful to have fixed elements of the tank or other objects attached to it, or a reflection of the elements above the tank. The flow of water at different speeds causes visible undulations in the shapes of these objects, while in the case of small elements floating on the water surface, it is possible to estimate the flow velocity on the basis of tracking their movement.

The experimental study was performed in a rainbow-trout farm located in Garnki in the northern part of Poland about 35 km from the Baltic Sea coast. The reference measurements were conducted using the electromagnetic water-flow meter MTF-10 and SonTek FlowTracker2 in the above-mentioned fish tank, as illustrated in Figure 1. Video acquisition took place using the color IP camera Dahua IPC-HFW2231T-ZS-27135-S2 typically applied in video surveillance and industrial monitoring systems. It is a varifocal IPC camera of the Lite series that records frames of video sequences with FullHD resolution (1920 $\times$ 1080 pixels) at the speed of 25 or 30 frames per second.

It has a two megapixel CMOS 1/2.8 inch Progressive Scan sensor with a sensitivity of 0.002 lux/F 1.5. It also has the option of using an additional illuminator in the form of 4 IR LEDs with a range of 60 m and an automatic ICR infrared filter. It has the built-in 2.7–13.5 mm lens with a viewing angle of 28–109 degrees and a motorized zoom with autofocus, which enables external, remote adjustment of the focal length level with the help of a built-in electric motor, and also automatically adjusts the sharpness level. The native format of data recording is Dahua's DAV, the camera also supports H.265+, H.265, H.264+ and MJPEG compression. The latter format, also referred to as Motion JPEG, may be used in the applications where a high compression ratio is not necessary, as in this case the video stream consists of a series of recorded frames compressed using a JPEG algorithm without using inter-frame information in the compression process. The housing has an IP67 tightness class, the camera supports 12 V DC or 48 V power supply using the Power over Ethernet interface (PoE 10/100 Base-T 802.3af).

Additional functionalities and advantages of the camera are the following technologies: 3D NR (noise reduction system that allows the image to be transparent in the event of changing signal levels, which protects the recordings against smudging), WDR (wide dynamic range—extended dynamic range, important in places with different lighting levels of the observed scene), BLC and HLC (back light compensation—important when the camera is directed towards a strong light source—and high lighting compensation to reduce the negative impact of point light sources, e.g., reflected in water), intelligent infrared lighting function and region of interest (RoI) limitation to minimize the size of saved files.

Due to some limitations of the local conditions, the camera was mounted directly above the tank (about 1 m above the water surface), enabling the observation of a selected part of the tank. One of the very important technical issues during the installation of the camera was also to ensure proper image sharpness. Another, quite obvious, limitation of the camera installation was the need to ensure the rigidity of the mount to prevent possible camera vibrations. In order to be able to effectively use vision techniques to estimate the water-flow velocity, it is necessary to ensure that the camera is stationary with respect to the housing of the rearing tank.

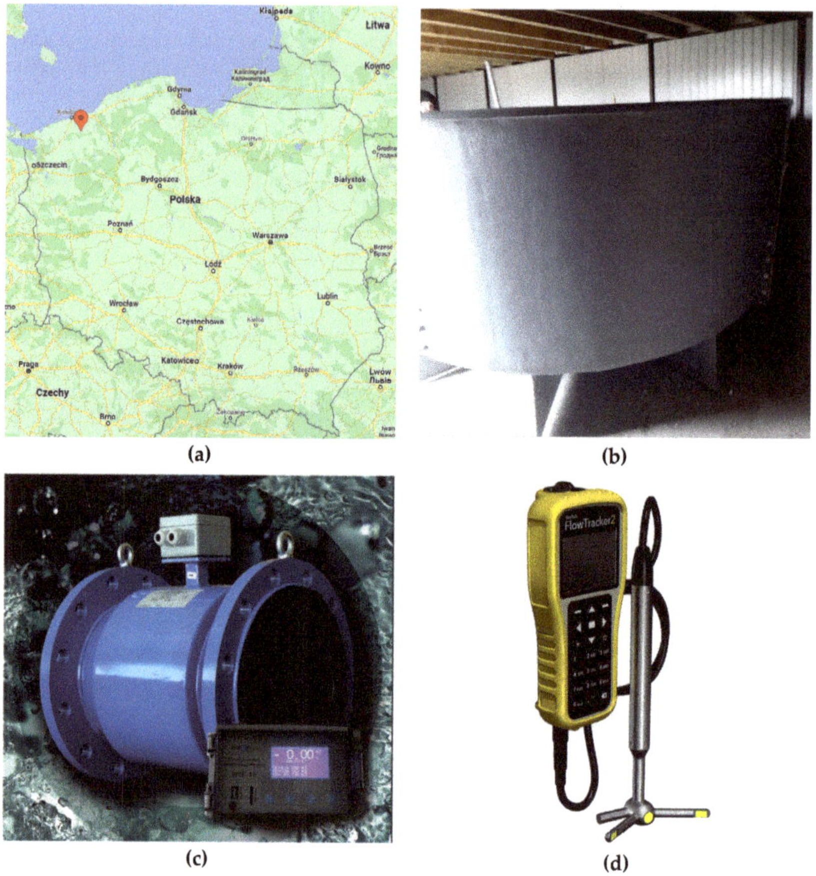

Figure 1. Location of the RAS fish farm and the equipment used in experiments: (**a**) location of Garnki, Poland shown on the map; (**b**) fish tank type SDK-RT 29-68 mounted on site; (**c**) electromagnetic water-flow meter MTF-10 (source: http://www.mtfflow.pl accessed on 13 September 2022); (**d**) SonTek FlowTracker2 Handheld-ADV (source: https://www.ysi.com/flowtracker2 accessed on 13 September 2022).

The set used in the research, consisting of a pump with an electromagnetic flow meter, allowed the use of flows (understood as the expenditure of water inflow) in the range of 1.5–5.0 dm^3 per second. Taking into account the need to stabilize the water flow in the entire reservoir, the recording of the video material was delayed in relation to the moment of setting the desired flow. The sample frames from the recorded video sequences obtained for various water flows are presented in Figure 2. Some changes are well visible even in the individual frames due to the presence of reflections of ceiling beams in the water. Some examples of the video frames captured in the presence of fish for various water-flow velocities are illustrated in Figure 3.

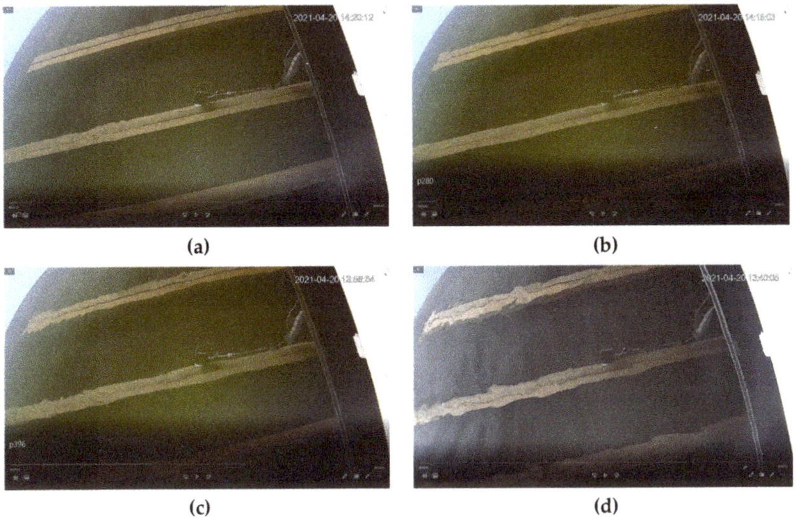

(a) (b) (c) (d)

Figure 2. Sample video frames acquired for various measured water-flow velocities: (**a**) 1.86 dm^3/s; (**b**) 2.80 dm^3/s; (**c**) 3.96 dm^3/s; and (**d**) 5.03 dm^3/s.

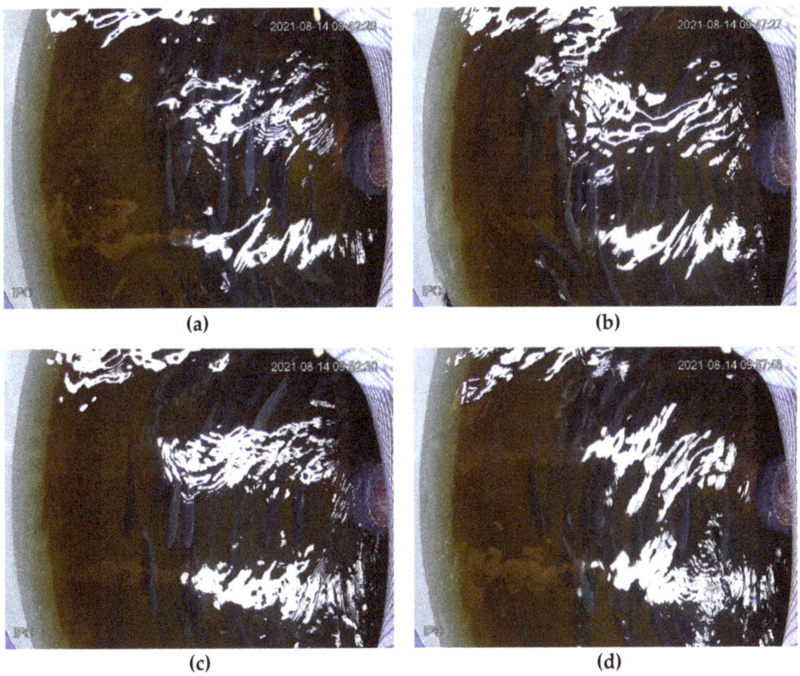

(a) (b) (c) (d)

Figure 3. Sample video frames acquired for various measured water flow velocities in the presence of fish: (**a**) 2.35 dm^3/s; (**b**) 2.99 dm^3/s; (**c**) 3.62 dm^3/s; and (**d**) 4.01 dm^3/s.

2.2. The Overview of the Applied Optical Flow Methods

The first of the considered approaches was the use of global entropy of the image and its local values determined for individual frames of the video sequence. According to the adopted assumption, for higher water-flow velocities, individual frames of the video sequence should be characterized by greater variability, and, thus, higher values of image entropy. However, the observed changes in entropy for individual frames of the video sequence did not allow to obtain satisfactory results, mainly due to the lack of use of inter-frame information. Therefore, subsequent studies focused on estimating the motion vectors between adjacent frames of the video sequence. For this purpose, it is possible to use several methods from the optical flow family [26].

To estimate the value of optical flows for individual frames of recorded video sequences, three different methods were used:

- Horn–Schunck (HS) method;
- Farneback (F) method;
- Lucas–Kanade (LK) method.

Determining the optical flow between two frames of a video sequence requires solving an equation with constraints:

$$I_x u = I_y v + I_t \tag{1}$$

where I_x, I_y and I_t are spatial and temporal (spatiotemporal) derivatives of image brightness, and u and v are the values of the horizontal and vertical optical flow (horizontal and vertical components of the vector), respectively.

The Horn–Schunck (HS) method is based on the assumption that the optical flow in the image is smooth over the entire surface of the image; hence, the velocity vector field $[u\ v]^T$ determined by this method should minimize the value of the equation:

$$E = \iint (I_x u + I_y v + I_t)^2 dx\, dy + \alpha \cdot \iint \left[\left(\frac{\partial u}{\partial x}\right)^2 + \left(\frac{\partial u}{\partial y}\right)^2 + \left(\frac{\partial v}{\partial x}\right)^2 + \left(\frac{\partial v}{\partial y}\right)^2 \right] dx\, dy \tag{2}$$

where the global coefficient α enables the selection of smoothness on the image surface, whereas the individual partial derivatives are determined in relation to the image pixel coordinates. Minimizing this expression in the HS method leads to the following relationships:

$$u_{x,y}^{k+1} = \bar{u}_{x,y}^{k} - \frac{I_x \cdot \left(I_x \bar{u}_{x,y}^{k} + I_y \bar{v}_{x,y}^{k} + I_t \right)}{\alpha^2 + I_x^2 + I_y^2}, \tag{3}$$

and

$$v_{x,y}^{k+1} = \bar{v}_{x,y}^{k} - \frac{I_y \cdot \left(I_x \bar{u}_{x,y}^{k} + I_y \bar{v}_{x,y}^{k} + I_t \right)}{\alpha^2 + I_x^2 + I_y^2}. \tag{4}$$

The vector $\left[u_{x,y}^k\ v_{x,y}^k\right]$ is an estimate of the flow velocity for a given image pixel with coordinates (x, y), whereas $\left[\bar{u}_{x,y}^k\ \bar{v}_{x,y}^k\right]$ contains its average values from the defined neighbourhood.

To determine the values of u and v flows, in the first step the gradient images I_x and I_y should be determined using the well-known Sobel convolution filter for a standard 3×3 pixels mask. The I_t value calculated as the difference between images is determined using the $[-1\ 1]$ mask, whereas the average velocity for each pixel (excluding it) is determined using a convolution with a mask typical for 4-connected neighborhood (two horizontal and two vertical neighbouring pixels). The determination of the values of flows u and v in subsequent steps is iterative. The default value of the parameter α is 1.

The Farneback (F) method assumes the generation of an image pyramid in which each level has a lower resolution compared to the previous level. Selecting a pyramid level greater than 1, the algorithm can track points at multiple levels of resolution, starting from the lowest level. An increase in the number of levels in the pyramid allows the algorithm to detect larger displacements of points between frames, however, increasing the number of necessary computations in this case. This is an example of a dense method that operates in the HSV color space (after the conversion from RGB). The flow vectors are also visualized in the HSV space, where the angle (hue) corresponds to the direction of the vector and the magnitude of the flow represented by the length of the vector is reflected by the value (V) component.

Motion tracking starts at the lowest resolution level and continues until the convergence is achieved. Point locations detected at this level are propagated as key points for the next level. Therefore, the Farneback method makes it possible to increase the tracking accuracy at each successive level. The decomposition of the pyramid enables the algorithm to detect large pixel movements for distances greater than the size of the neighborhood under consideration. The default number of levels in the pyramid is 3 with the scale factor 0.5 (split into 4 parts). By default, three iterations of the solution search are performed for each level with the default neighborhood area of 5 pixels. The convolutional Gaussian filter with the default 15×15 pixels mask is used as the averaging (smoothing) filter.

The most popular Lucas–Kanade (LK) method belongs to the group of sparse methods; hence, full calculations are performed only for selected points of the image. In this method, the image is divided into fragments in which a constant flow velocity is assumed. To find a solution, a weighted least-squares fit of the optical flow limitation equation to the constant model $[u\ v]^T$ for each Ω section is performed by minimizing the expression:

$$\sum_{x \in \Omega} W^2 \left(I_x u + I_y v + I_t \right)^2 \tag{5}$$

where W denotes a centrally weighted window function. The solution has the following form:

$$\begin{bmatrix} \sum \left(W^2 I_x^2 \right) & \sum \left(W^2 I_x I_y \right) \\ \sum \left(W^2 I_x I_y \right) & \sum \left(W^2 I_y^2 \right) \end{bmatrix} \cdot \begin{bmatrix} u \\ v \end{bmatrix} = - \begin{bmatrix} \sum \left(W^2 I_x I_t \right) \\ \sum \left(W^2 I_y I_t \right) \end{bmatrix}. \tag{6}$$

Similarly to in the HS method, the I_t value is determined between video frames using the $[-1\ 1]$ mask. To determine the values of flows u and v, the following actions are performed:

- Determining the values of the directional gradients I_x and I_y using the convolutional filter with the mask $[-1/12\ \ 2/3\ \ 0\ \ -2/3\ \ 1/12]$ and its transposed version, respectively;
- Calculation of the differential I_t value between the frames;
- Smoothing the gradients I_x I_y and I_t using the 5×5 pixels convolutional filter with the mask type $[1/16\ \ 1/4\ \ 3/8\ \ 1/4\ \ 1/16]$;
- Solving the system of two linear equations for each pixel:

$$A = \begin{bmatrix} a & b \\ c & d \end{bmatrix} = \begin{bmatrix} \sum \left(W^2 I_x^2 \right) & \sum \left(W^2 I_x I_y \right) \\ \sum \left(W^2 I_x I_y \right) & \sum \left(W^2 I_y^2 \right) \end{bmatrix} \tag{7}$$

using the eigenvalues of the matrix A determined as $\lambda_1 = \frac{a+c}{2} + \frac{\sqrt{4b^2+(a-c)^2}}{2}$ and $\lambda_2 = \frac{a+c}{2} - \frac{\sqrt{4b^2+(a-c)^2}}{2}$. Both those values are compared to the threshold τ applied for the reduction in the noise effect. If at least the threshold value τ is obtained by both eigenvalues, the system of equations can be solved by Cramer's method. When both eigenvalues are less than the threshold τ, the optical flow is zero, whereas if only $\lambda_1 \geq \tau$ (and $\lambda_2 < \tau$), the matrix A is singular and the gradient flow is normalized to calculate u and v values. The default threshold value is $\tau = 0.0039$.

One of the possible extensions of the LK method is the use of the additional temporal filtering based on the application of the DoG filter, leading to the Lucas–Kanade derivative of Gaussian (LKDoG) method. Nevertheless, during the conducted initial experiments, this method led to significantly worse results and, therefore, further experiments focused on the application of three above-described algorithms, the implementation of which is available, e.g., in MATLAB® and OpenCV library.

The illustration of the simplified flowchart of the method proposed in the paper is shown in Figure 4.

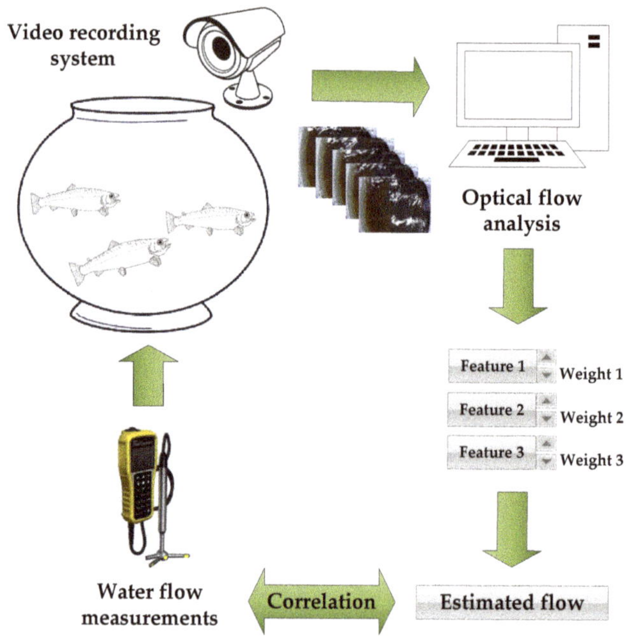

Figure 4. The simplified flowchart illustrating the idea of the proposed approach.

2.3. Experiments

In the conducted experimental studies, the possibility of using the above optical flow methods was verified in order to validate the features of video sequences based on various methods of determining optical flows, characterized by the highest possible compliance with the values of water flows measured with a water-flow meter. For this purpose, a set of video sequences was recorded for fixed flow values, and then the optical flow values for each frame, as well as their selected parameters, were calculated.

Due to the assumption of a constant value of the water flow in the tank during the recording of individual video sequences, as well as the rigid mounting of the camera (fixed in relation to the tank housing), it was assumed that the time-averaged and median values determined for individual pixels would be subject to analysis. This assumption was made despite the existence of certain differences between the values of optical flows represented by the designated motion vectors for individual frames, although leading to reliable results as presented later.

As the optical flow methods are based on the calculation of local gradients, global changes in brightness have no effect on the obtained results. An opposite situation can only take place in the case of very strong lighting or darkness, which is considered an emergency state. The method proposed in the article was developed primarily for fish breeding farms located in closed halls, where it is possible to maintain controlled lighting conditions. Due to the fact that intensive fish fattening, justified for economic reasons, usually requires the use of continuous 24-h lighting to intensify food intake by fish, the lighting used in this case is sufficient for the proper functioning of the developed method.

3. Discussion of the Experimental Results

Calculation of the individual flow maps was made with the use of the three methods considered in the paper: Horn–Schunck, Farneback and Lucas–Kanade. Since all calculations were made for the recorded video sequences, the results obtained for individual pixels were averaged. Sample results obtained with the use of the three above-listed methods for two exemplary water-flow velocities are illustrated in Figure 5. All the presented flow maps demonstrate the highest values of detected optical flows for the reflections of several elements located above the tank visible on the water surface (wooden roof boards, electric extension cord, camera). Therefore, further analyses related to determining the flow characteristics were carried out for the central parts of the visible area. Nevertheless, it is worth noting that the presence of reflections on the water surface of solid elements of the structure located above the tank is in fact beneficial for the methods applied to determine the optical flows.

As noticed on the presented flow maps, the flow rates shown in the images did not always follow expectations, which is well-visible, especially for the LK method. However, in this case, the application of the Farneback method yielded water-flow estimates more in line with the predictions. To improve the accuracy of the flow estimation, providing a highly linear correlation with the velocities measured with an electromagnetic flow meter, a set of features determined based on the obtained flow-velocity maps for the central parts of the scene was selected. The goal of the application of these features is the efficient utilization of the visual feedback for the detection of changes in the water-flow values based on the continuous analysis of the video sequence. Thanks to such an approach, the detection of some failures of, e.g., a pump or flow meter (significant non-compliance of the value determined using the video sequence analysis with the values from the flow meter) would be possible. Ultimately, after calibrating the system, it is also possible to use the proposed vision approach for fish-farming tanks in situations when an installation of a water-flow meter is impossible.

The achieved relationship, based on a non-linear combination of the three features with the fitting curves obtained for the linear and quadratic functions, is shown in Figure 6. The features used in this combination, which ensure the Pearson's linear correlation coefficient greater than 0.9, are:

- Squared entropy of the motion vectors map for the Farneback method (Ent_F^2);
- Standard deviation of the motion vector map for the Lucas–Kanade method (std_{LK});
- Variance in the local entropy of the motion vector map for the Lucas–Kanade method ($varEnt_{LK}$).

The final formula combining the three above features may be obtained using the optimization of the weighting coefficients, finally leading to the following expression:

$$Est_{flow} = A \cdot \left(Ent_F^2 - 12 \cdot std_{LK} + varEnt_{LK} \right) + B \quad (8)$$

where the parameters $A = 4.5$ and $B = -0.4$ depend on the spatial configuration of the camera and the tank and might be changed according to needs.

The dependence presented on the graph is approximately linear, however, one can notice an increase in the flow estimation error with the increase in its value. For this reason, the application of the proposed method in real conditions may be limited to lower flow values, for which the linearity of the relationship between the estimated and measured values is easily noticeable. However, additional verification and calibration is required if a different type of tank is used.

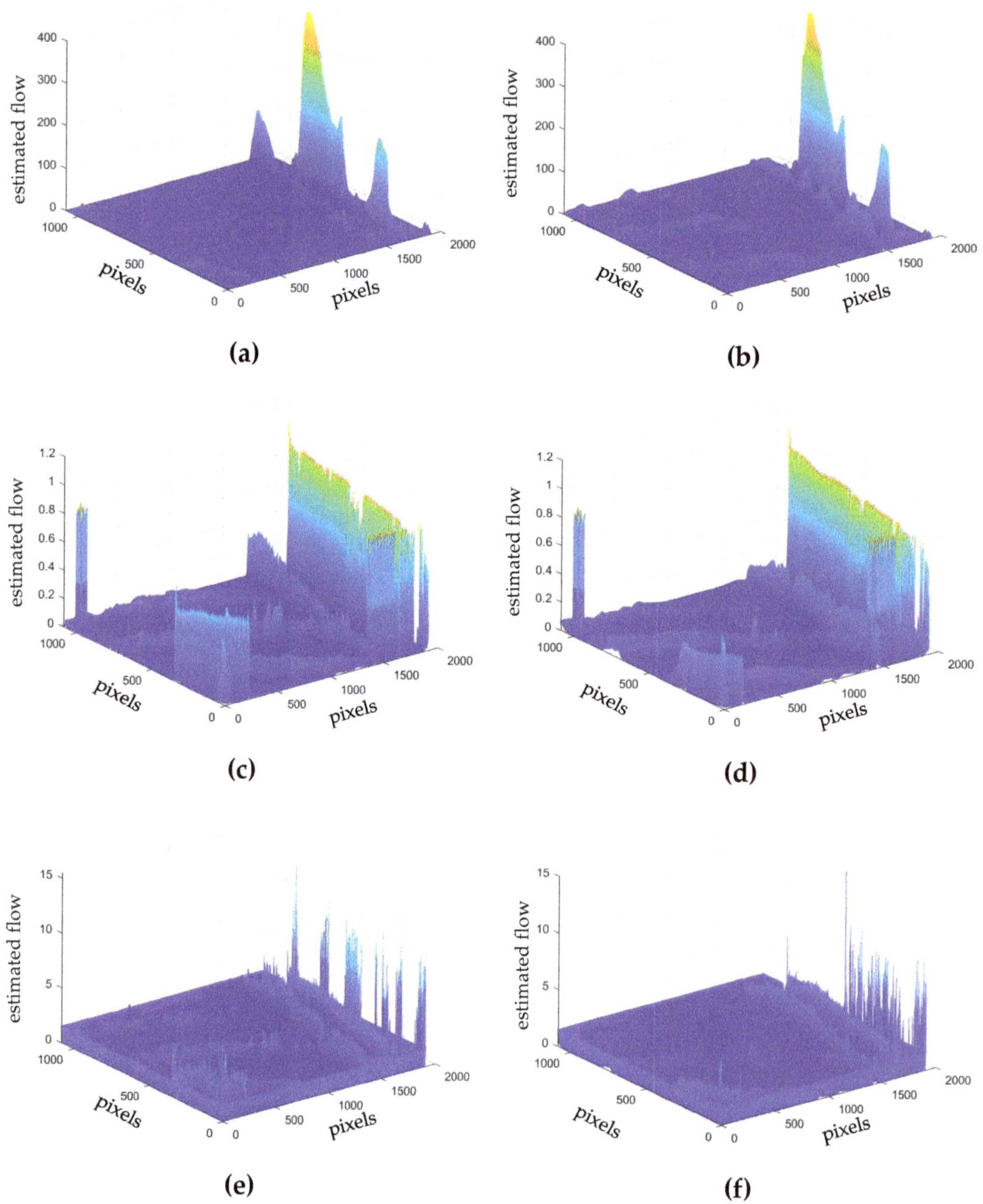

Figure 5. Sample averaged optical flows obtained for two different water flow velocities: (**a**) Farneback method for 2.17 dm^3/s; (**b**) Farneback method for 5.03 dm^3/s; (**c**) HS method for 2.17 dm^3/s; (**d**) HS method for 5.03 dm^3/s; (**e**) LK method for 2.17 dm^3/s; and (**f**) LK method for 5.03 dm^3/s.

An additional factor affecting the results obtained in the video feedback loop is the presence of fish; however, considering the fact that most of the time the fish do not move significantly, it is possible to detect and filter out the detected sudden movements of the fish. As verified experimentally by some other researchers [25,27,28], the presence of fish may decrease the water velocity by 25–30% in comparison to the tank without fish, affecting the flow pattern. However, it has been confirmed that trends of mean velocity in the radial direction remain unchanged [25]. Therefore, as verified for some video sequences acquired in the presence of fish, the method proposed in the paper also leads to similar conclusions and results in the presence of fish biomass, assuming the necessary calibration, taking into account the expected velocity decrease.

Nevertheless, it is also worth noting that a typical water flow used in fish farming for such types of tanks does not exceed 3.7 dm^3/s; hence, the proposed method may be successfully applied for the typical velocity range from 1.5 to about 3.7 dm^3/s, ensuring the Pearson linear correlation coefficient equal to 0.9868.

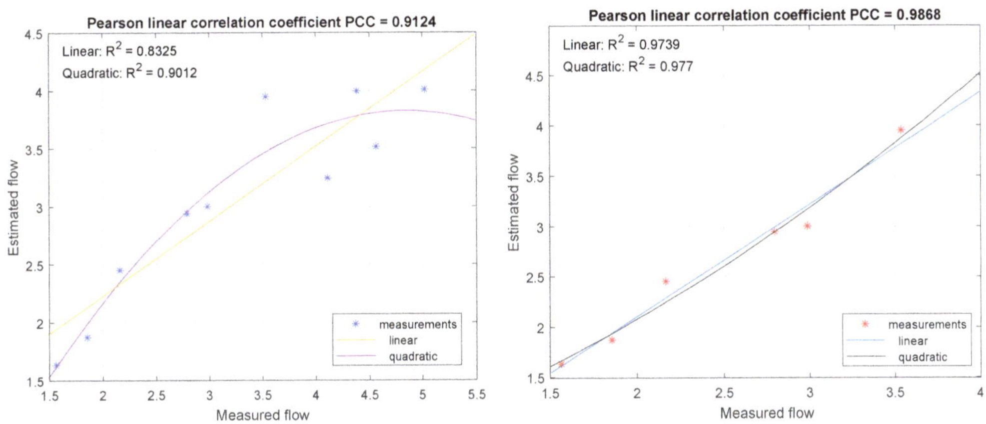

(**a**) (**b**)

Figure 6. The relationships between the uncalibrated estimates achieved using the proposed method and the velocities measured using the water-flow meter: (**a**) for the whole range 1.5–5.0 dm^3/s; (**b**) for the typical range 1.5–3.7 dm^3/s.

For further verification of the universality of the proposed solution in fish-farming tanks, additional experiments were performed after recording several hours of video sequences using five water-flow velocities in the presence of fish, as illustrated in Figure 3. However, the proposed method requires the additional calibration in the presence of fish and some changes in the parameters, also using a different camera location. In such cases, very good experimental results were also obtained applying the fourth root of the entropy of the motion vectors map for the Lucas–Kanade method ($Ent_{LK}^{0.25}$) instead of the squared entropy of the Farneback motion vectors map (with $A = 30$ and $B = -55.5$). Additionally, in this case, there is no necessity to use the Farneback method at all, since all three features originate from the Lucas–Kanade method. Therefore, an additional advantage of this simplified approach is the decrease in the number of computations, preserving relatively high correlation with measurement results (0.9847 in the experiments illustrated in Figure 7). Similarly, considering the four measurements obtained for the water flows from 2.35 to 4.01 dm^3/s, a high linearity of the obtained relationship may be noticed as well.

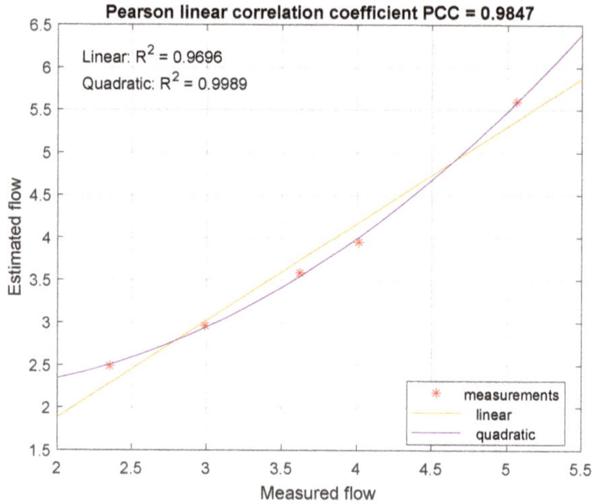

Figure 7. The relationships between the uncalibrated estimates achieved using the proposed simplified approach utilizing the Lucas–Kanade method and the velocities measured using the water-flow meter in the presence of fish.

4. Conclusions

The video feedback method presented in the paper makes it possible to monitor the water flow in the RAS fish-farming tanks and even avoid the use of water flow meters in some configurations. The possibility of the detection of potential failures of the water pumps or water-flow meters makes the proposed solution not only an interesting alternative to typical flow measurements but also a relatively cheap addition due to the increasing availability of affordable high-resolution video cameras.

The proposed solution may also limit the necessity of mounting some underwater sensors which may injure individual fish. The correctness of the water-flow estimation, verified for typical velocities, also allows its application in some other tanks after the necessary calibration of the algorithm's parameters. As the proposed method is based on the analysis of the local gradients, it is not sensitive to typical changes in lighting and weather conditions.

Further research on this topic will concentrate on the verification of sensitivity-to-lighting conditions as well as some other constraints, to achieve the higher universality of the proposed approach, allowing it to work in some unexpected conditions. Another direction of research will be related to extensions related to the combination of the proposed method with fish-counting and fish-tracking methods.

Author Contributions: Conceptualization, P.L., K.O., A.T. and K.F.; methodology, P.L. and K.O.; software, K.O.; validation, D.A., D.N., A.T. and K.F.; formal analysis, P.L. and K.O.; investigation, P.L. and K.O.; resources, A.T., A.K.-O. and K.F.; data curation, P.L. and K.O.; writing—original draft preparation, P.L. and K.O.; writing—review and editing, K.O. and K.F.; visualization, D.A., D.N., A.T. and K.O.; supervision, K.F.; project administration, A.K.-O. and K.F.; funding acquisition, K.F. All authors have read and agreed to the published version of the manuscript.

Funding: The research was conducted within the project no 00002-6521.1-OR1600001/17/20 financed by the "Fisheries and the Sea" program.

Institutional Review Board Statement: Not applicable.

Informed Consent Statement: Not applicable.

Data Availability Statement: Not applicable.

Conflicts of Interest: The authors declare no conflict of interest. The funders had no role in the design of the study; in the collection, analyses, or interpretation of data; in the writing of the manuscript; or in the decision to publish the results.

Abbreviations

The following abbreviations are used in this manuscript:

AFMV	Airborne feature matching velocimetry
BLC	Back light compensation
CMOS	Complementary metal-oxide-semiconductor
CNN	Convolutional neural network
DC	direct current
FPGA	Field-programmable gate array
HLC	High light compensation
ICR	Infrared cutfilter removal
IoT	Internet of Things
IR	Infrared
JPEG	Joint Photographic Experts Group
LED	Light-emitting diode
MJPEG	Motion JPEG
PoE	Power over Ethernet
PIV	Particle image velocimetry
RAS	Recirculating Aquaculture System
RoI	Region of interest
WDR	Wide dynamic range

References

1. Liu, Y.; Ma, X.; Shu, L.; Hancke, G.P.; Abu-Mahfouz, A.M. From Industry 4.0 to Agriculture 4.0: Current Status, Enabling Technologies, and Research Challenges. *IEEE Trans. Ind. Inform.* **2021**, *17*, 4322–4334. [CrossRef]
2. Spampinato, C.; Chen-Burger, Y.H.; Nadarajan, G.; Fisher, R.B. Detecting, tracking and counting fish in low quality unconstrained underwater videos. In Proceedings of the Third International Conference on Computer Vision Theory and Applications—Volume 1: VISAPP, (VISIGRAPP 2008), Funchal, Madeira, Portugal, 22–25 January 2008; SciTePress: Setúbal, Portugal, 2008; pp. 514–519. [CrossRef]
3. Forczmański, P.; Nowosielski, A.; Marczeski, P. Video Stream Analysis for Fish Detection and Classification. In *Soft Computing in Computer and Information Science, Advances in Intelligent Systems and Computing*; Wiliński, A., ElFray, I., Pejaś, J., Eds.; Springer International Publishing: Cham, Switzerland, 2015; Volume 342; pp. 157–169. [CrossRef]
4. Han, F.; Yao, J.; Zhu, H.; Wang, C. Underwater Image Processing and Object Detection Based on Deep CNN Method. *J. Sensors* **2020**, *2020*, 6707328. [CrossRef]
5. Spampinato, C.; Giordano, D.; Salvo, R.D.; Chen-Burger, Y.H.J.; Fisher, R.B.; Nadarajan, G. Automatic fish classification for underwater species behavior understanding. In Proceedings of the First ACM International Workshop on Analysis and Retrieval of Tracked Events and Motion in Imagery Streams—ARTEMIS'10, Firenze, Italy, 29 October 2010; ACM Press: New York City, NY, USA, 2010; pp. 45–50. [CrossRef]
6. Papadakis, V.M.; Papadakis, I.E.; Lamprianidou, F.; Glaropoulos, A.; Kentouri, M. A computer-vision system and methodology for the analysis of fish behavior. *Aquac. Eng.* **2012**, *46*, 53–59. [CrossRef]
7. Zhao, J.; Gu, Z.; Shi, M.; Lu, H.; Li, J.; Shen, M.; Ye, Z.; Zhu, S. Spatial behavioral characteristics and statistics-based kinetic energy modeling in special behaviors detection of a shoal of fish in a recirculating aquaculture system. *Comput. Electron. Agric.* **2016**, *127*, 271–280. [CrossRef]
8. Li, D.; Wang, Z.; Wu, S.; Miao, Z.; Du, L.; Duan, Y. Automatic recognition methods of fish feeding behavior in aquaculture: A review. *Aquaculture* **2020**, *528*, 735508. [CrossRef]
9. Zhou, C.; Lin, K.; Xu, D.; Chen, L.; Guo, Q.; Sun, C.; Yang, X. Near infrared computer vision and neuro-fuzzy model-based feeding decision system for fish in aquaculture. *Comput. Electron. Agric.* **2018**, *146*, 114–124. [CrossRef]
10. An, D.; Hao, J.; Wei, Y.; Wang, Y.; Yu, X. Application of computer vision in fish intelligent feeding system—A review. *Aquac. Res.* **2020**, *52*, 423–437. [CrossRef]
11. Lech, P.; Okarma, K.; Korzelecka-Orkisz, A.; Tański, A.; Formicki, K. Monitoring the Uniformity of Fish Feeding Based on Image Feature Analysis. In Proceedings of the Computational Science—ICCS 2021, Krakow, Poland, 16–18 June 2021; Paszynski, M., Kranzlmüller, D., Krzhizhanovskaya, V.V., Dongarra, J.J., Sloot, P.M., Eds.; Lecture Notes in Computer Science; Springer International Publishing: Cham, Switzerland, 2021; pp. 68–74. [CrossRef]

12. Xiao, G.; Feng, M.; Cheng, Z.; Zhao, M.; Mao, J.; Mirowski, L. Water quality monitoring using abnormal tail-beat frequency of crucian carp. *Ecotoxicol. Environ. Saf.* **2015**, *111*, 185–191. [CrossRef] [PubMed]
13. Angani, A.; Lee, C.B.; Lee, S.M.; Shin, K.J. Realization of Eel Fish Farm with Artificial Intelligence Part 3: 5G based Mobile Remote Control. In Proceedings of the 2019 IEEE International Conference on Architecture, Construction, Environment and Hydraulics (ICACEH), Xiamen, China, 20–22 December 2019; IEEE: Piscataway, NJ, USA, 2019; pp. 101–104. [CrossRef]
14. Angani, A.; Lee, J.C.; Shin, K.J. Vertical Recycling Aquatic System for Internet-of-Things-based Smart Fish Farm. *Sensors Mater.* **2019**, *31*, 3987–3998. [CrossRef]
15. Zhang, Q.; Ren, X.; Liu, C.; Shi, X.; Gui, J.; Bi, C.; Xue, B. The Influence Study of Inlet System in Recirculating Aquaculture Tank on Velocity Distribution. In Proceedings of the 29th International Ocean and Polar Engineering Conference, Honolulu, HI, USA, 16–21 June 2019; Paper No. ISOPE-I-19-332.
16. Oca, J.; Masalo, I. Flow pattern in aquaculture circular tanks: Influence of flow rate, water depth, and water inlet & outlet features. *Aquac. Eng.* **2013**, *52*, 65–72. [CrossRef]
17. Khalid, M.; Pénard, L.; Mémin, E. Optical flow for image-based river velocity estimation. *Flow Meas. Instrum.* **2019**, *65*, 110–121. [CrossRef]
18. Cao, L.; Weitbrecht, V.; Li, D.; Detert, M. Airborne Feature Matching Velocimetry for surface flow measurements in rivers. *J. Hydraul. Res.* **2020**, *59*, 637–650. [CrossRef]
19. Sirenden, B.H.; Mursanto, P.; Wijonarko, S. Galois field transformation effect on space-time-volume velocimetry method for water surface velocity video analysis. *Multimed. Tools Appl.* **2022**, 1–23. [CrossRef]
20. Adrian, R.J. Particle-Imaging Techniques for Experimental Fluid Mechanics. *Annu. Rev. Fluid Mech.* **1991**, *23*, 261–304. [CrossRef]
21. Guler, Z.; Cinar, A.; Ozbay, E. A New Object Tracking Framework for Interest Point Based Feature Extraction Algorithms. *Elektron. Ir Elektrotechnika* **2020**, *26*, 63–71. [CrossRef]
22. Yagi, J.; Tani, K.; Fujita, I.; Nakayama, K. Application of Optical Flow Techniques for River Surface Flow Measurements. In Proceedings of the 22nd IAHR APD Congress, Sapporo, Japan, 14–17 September 2020; International Association for Hydro-Environment Engineering and Research-Asia Pacific Division: Beijing, China, 2020.
23. Ammar, A.; Fredj, H.B.; Souani, C. An efficient Real Time Implementation of Motion Estimation in Video Sequences on SOC. In Proceedings of the 18th International Multi-Conference on Systems, Signals and Devices (SSD), Monastir, Tunisia, 22–25 March 2021; IEEE: Piscataway, NJ, USA, 2021; pp. 1032–1037. [CrossRef]
24. Wu, H.; Zhao, R.; Gan, X.; Ma, X. Measuring Surface Velocity of Water Flow by Dense Optical Flow Method. *Water* **2019**, *11*, 2320. [CrossRef]
25. Gorle, J.; Terjesen, B.; Mota, V.; Summerfelt, S. Water velocity in commercial RAS culture tanks for Atlantic salmon smolt production. *Aquac. Eng.* **2018**, *81*, 89–100. [CrossRef]
26. Sirenden, B.H.; Arymurthy, A.M.; Mursanto, P.; Wijonarko, S. Algorithm Comparisons among Space Time Volume Velocimetry, Horn-Schunk, and Lucas-Kanade for the Analysis of Water Surface Velocity Image Sequences. In Proceedings of the International Conference on Computer, Control, Informatics and its Applications (IC3INA), Tangerang, Indonesia, 23–24 October 2019; IEEE: Piscataway, NJ, USA, 2019; pp. 47–52. [CrossRef]
27. Timmons, M.B.; Summerfelt, S.T.; Vinci, B.J. Review of circular tank technology and management. *Aquac. Eng.* **1998**, *18*, 51–69. [CrossRef]
28. Plew, D.R.; Klebert, P.; Rosten, T.W.; Aspaas, S.; Birkevold, J. Changes to flow and turbulence caused by different concentrations of fish in a circular tank. *J. Hydraul. Res.* **2015**, *53*, 364–383. [CrossRef]

applied sciences

MDPI

Article

The Impact of the 4.0 Paradigm in the Italian Agricultural Sector: A Descriptive Survey

Federico Angelo Maffezzoli *, Marco Ardolino and Andrea Bacchetti

RISE, Laboratory of Research and Innovation for Smart Enterprises, Department of Mechanical and Industrial Engineering, University of Brescia, 25121 Brescia, Italy

* Correspondence: f.maffezzoli@unibs.it

Abstract: This paper investigates how much Italian farms are involved in the so-called "Agriculture 4.0" (Agri 4.0) journey. The paper focuses on analyzing the knowledge and adoption levels of specific 4.0-enabling technologies while also considering the main benefits and obstacles. A descriptive survey was carried out on a total of 670 respondents related to agricultural companies of different sizes. The findings from the survey demonstrate that Italian farms are in different positions in their journey toward the Agri 4.0 paradigm, mainly depending on their size in terms of revenues and land size. Furthermore, there are strong differences concerning both the benefits and obstacles related to the adoption of the Agri 4.0 paradigm, here depending on the technology adoption level. Regarding future research, it would be interesting to carry out the same study in other countries to make comparisons and suitable benchmark analyses. Although scholars have debated about the adoption of technologies and the benefits related to the Agri 4.0 paradigm, to the best of the authors' knowledge, no empirical surveys have been carried out on the adoption level of digital solutions in agriculture in specific countries.

Keywords: Agriculture 4.0; smart agriculture; digitalization; descriptive survey; digital technologies

Citation: Maffezzoli, F.A.; Ardolino, M.; Bacchetti, A. The Impact of the 4.0 Paradigm in the Italian Agricultural Sector: A Descriptive Survey. *Appl. Sci.* **2022**, *12*, 9215. https://doi.org/10.3390/app12189215

Academic Editors: José Miguel Molina Martínez, Ginés García-Mateos and Dolores Parras-Burgos

Received: 29 July 2022
Accepted: 6 September 2022
Published: 14 September 2022

1. Introduction

In the coming decades, the world will face major issues that will have massive effects on the agricultural sector. Three main challenges are on the horizon: (1) The world population is set to increase. It is estimated that the human population will reach 9 billion people by 2050, increasing the demand for food by 70%, and water consumption in agriculture is expected to increase by 41% (the sector is already responsible for the consumption of almost 70% of the fresh drinking water on the planet) [1]. (2) In the medium term, climate change will profoundly affect the extent of arable land worldwide [2]. (3) The aging population in developed economies will soon bring the need to automate and digitize the agriculture sector [3].

Agriculture is a fundamental part of all economies in the world and, like all key sectors, is involved in the Fourth Industrial Revolution. The evolution of the primary sector toward digitalization is not dictated by an overall trend but aims to address the main macro issues in the years and decades to come, such as the need to make crops more efficient and effective and to evolve in an environmentally sustainable way. The strong link between sustainability and digital innovation is not limited to the primary sector but involves all major economic ones. From this approach, the phenomenon of Agriculture 4.0 (from now on, Agri 4.0) derives from the broader theme of Industry 4.0, which is considered to have huge potential in providing digital solutions to address the main problems encountered by traditional agriculture, enabling support for farmers to make faster decisions, achieve higher process efficiency, and have the ability to take timely action to meet market demands [4]. The literature sometimes also refers to this emerging phenomenon as "smart agriculture," basically taking its cue from the concept of smart manufacturing, which is already widely used in industry [5]. In other cases, scholars have used the term "smart farming" [6,7] or "digital farming" [8]. All these terms can be seen as synonyms, so for the current paper, the term Agri 4.0 will be used for simplicity purposes.

Scholars have focused on how digital technologies impact the agricultural sector [9,10] and how the diffusion of the Agri 4.0 paradigm can transform production processes and business strategy [4].

Although this paradigm has been investigated in the literature, presenting concrete examples of categorization of the possible benefits, obstacles, and dedicated digital technologies, there is no pervasive study focusing on the knowledge of digital solutions in agriculture and their degree of utilization. Moreover, the scientific literature presents no contributions when it comes to surveying the state of knowledge of the solutions among farmers and their degree of use, as well as the impacts received in using these solutions, both in general terms and specifically in the Italian context. In addition, research on Agri 4.0 neglects the use of empirical methods, such as empirical surveys, to develop scientific results from information from farmer practitioners. The few empirical surveys carried out by scholars have tended to focus on other drivers or on a single aspect throughout the whole questionnaire, such as the ones by Bolfe [11] and Chuang [12].

In an attempt to fill the above-mentioned literature gaps, the following research questions have been formulated:

RQ1: What is the level of awareness of Agri 4.0 solutions among farm enterprises?

RQ2: What is the level of adoption of Agri 4.0 solutions?

RQ3: What are the main benefits perceived in adopting Agri 4.0 solutions?

RQ4: What are the main challenges perceived in adopting Agri 4.0 solutions?

The research questions were set based on a reference scheme developed by the authors, which is presented in Figure 1.

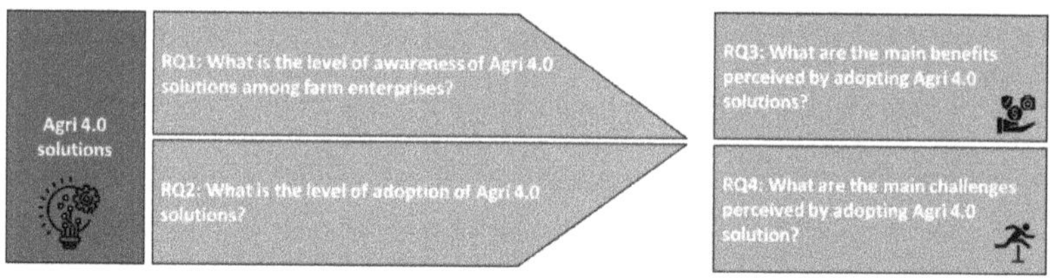

Figure 1. Reference scheme.

In particular, RQ1 and RQ2 aim at investigating the technological issues concerning Agri 4.0, while RQ3 and RQ4 investigate the effects in terms of the benefits and obstacles of the previous research questions.

Therefore, the present paper addresses the Agri 4.0 paradigm, aiming to gather evidence from the current state-of-the-art in the Italian agricultural context. Based on a descriptive survey completed by 670 respondents, the current paper aims to understand the degree of penetration of the phenomenon, covering many different open points of the paradigm and addressing these in multiple dimensions (distinctive solutions knowledge and utilization rate, benefits, and challenges).

The current paper concentrates on the Italian context. This choice was driven by the fact that, given the composition of the research group, the number of companies that could be involved was larger and because the Italian agriculture system is first in agriculture in Europe based on added value and third based on gross saleable production. Italy is also the world's leading producer of wine by volume and leading European producer of vegetables by value [13].

The present study also provides a systematization of the technological solutions adopted within the Agri 4.0 paradigm. Finally, the current paper provides a rationalization of the benefits and obstacles related to the implementation of the aforementioned digital technological solutions in the primary sector.

The current article is structured as follows: Section 2 gives an overview of the paradigm and presents the Agri 4.0 studied solutions. Section 3 describes the research methodology used, which is followed by Section 4, in which the four main thematic analyses are discussed. Next, Section 5 discusses the results, providing the research implications of the study and proposals for future research agendas in Agri 4.0.

2. Literature Review

2.1. Agri 4.0: Phenomenon and Paradigm Definition

The concept of Agri 4.0 encompasses several different scientific domains, some of which are directly related to land cultivation (water control, crop cultivation, harvesting, etc.), while others are an expansion of the agricultural area to different disciplines, such as engineering, economics, management, and so forth. Advances in different areas of the information and communication technology (ICT) domain, combined with the need to improve agricultural productivity, both for food safety and environmental impact issues, have determined the research area for Agri 4.0. Therefore, Agri 4.0 is derived from the broader concept of Industry 4.0 [9], which aims to define the integration of different technologies (such as Internet of Things (IoT), artificial intelligence, cloud computing, etc.) to automate cyber-physical tasks and processes, allowing for better planning and control of agricultural systems. The relationship of this concept with that of the Industry 4.0 paradigm, that is, the adoption of digital technologies to support the processes of manufacturing companies, is clear.

As reported in the literature, reducing input costs and increasing productivity seem to be the driving forces behind the progress in agriculture. However, the importance of sustainability should not be overlooked, a concept that has emerged as a major issue across the spectrum of human activities. Therefore, one of the goals of Agri 4.0 is to minimize the environmental impact of agricultural activities [14]. Thus, the implementation of Agri 4.0 solutions implies the possibility of farms achieving certain goals and benefits.

2.2. Enabling Digital Solutions for Agri 4.0

Taxonomies to group digital solutions in Agri 4.0 have already been presented in the literature. In particular, some interesting solutions are the ones presented by Lezoche et al. [9] and Liu et al. [10]; in both studies, the authors have presented an interesting categorization and description of the main technologies to be considered in Agri 4.0.

On the other hand, the current study focuses on solutions rather than technologies (i.e., different technologies can be part of the same type of solution); therefore, drawing on information and insights arising from the literature, five different clusters are presented: decision support system software; monitoring systems; systems for precision activities; mapping systems; and autonomous systems. The full list is presented in Table 1.

Table 1. List of the Agri 4.0 solutions considered.

Cluster of Technology Solutions	Solutions	References
Decision support system software	Business management software	[15,16]
	Decision support system (DSS)	
Monitoring systems	Agricultural machinery and equipment	[17,19]
	Crop and soil	
	Enterprise infrastructure	
	Indoor cultivation	

Table 1. *Cont.*

Cluster of Technology Solutions	Solutions	References
Systems for precision activities	Precision irrigation systems	[9,20]
	Variable rate distribution system	
	On-field treatment with drones	
Mapping systems	Satellite technologies	[18,21]
	Mapping equipment installed on machinery	
	Mapping drones	
Autonomous systems	Robot for field activities	[10,22]
	Satellite guidance	

(a) **Decision support system software:** This type of software facilitates the decision-making process by helping prioritize goals, evaluate alternatives, and simulate outcomes. Within this category, there are two key reference solutions: (a) Business management software helps in automating the management processes within companies. In particular, this software is useful for various business functions, such as agricultural production, warehouse management, and accounting [15]. (b) Decision support systems (DSS) are computer tools that use data and mathematical models to support the decision maker, here being the farmer [16].
(b) **Monitoring systems:** Monitoring systems use different technologies, such as smart sensors and pervasive connectivity, to monitor different areas of a farm [17,18]. The Agri 4.0 paradigm diffusion strongly relies on the development of innovative technologies such as sensors, the IoT, and Big Data [19]. Therefore, this cluster is divided into four different application areas that cover the main areas to be monitored within an agricultural company: agricultural machinery and equipment domain, crop and soils monitoring, enterprise infrastructure, and indoor cultivation.
(c) **Systems for precision activities:** The systems for precision activities enable the targeted use of various agricultural inputs [9,20]. Specifically, in this cluster, it is possible to identify three solutions: precision irrigation systems, variable rate distribution systems, and on-field treatment with drones.
(d) **Mapping systems:** Land mapping is a fundamental activity of Agri 4.0. Thanks to the knowledge of the spatial variability of soil properties, farm potential in terms of quality, quantity, and yield can be optimized [18,21]. In this regard, the three solutions refer to satellite technologies, mapping equipment installed on machinery, and mapping through drones.
(e) **Autonomous systems:** The use of increasingly advanced IT (information technology) and OT (operation technology) leads the agriculture industry to use autonomous systems both in terms of moving machines during operations and in terms of deciding on the activities to be performed within the fields [10,22]. The main solutions of this cluster are robots for field activities and satellite guidance.

2.3. *Agri 4.0: Benefits and Obstacles*

A long list of potential benefits can be listed under different economic areas, as well as environmental benefits with a reduction of pollutants [23,24] and social benefits with positive effects on the well-being of the workers involved and on society in general [25]. At the same time, there are also criticalities involved in implementing new systems, especially digital ones, in contexts such as the agricultural sector. For those who decide to implement innovative systems, there can be challenges of a technological, economic, and implementation nature, as well as those arising from corporate culture and organizational issues [26,27].

The benefits investigated can be categorized into four clusters, which have been identified according to the triple bottom line (TBL; that is, people, planet, and profit) principles [28]. The first two clusters (effectiveness and efficiency) refer to the profit or bottom line. The next two are environmental and social benefits. From these four clusters, a set of 14 benefits was proposed. A full list of the benefits and references is presented in Table 2.

Table 2. List of benefits.

TBL Cluster	Cluster of Benefit	Benefits	References
Profit	Effectiveness	Product quality increase	[24,29]
		Yield increase	
		Soil quality increase	
		Increase in selling price	
	Efficiency	Less water consumption	[30,31]
		Less technical input consumption	
		Less machinery usage	
		Simplification in the cultivation decisions to be made	
		General cost reduction	
Planet	Environmental benefits	Air pollution (CO_2, N_2O, …) decrease	[23]
		Water pollution decrease	
People	Social Benefits	Work safety increase	[25,32,33]
		Reduced time spent on bureaucratic tasks	
		Physical labor reduction	

(a) **Effectiveness benefits:** The benefit cluster related to the economic part of the TBL. Operational effectiveness encompasses the practices employed to maximize resources and deliver high-quality results [24,29]. Here, the authors investigated four different benefits related to effectiveness: higher product quality, yield increase, better soil quality, and an increase in the selling price of goods produced.

(b) **Efficiency benefits:** The benefit cluster related to the economic part of the TBL. Reducing the consumption of productive inputs and, thus, the associated costs is critical because it is known that a firm that has lower cumulative costs to perform all value-generating activities than its competitors has a cost advantage [30,31]. Here, the authors investigated five different benefits related to efficiency: less water consumption, less technical input consumption, less machinery usage, simplification in the cultivation decision to be made, and general cost reduction.

(c) **Environmental benefits:** Sustainability from an environmental perspective is another benefit that can be achieved through the use of 4.0 solutions in agriculture [23]. Reducing the use of pollutants (such as agrochemicals and various fertilizers) increases soil quality, but from a purely environmental standpoint, there are real effects on air pollution decrease (CO, NO, etc.) and decreases in water pollution.

(d) **Social benefits:** The adoption of Agri 4.0 techniques has the potential to increase farmers' quality of life in terms of increased work safety and decreased work stress. The controllable work environment in a plant factory is much more desirable than field cultivation, which requires a lot of physical energy to complete [32,33]. Specifically, the benefits under this cluster are three: increase in work safety, reduced time spent in bureaucratic activities, and a reduction in physical labor [25].

For 'obstacles,' four main clusters have been identified. The clusters cover the main areas of challenge when introducing a technological evolution in a certain environment:

(a) **Technological challenges:** This cluster refers to technical and technological issues related to the implementation of 4.0 solutions in agriculture [34]. Technical barriers can be limited or without interoperability (the data collected cannot be reused and different solutions do not work together) and lacking connectivity. However, it is also important to mention the limited or absent flexibility in the sense that the provided solution works only under optimal operating conditions (primarily weather) [27].
(b) **Economic challenge:** The implementation of innovative solutions and technologies in every field leads to a significant economic effort by the company that intends to adopt an innovative path. For this reason, it takes into consideration the economic return of the investment made in 4.0 solutions [35]. The challenge in question is the low investment return rate.
(c) **Implementation challenges:** The skills needed to properly implement 4.0 solutions, especially in companies in the primary sector, inevitably lead to the challenge of usage difficulty. Subsequently, the challenge connected to the first one is insufficient assistance because many companies can face obstacles that are difficult to overcome without an appropriate implementation assistance path [26].
(d) **Cultural and organizational challenge:** Because of the introduction of digitalization in agriculture, the 4.0 revolution in all economic sectors will require a new set of skills related to the introduction of digital solutions in companies [25]; for this reason, the challenge presented in the survey is the lack of key digital skills in the farm.

Out of these clusters, seven different obstacles could be derived. The full list of challenges and references is presented in Table 3.

Table 3. List of challenges.

Cluster of Challenges	Challenges	References
Technological challenge	Limited or absent interoperability	[27,34]
	Limited or absent flexibility	
	Lack of connectivity	
Economic challenge	Low return of investment	[35]
Implementation challenge	Insufficient assistance	[26]
	Usage difficulty	
Cultural and organizational challenge	Lack of key digital skills in the farm	[25]

3. Methodology

Survey research is useful for obtaining information about a specific phenomenon concerning large populations, allowing for an adequate level of accuracy [36,37]. The current research adopts descriptive survey research because the objective is to understand the significance of a phenomenon and describe its occurrence in a population [38,39]. Indeed, descriptive surveys are highly valuable for gathering data from diverse populations because the researcher can extract the attitudes and features of respondents accurately [40]. Moreover, it is possible to provide an effective "picture" of the phenomenon being investigated from which evidence can be drawn. Thus, a descriptive survey is a convenient method when knowledge of a phenomenon is not too poorly underdeveloped, the variables and context can be described in detail, and the objective is to understand to what extent a given relationship is present. The intent of descriptive surveys is not necessarily to determine causal relationships, but they do provide an effective method for investigating a representative sample and enabling data regarding particular issues to be collected, which may be used to form the basis of decision-making activities in the future [41].

Therefore, the primary research objective is not theory development but rather the investigation of the impacts of the Agri 4.0 paradigm in the Italian primary sector by describing the knowledge levels, achieved benefits, and perceived challenges.

To obtain the above-mentioned objectives, a survey research process consisting of three steps was adopted: survey design, pilot testing, and data collection and analysis.

3.1. Survey Design and Pilot Testing

The questionnaire was characterized by 18 mixed open and closed questions, and it was structured into four sections. The first section aimed to collect general information and a registry about the company and respondents. The second section asked about the level of knowledge for each solution proposed; the description of each technological solution was provided through a "link" button to the respondents to provide the same interpretation of technology meaning and avoid bias related to ambiguous questions. The third section inquired about the company's perceptions of the benefits of Agri 4.0. Finally, the fourth section investigated the challenges and obstacles in adopting the Agri 4.0 paradigm.

To reach the highest number of respondents, a web survey was administered for conducting the research [42]. The trend of conducting surveys online has grown in recent times because they can offer many benefits over paper-based surveys. Indeed, with respect to face-to-face and e-mail surveys, web surveys do not require the manual transfer of responses into a database; the cost is minimal compared with other means of distribution, and greater anonymity is guaranteed, helping to avoid interviewer biases [42]. Online survey research can also allow researchers to isolate specific groups of participants who share common features [43].

Subsequently, to test possible question bias, translation accuracy, and the logical flow of the survey, pilot testing was performed before survey distribution [44]. In the first step, a group of colleagues was involved to check the readability and help pinpoint whether the questionnaire was within the study objectives. After refining the survey, the second step then involved sending the questionnaire to seven beta-tester companies to get feedback from them for further possible improvement. The pilot testing helped assess the content of the questionnaire and guaranteed the validity for the official launch.

Concerning the survey sample, the unit of analysis refers to Italian agricultural companies and farms. Moreover, this research involves all types of agricultural companies—except livestock farms—with no limits concerning their size and cultivation sector. The survey was carried out from January to October 2021, and repeated waves of reminders and recall activities were conducted with the support of the main Italian agricultural associations. The analysis started with a total number of 1273 responses before eliminating incomplete responses, duplicate responses, and test responses conducted internally by the team. As a result, a sample of 670 companies was validated. The survey respondent is the owner of the company to which the questionnaire was sent (or the decision maker in their place), who, therefore, has an overview of their farm.

3.2. Sample Description

Table 4 shows the company size of the cluster. Because there is currently no specific classification for farm size in the primary sector, it was decided to develop five customized clusters. Indeed, it was considered misleading to use the classical criteria related to manufacturing enterprises because of the great diversity in terms of turnover between the sectors. It should be noted in the table that most of the sample, 70%, is below EUR 250,000 in turnover. Only the remaining 30% are above this threshold, of which 17% have a turnover of over half a million euros.

Table 4. Revenue clusters distribution in the sample.

		Number of Farms	(%)
Revenue Cluster	A. <EUR 30,000	147	22
	B. between EUR 30,000 and 100,000	194	28
	C. between EUR 100,000 and 250,000	133	20
	D. between EUR 250,000 and 500,000	84	13
	E. >EUR 500,000	112	17
	Total	670	100

For a more complete analysis and because of the peculiarities of the sector under study, an additional proxy for the size of the sample companies was used (Table 5), that is, cultivated hectares. In this case, there is no clear definition of the classes to be considered.

Table 5. Land size cluster distribution in the sample.

		Number of Farms	(%)
Hectares Cluster	A. <10	192	29%
	B. 10–20	112	17%
	C. 20–50	148	22%
	D. >50	218	32%
	Total	670	100

Figure 2 represents the Italian distribution of farm locations; to have data with the correct granularity, the data are represented by province rather than by region. Here, the distribution of the sample subject ranges over the entire country, demonstrating a very important capillarity. In detail, there are 178 companies in Southern Italy, 96 in the center, and 396 belonging to Northern Italy.

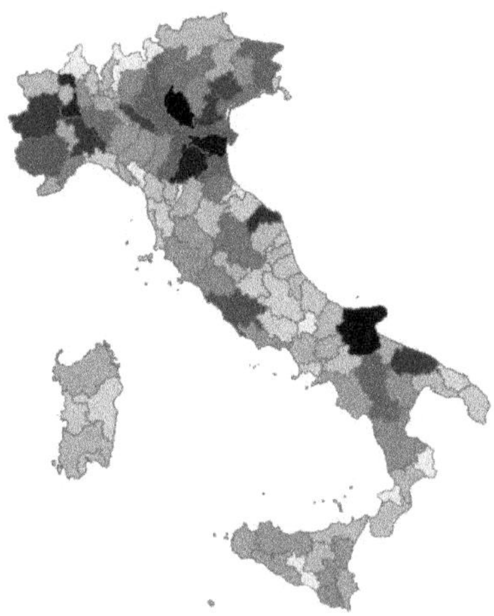

Figure 2. Geographical distribution of companies in the sample.

As a final representative analysis of the reference sample, Table 6 shows the distribution of prevalent cultivation. The classification method presented was developed following two interviews with experts in the field (agronomists), who indicated the categories listed in the table. The sample is highly heterogeneous in this respect as well, reinforcing the generalization of the analyses and considerations made in the current study.

Table 6. Pareto distribution of prevalent cultivation.

Prevalent Cultivation	Sample Pareto Distribution (%)
(a) Cereals	37.4
(b) Vineyards	19.7
(c) Forage	8.2
(d) Olive groves	7.9
(e) Vegetables	5.5
(f) Pome fruit trees	4.7
(g) Stone fruit orchards	4.6
(h) Industrial crops	3.3
(i) Leguminous	2.1
(j) Flowers and ornamental plants	1.6
(k) Nursery tree	1.6
(l) Other arboretums	1.4
(m) Potatoes	1.3
(n) Citrus groves	0.5
(o) Other	0.3
Total	100

3.3. Variable Definition and Measure

Table 7 shows the variables used in the survey. The variable "Agri 4.0 solutions knowledge level" evaluates the degree of knowledge of the various solutions proposed. Four options are considered: "I have never used this solution, and I am not familiar with it," "I have never used this solution, but I know it," "I do not currently use this solution but have used it in the past," and "I currently use this solution." A variable implicitly connected to the one just described is "Agri 4.0 solutions adoption," in which the answer "I currently use this solution" was used to represent the results.

Table 7. Variable definition and criteria.

Variable	Type	Nr. of Levels/Clusters	Levels
Company size (revenues)	Ordinal	5	A. Less than EUR 30,000; B. between EUR 30,000 and 100,000; C. between EUR 100,000 and 250,000; D. between EUR 250,000 and 500,000; E. over EUR 500,000
Company size (land)	Ordinal	5	A. Lower than 10 hectares; B. Between 10 and 20 hectares; C. Between 20 and 50 hectares; D. Over 50 hectares
Agri 4.0 solutions knowledge level	Ordinal	4	I am not familiar with the solution; I am a little familiar with the solution; I am familiar with the solution at a theoretical level; I am familiar with the solution at a practical level
Agri 4.0 solution adoption	Ordinal	2	I use the solution; I do not use the solution
Benefits	Ordinal	5	Null; Low; Middle; High; Very High
Challenges	Ordinal	5	Null; Low; Middle; High; Very High

To identify the enabling solutions, benefits, and obstacles related to the Agri 4.0 paradigm, no systematic analysis was carried out, but a narrative literature review was conducted. This type of analysis, which is widely used in studies related to the medical sciences [45], does not involve following a strict protocol or specific standards but still allows for the identification of the main studies describing a problem of interest [46]. Concerning the identification of the articles to be analyzed, the Scopus and Web of Science databases were surveyed using strings formulated from the keywords related to agriculture and digitalization ("Smart Agrifood," "Smart Agriculture," "Smart Farming," "Agrifood 4.0," "Agriculture 4.0," "Farming 4.0," "Internet of Farming," "Digital Agrifood," "Digital Agriculture," "Digital Farming," "Precision Agriculture," and "Precision Farming"). The set of enabling solutions, benefits, and obstacles have already been presented in Section 2.

4. Results

4.1. RQ1: What Is the Level of Awareness of Agri 4.0 Solutions among Farm Enterprises?

The first highlight of the analysis derives from the investigation of the current degree of knowledge of Agri 4.0 solutions within the sample considered. Figure 3 summarizes the results. The level of awareness was measured using a 4-point scale, from a low to a high level of awareness of the solutions, specifically (a) I am not familiar with the solution, with no awareness of solutions existence; (b) I am a little familiar with the solution, meaning having only marginally heard of the solution; (c) I am familiar with the solution at a theoretical level, meaning having a good level of theoretical knowledge; and (d) I am familiar with the solution at a practical level, meaning knowing the solution and having knowledge of practical examples in the field.

Figure 3 shows all the solutions proposed within the questionnaire, ordering them from the most to least known. Another important aspect to consider is the statistical distribution of the number of solutions deeply known by the respondents (counting only answers in which the solution is familiar to the respondents). The distribution depicted in Table 8 represents the number of times a certain number of solutions is known at the same time, presenting the percentage over the entire sample and number of respondents.

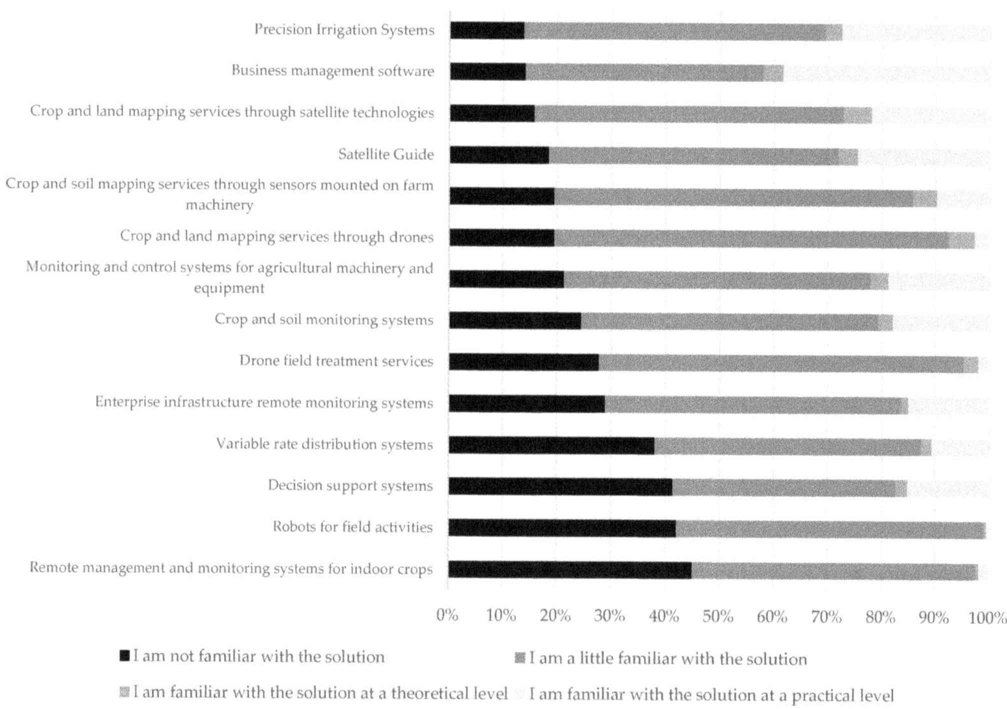

Figure 3. Agri 4.0 solutions awareness level.

Table 8. Statistical distribution of the number of digital solutions known.

Number of Digital Solutions Known Simultaneously	Respondents (%)	Respondents (Nr.)
0	28.4	190
1	17.3	116
2	13.7	92
3	11.2	75
4	9.6	64
5	7.5	50
6	4.2	28
7	3.4	23
8	2.1	14
9	1.3	9
10	0.7	5
11	0.3	2
12	0.3	2
Total:	100	670

The table clearly shows that the number of respondents claiming to know more solutions decreases as the number of known solutions increases.

The most well-known solutions within the given answer set are by far precision irrigation systems and business management software, followed by two technological solutions that share a similar technological basepoint, i.e., crop and land mapping services through satellite technologies and satellite guides. In this case, the management software solution is the most well known in practice, demonstrating that it is the solution most likely to be implemented by companies.

Crop and land mapping though drones deserves a separate discussion, which, despite being in the middle of the ranking for awareness, is one of the least known at the practical level, with only 3% of the respondents indicating that they had seen a practical example of this type of solution. A similar argument can be made for robots for field activities, of which not a single respondent claimed to have any practical knowledge, and remote management and monitoring for indoor crops, as the two least well-known solutions of the solution set.

The level of awareness of the solutions identified in the current study can be correlated with the descriptive variables of the analysis used as control variables to check for the presence of trends and patterns. To calculate the level of knowledge, scores were assigned from 0 to 3 in ascending order, here based on the answers given to the question about the level of awareness. To determine the level of awareness for each respondent, the sum of the level for each solution was divided by the maximum obtainable.

Figure 4 shows an increased pattern of awareness level related to hectare size of the farm, with the cluster of largest companies having a higher average (45%), median (45%), inferior quartile (33%) and major quartile (55%) than any other cluster. Furthermore, an increasing trend in the awareness of Agri 4.0 solutions is evident with respect to the size of the land worked.

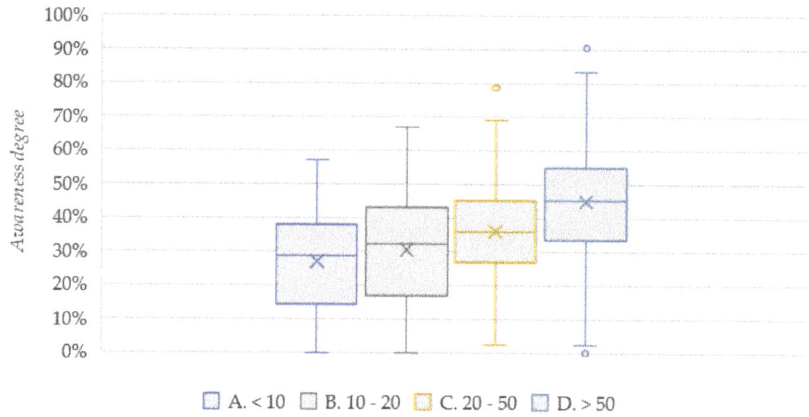

Figure 4. Boxplot graph of awareness level depending on the size of the company (land size).

This trend is further verified and reinforced by the analysis of the relationship between awareness level and turnover (Figure 5), in which it is possible to see how the level of awareness increases with an increase in the revenue cluster. The boxplot graph is particularly significant because each element (minimum, maximum, inferior quartile, major quartile, mean, and median) of the larger revenue class is greater than each element of the smaller revenue class. The boxplot shows that, on average, companies with a turnover of more than EUR 500,000 are currently using half (48%) of the solutions proposed in the survey.

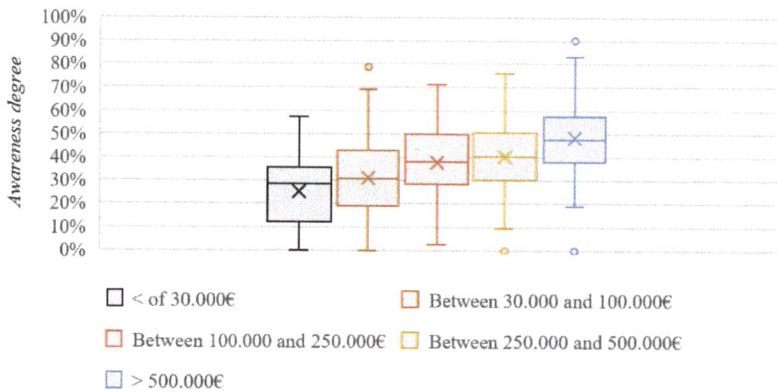

Figure 5. Boxplot graph of awareness level depending on company size (revenues).

4.2. RQ2: What Is the Level of Adoption of Agri 4.0 Solutions?

For each Agri 4.0 solution, the respondents were asked to specify whether they used the solution or not, allowing for the identification of adopters and nonusers.

Comparing the level of awareness versus the level of adoption, as expected, the rate of awareness increases for those using Agri 4.0 solutions compared with those who do not utilize any of the solutions.

Figure 6 links the first two research questions, highlighting higher awareness of different Agri 4.0 solutions among the respondents who used at least one solution compared with those who did not use any.

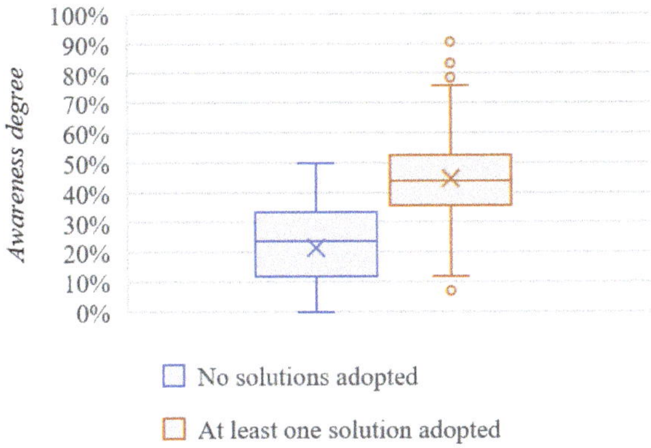

Figure 6. Boxplot graph of awareness level and utilization of 4.0 solutions.

The level of adoption can also be analyzed by comparing it against some control variables relative to the surveyed sample. First, it was examined whether there was a link between the size of companies and rate of utilization of technological solutions.

To assess whether the data and analyses had statistical significance or not, a chi-square test was performed, here measuring the *p*-value. Typically, its value is a very small number, close to zero. Here, the *p*-value is the assigned level of significance (i.e., a measure of evidence against the null hypothesis) and, to be statistically significant, this value must be less than 0.05. A significant association was found between revenue size cluster and utilization level of Agri 4.0 solutions (Table 9), in which the Pearson's χ^2 test *p*-value was very low (3.48×10^{-19}, ensuring the significance of the analysis.

Table 9. Pearson's χ^2 test for adoption level and revenue clusters.

	Agri 4.0 Solution Adoption Level	**At Least One Solution Adopted**	**No Solutions Adopted**
Revenue Cluster	A. <EUR 30,000	47	100
	B. between EUR 30,000 and 100,000	98	96
	C. between EUR 100,000 and 250,000	97	36
	D. between EUR 250,000 and 500,000	63	21
	E. >EUR 500,000	91	21

Pearson's χ^2 test: *p*-value = 3.48×10^{-19} (significant).

The growing trend in the level of adoption depends on the size (in terms of turnover) of the companies. This trend is further confirmed by the boxplot presented in Figure 7. To calculate the levels of adoption in the boxplot graphs, the sum of the usage responses for each respondent for the various solutions was analyzed and then divided by the total number of proposed solutions.

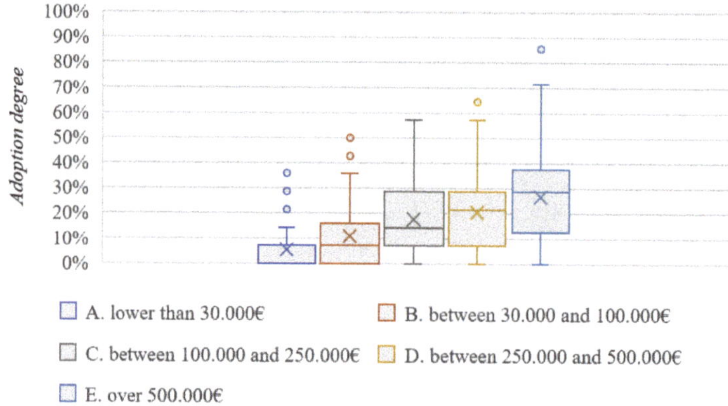

Figure 7. Boxplot graph of the adoption rate depending on company size (revenue).

The association is clear in Figure 7, in which, from the lowest to the highest revenue class, all significant adoption-level metrics increase. The sample presents an average of 5% from the smallest class of revenue to an average of 27%, also taking into consideration the fact that the sample shows a maximum percentage of adoption level that goes from 14% to 71% for the most significant turnover class.

The analysis represented in Figure 7 indicates that companies with higher turnover not only have more knowledge of the available solutions, but also a higher degree of use, perhaps because of the greater capacity to spend resources on these solutions.

The trend shown above is also confirmed when using the size of the cultivated area as a proxy for farm size. This can be prooved by the strong association between these two variables (utilization rate–size in hectares) with a Pearson's χ^2 test p-value equal to 4.618×10^{-15} (significant).

Table 10 shows the increase of utilizers as the farm's size increases. At the same time, the number of farms not using 4.0 solutions drops, resulting in a strong relationship between these two variables. Moreover, from a visual perspective (Figure 8), the boxplot graph helps in seeing the main message of this analysis: the utilization values increase as the number of hectares increases, but it is interesting to note the strong increase from 50 hectares onwards.

Table 10. Pearson's χ^2 test for adoption level and land size cluster.

	Agri 4.0 Solution Adoption Level	At Least One Solution Adopted	No Solutions Adopted
Hectares Cluster	A. <10	57	53
	B. 10–20	26	56
	C. 20–50	46	66
	D. >50	94	54

Pearson's χ^2 test: p-value = 4.618×10^{-15} (significant).

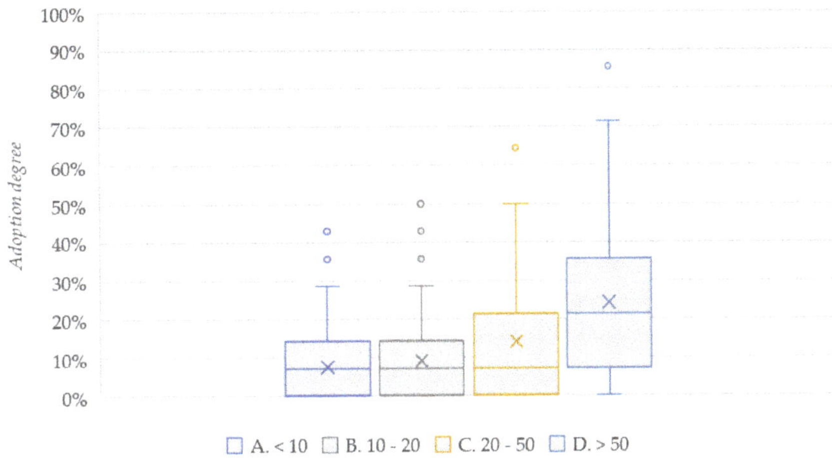

Figure 8. Boxplot graph of the utilization rate depending on company size (land size).

Subsequently, the focus of the analysis shifted to another important control variable in the questionnaire: the respondent's educational qualification (whether agricultural or not).

In Table 11, it is possible to see the strong relationships between the degree of utilization and type of education received by the business owner. The Pearson's chi-square test *p*-value results in a very small value, ensuring the statistical significance of the analysis.

Table 11. Pearson's χ^2 test for adoption level and type of education.

	Agri 4.0 Solutions Adoption Level	At Least One Solution Adopted	No Solutions Adopted
Agricultural education	A. No	205	194
	B. Yes	191	80

Pearson's χ^2 test: p-value = 7.98×10^{-7} (significant).

The graphical relationship of the effect that the control variable has on the degree of adoption is depicted in Figure 9. The subgroup of respondents with an educational background in agriculture presents a greater degree of adoption of Agri 4.0 solutions than the subgroup without this type of background. This can be seen in all aspects of the boxplot, from the minimum to the maximum, as well as for the interquartile range (0–21% vs. 7–29%), the mean (13% vs. 18%), and the median (7% vs. 14%).

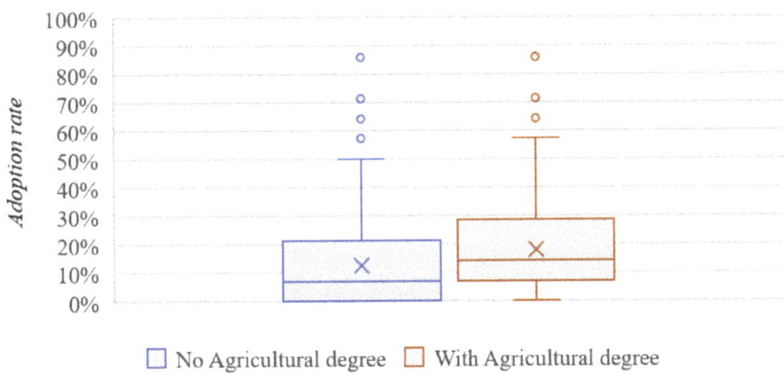

Figure 9. Boxplot graph of adoption level and type of education.

4.3. RQ3: What Are the Main Benefits Perceived in Adopting Agri 4.0 Solutions?

Figure 10 shows a boxplot that compares the benefit (divided into 14 different classes of benefit) obtained from the implementation of 4.0 solutions by users with the expected benefit by those who are not currently using any of the 4.0 solutions proposed in the questionnaire. This analysis highlights how the expectations of nonusers exceed the reality of users in terms of the level of benefit.

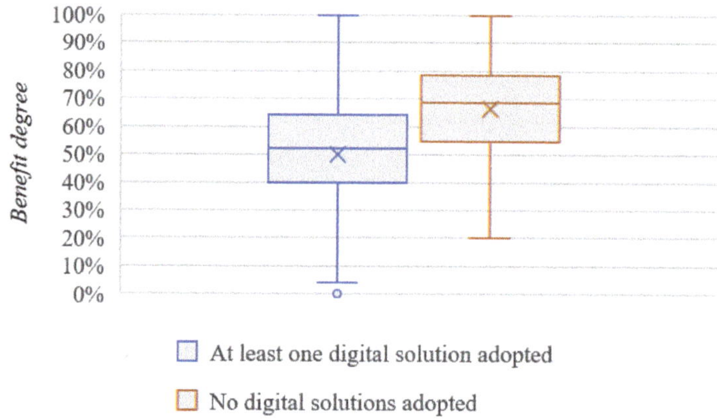

Figure 10. Boxplot graph of the level of benefit between users and nonusers.

This result deserves a more specific analysis; in Figure 10, the average of the benefit obtained and that expected from the two types of different actors, here unpacked in the 14 different obtainable benefits, is visualized. Figure 10 also represents the average benefit perceived by large users, which means those users who are currently operating many solutions (above eight different solutions).

Figure 11 represents what was previously summarized in Figure 10, providing more detail regarding each benefit presented to the respondents. It is interesting to note that, for all benefits (except for the benefit of reducing water consumption), the respondents who are users of 4.0 solutions present an average level of benefit lower than the average benefit that nonusers expect. In particular, it is possible to see how this gap is wider for "increase in sales price," which reports a rather low value (1.7 average value) for users while showing a potentially higher benefit for nonusers. In general, the benefits that have brought the most benefit to the sample under analysis are "lower consumption of technical inputs," "lower water consumption," and "soil quality improvement." However, it is also interesting to notice an upward trend. As previously stated, the average of the users is clearly lower than the expected benefit of the nonusers, but it is also true that the average of the large users (in this case, those respondents declaring that they use eight or more different solutions) increases considerably to the level of the expectations. A takeaway from this trend is that to reach (at least at the level of the average) the level of benefit expectation, it is necessary to use several solutions in parallel to exploit the joint work to achieve the desired benefits.

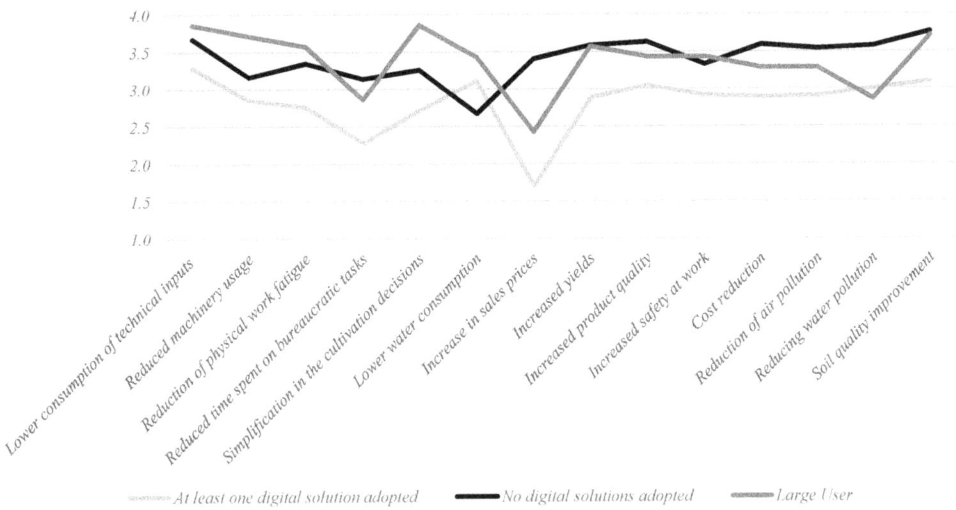

Figure 11. Benefits of Agri 4.0 solutions.

To gain a better understanding of the differences between users and nonusers, a ranking of benefit levels for users and expectations for nonusers was drawn up in descending order to identify the relative position of each benefit in these two lists. The results are shown in Table 12. The columns of "position" represent the relative position for users' benefits, and, in brackets, the position difference for nonusers' expected benefits is given.

Table 12. Relative position of benefits.

Benefit	Position: Users	Position: Nonusers
Lower consumption of technical inputs	1	2 (−1)
Lower water consumption	2	14 (−12)
Soil quality improvement	3	1 (+2)
Increased product quality	4	3 (+1)
Reducing water pollution	5	6 (−1)
Increased safety at work	6	10 (−4)
Reduction of air pollution (CO_2, N_2O, …)	7	7 (-)
Cost reduction	8	4 (+4)
Increased yields	9	5 (+4)

Table 12. *Cont.*

Benefit	Position: Users	Position: Nonusers
Lower consumption of technical inputs	1	2 (−1)
Lower water consumption	2	14 (−12)
Cost reduction	8	4 (+4)
Increased yields	9	5 (+4)
Reduced machinery usage	10	12 (−2)
Reduction of physical work fatigue	11	9 (+2)
Simplification in the cultivation decisions to be made	12	11 (+1)
Reduced time spent on bureaucratic tasks	13	13 (-)
Increase in sales prices	14	8 (+6)

The message that emerges from Table 12 is indicative of whether the various benefits perceived by users are in line (at least from the point of view of relative position) with expectations or not. Maintaining this approach but aggregating the benefits by cluster, we find interesting results.

As depicted in Figure 12, it is possible to notice that the "people" and "planet" clusters have an average position in line between the two samples. The "profit" cluster is a different matter. In this case, the analysis should be divided into subclusters of efficiency and effectiveness. In the first one, the perceived benefit is higher than the expected one, demonstrating the usefulness of 4.0 solutions in this area, while the relative position of the effectiveness subcluster is lower than expected (on average 2.2 positions lower in the ranking).

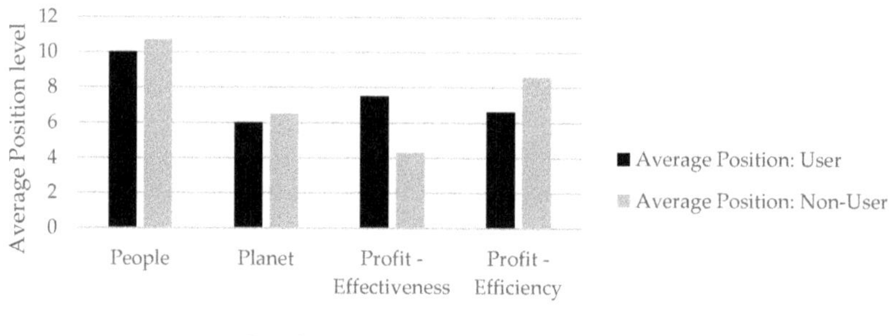

Figure 12. Aggregate positioning of benefits.

4.4. RQ4: What Are the Main Challenges Perceived in Adopting Agri 4.0 Solutions?

The analysis aimed to describe the barriers to implementing 4.0 systems in agriculture and expected difficulty in overcoming these barriers. In this paragraph, the aim is to replicate the structure of analysis presented in the previous research question, replicating the same type of analysis to identify analogies between the two research questions.

As a first analysis, Figure 13 represents the level of challenge declared by respondents, dividing the sample between users and nonusers, with users defining their perceived level and nonusers defining their expectations of the proposed challenge. The analysis of the boxplot depicted in Figure 13 contrasts with the analysis seen for benefits. In this case, on average, users experience a lower level of obstacles than nonusers. However, the analysis is at an aggregate level, and one cannot see the obstacles one by one. For this reason, the analysis of the level per obstacle has also been replicated (Figure 14).

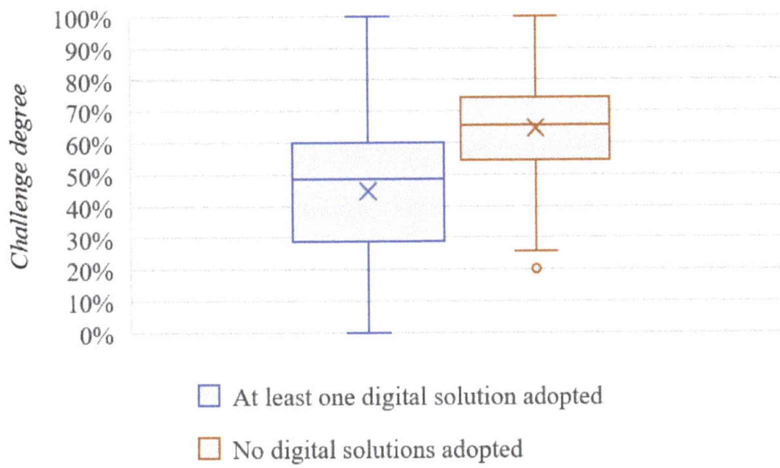

Figure 13. Boxplot graph of level of challenge between users and nonusers.

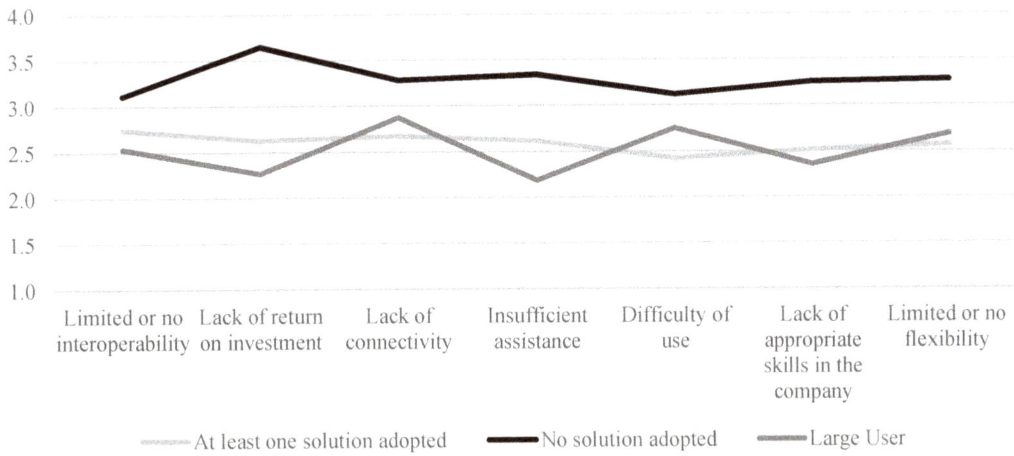

Figure 14. Challenges with Agri 4.0 solutions.

In this case, unlike the analysis carried out for the benefits, there is not the same trend, and the analysis depicted in Figure 14 contrasts with the findings of the previous research question. In this case, it is significant that each of the barriers has a lower challenge level found by users compared with the expectations of nonusers. In addition, the trend for those who use many different solutions in parallel does not lead to a particular increase or decrease in the challenge level for each obstacle proposed, thus identifying a constant trend in the challenge level as the number of solutions used increases but with an increase in the variability of the level per item, as can be seen in Figure 14.

To better understand the challenge level and relationship for each item in the list between users and nonusers, a ranking of the items from the highest level of perceived challenge to the lowest level was again drawn up (Table 13).

Table 13. Relative position of challenges.

Challenge	Position: Users	Position: Nonusers
Limited or no interoperability	1	1 (-)
Lack of connectivity	2	5 (−3)
Lack of return on investment	3	7 (−4)
Insufficient assistance	4	6 (−2)
Limited or no flexibility	5	4 (+1)
Lack of appropriate skills in the company	6	3 (+3)
Difficulty of use	7	2 (+5)

In contrast to the same table presented for benefits, in this case, a higher position corresponds to a more serious problem for respondents. The first important consideration that is possible to see from Table 13 is that limited or no interoperability is at the top of both the problems encountered by users and expectations of respondents who do not use any Agriculture 4.0 solution, demonstrating the centrality of the issue for Agri 4.0 and, more generally, in the 4.0 paradigm. Furthermore, in this case, it is interesting to compare the clusters, as carried out in the benefits, and compare the relative position in the case of user response and expectations of nonusers.

As presented in Figure 15, in this case, economic obstacles hold a higher position, so the perceived problem is greater than the expectations of nonusers; here, even four positions differ between the two types of respondents. As far as the technology category is concerned, the position is relatively stable between the two samples. It is also interesting to note that technological challenges rank first among the problems encountered by users. Less serious than expected are cultural and organizational challenges and implementation problems, both of which have a lower relative position than expected for these clusters.

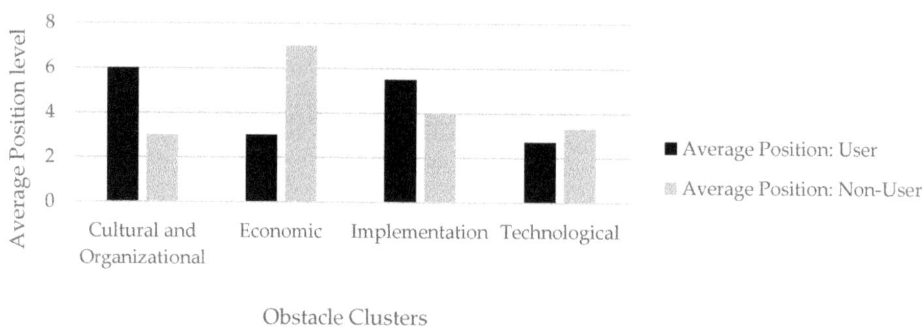

Figure 15. Aggregate positioning of obstacles.

5. Discussion and Conclusions

The current study aimed to map the state-of-the-art in Agri 4.0 within Italian farms through a descriptive survey, here adopting the perspectives of the awareness and adoption level, understanding which benefits users value the most and the differences between nonuser clusters, as well as identifying the critical factors and challenges that impact a company's adoption level regarding Agri 4.0 solutions. A large sample of 670 agricultural companies in Italy was analyzed. In particular, the digital solutions presented to respondents refer to five different clusters: DSS software, monitoring systems, systems for precision activities, mapping systems, and autonomous systems. In addition, this study considered the benefits by clustering the specific items by referring to the TBL, that is, economic, social, and environmental benefits, while analyzing several kinds of challenges: technological, economic, implementation, and cultural and organizational.

At a general level, our study shows that Italian farms display a heterogeneously distributed level of knowledge of the proposed solutions, but few farms know more than one solution in depth. Moreover, the current study shows that some control variables influence the level of awareness more than others; as turnover and cultivable area increase, an increase in the average level of awareness of Agri 4.0 solutions can be seen. The first is the level of awareness of each digital solution, which is still not pervasive over all the different solutions identified. Extensive knowledge of the solutions is still far from common. Above all, it is possible to see that the percentage of those who claim to know examples of practical implementation is low for each solution. The other important point is the degree of adoption. Here, the key message is that the average level of adoption increases as the turnover and size of the arable land increase. The level of maturity, therefore, is still low on average, and the market is not very dynamic if smaller companies are more out of the change process. In fact, smaller companies have less capacity to invest, and in line with the result of the barriers to adoption (which puts the economic problem at the top), this leads to a greater shift in adoption toward larger companies. At the same time, a similar increasing trend can be noticed in the degree of adoption. Although the average penetration rate is not particularly high, companies that have embarked on the journey to Agri 4.0 transformation have generally perceived lower barriers than companies that have yet to begin this journey. Finally, the present article has also investigated the benefits and potential obstacles to implementing Agri 4.0 solutions. The analysis shows that the main benefits perceived by the user are the reduction of technical inputs and water, which, in turn, benefit the entrepreneur economically but can also be said to have a positive impact on the environment. It is also interesting to note that the main problem encountered is the limited, even lack of, interoperability between 4.0 systems in the field. This obstacle is the point at which the actors and technology providers of the Agri 4.0 value chain must focus on to extract the maximum value from the digitalization of agricultural systems.

5.1. Research and Managerial Implications

For the current study, there are several implications, both for scholars and for practitioners.

The first aim of the proposed study was to provide evidence in a developed economy market of the state of adoption of Agri 4.0, here trying to define through concrete numbers the state of adoption of the paradigm in Italy, which can be representative also of other European economies. As defined above, there is no study analyzing the state of penetration of Agri 4.0 in Italy or Europe. Hence, the current study paves the way for scholars to pursue empirical research regarding the paradigm and state of the art. Some of the key insights of the proposed study are that Agri 4.0 is gaining more momentum, mainly because of the continuing need to be more sustainable, efficient, and using increasingly circular means while using digital leverage. However, within the main applicable solutions, it appears there are different levels of awareness that make up the digital solutions because some solutions are probably not yet mature enough to fit the needs of farmers.

The same applies even more so to the level of adoption because, on paper, the expected benefits are far greater than those perceived. This seems to me an important implication, and there is probably a mismatch between practical application and theory. A further implication is that the more solutions are used simultaneously, the more there is an alignment between expected and actual benefits.

The results of the exploratory survey presented in the current article provide several insights that can be useful for professionals working in the agricultural sector, technological suppliers of Agri 4.0 solutions, and public administration decision makers. First, it is clear that the approach to the digitalization of agricultural processes is currently possible for all companies, regardless of size, here in terms of revenue and arable land, even though an increasing trend is noticed in Figures 7 and 8.

The identification of the most known solutions within the sample leads to two possible implications for practitioners: (1) from the point of view of public institutions, it helps us to understand which solutions or clusters of solutions should be invested in from a communication and knowledge point of view as a way to inform potential users of the potential of these solutions; (2) it helps the companies providing the different solutions to identify the most well-known ones and, ultimately, which solutions can be used the most (at least in the short term).

In addition, the benefit analysis has shown that the average benefit among users is lower than the expectations of nonusers, but that, for those who use a large number of solutions in parallel, the average benefit per solution is similar to the level of expectations. At the same time, challenge expectations are higher than the challenges experienced by users. Furthermore, the results highlight that, for both nonuser and user expectations, the technological obstacle (particularly from the point of view of interoperability between different systems and lack of connectivity in the fields) is the worst and, therefore, the most important to pay attention to, particularly from the point of view of policy-makers, who must focus on these aspects to entice and channel investment from farms that have not yet invested in Agri 4.0. In fact, one of the main difficulties that can undermine the success of a digitalization strategy in agriculture is the risk of not being able to connect the new technologies with the infrastructure already in place on the farm or even with other solutions in parallel. To overcome this constraint, it is important to develop an integration strategy plan that allows for effectively linking not only the solutions to each other, but also the people who must be properly trained and whose skills must be properly aligned. In this way, it is possible to properly implement Agri 4.0.

5.2. Limitations and Future Research Directions

As with any other research, the current study also comes with limitations. First, the sample investigated in the present study is not perfectly aligned with the current Italian agricultural context in terms of revenues and land size. Indeed, the sample differs from the Italian landscape, which is smaller in terms of the size of arable land and turnover [47]. Thus, there is still extensive room for improvement. Moreover, the current study focused only on Italian agricultural companies, which may limit the generalization of the results. Despite this, however, it is necessary to specify that the Italian agricultural sector is one of the best performers in the Italian economy, with one of the highest added value to the gross domestic product (GDP) in Europe.

A possible limitation lies in comparing the benefits and barriers from two "parallel" clusters, potentially creating bias. To overcome this problem, it would be interesting to perform a longitudinal study as a follow-up. Here, the current survey could be repeated in a few years, comparing the cluster that is not currently adopting any solution and evaluating their evolution over time; this can be done mainly to compare the evolution from the point of view of adoption and analyze what the new users see as the benefits and barriers compared with initial expectations.

Another interesting future direction of work could be to compare the level of awareness and adoption with other countries, both with a similar sector structure (such as Spain or France) and others that are among the early adopters of Agri 4.0 and precision farming (such as the Netherlands), to carry out specific benchmark analyses. Further areas of research can be derived from adopting the same research also in companies from another sector, such as breeders of livestock for meat production and meat, eventually comparing the differences between the internal branches of the agricultural sector.

Author Contributions: Conceptualization, F.A.M., M.A. and A.B., methodology, M.A. and F.A.M.; validation, F.A.M., M.A. and A.B.; formal analysis, F.A.M.; investigation, F.A.M., A.B.; data curation, F.A.M.; writing—original draft preparation, F.A.M.; writing—review and editing, F.A.M., M.A. and A.B.; supervision, A.B. All authors have read and agreed to the published version of the manuscript.

Funding: This research received no external funding.

Institutional Review Board Statement: Not applicable.

Informed Consent Statement: Informed consent was obtained from all subjects involved in the study.

Data Availability Statement: Not applicable.

Acknowledgments: This paper was inspired by the activities of the Smart Agri-Food Observatory, an industry–academia community aimed at developing knowledge and innovation in the primary sector and impact of the digital revolution in agriculture (www.osservatori.net, (accessed on 4 September 2022)).

Conflicts of Interest: The authors declare no conflict of interest.

References

1. Mohammad Fakhrul Islam, S.; Karim, Z. World's Demand for Food and Water: The Consequences of Climate Change. In *Desalination—Challenges and Opportunities*; Hossein Davood Abadi Farahani, M., Vatanpour, V., Hooshang Taheri, A., Eds.; IntechOpen, 2020; ISBN 978-1-78984-738-3.
2. Sott, M.K.; Furstenau, L.B.; Kipper, L.M.; Giraldo, F.D.; Lopez-Robles, J.R.; Cobo, M.J.; Zahid, A.; Abbasi, Q.H.; Imran, M.A. Precision Techniques and Agriculture 4.0 Technologies to Promote Sustainability in the Coffee Sector: State of the Art, Challenges and Future Trends. *IEEE Access* **2020**, *8*, 149854–149867. [CrossRef]
3. Abeliansky, A.; Prettner, K. Automation and Demographic Change. *SSRN Electron. J.* 2017. [CrossRef]
4. Yadav, V.S.; Singh, A.R.; Raut, R.D.; Mangla, S.K.; Luthra, S.; Kumar, A. Exploring the application of Industry 4.0 technologies in the agricultural food supply chain: A systematic literature review. *Comput. Ind. Eng.* **2022**, *169*, 108304. [CrossRef]
5. Mazzetto, F.; Gallo, R.; Sacco, P. Reflections and Methodological Proposals to Treat the Concept of "Information Precision" in Smart Agriculture Practices. *Sensors* **2020**, *20*, 2847. [CrossRef]
6. Jayaraman, P.; Yavari, A.; Georgakopoulos, D.; Morshed, A.; Zaslavsky, A. Internet of Things Platform for Smart Farming: Experiences and Lessons Learnt. *Sensors* **2016**, *16*, 1884. [CrossRef]
7. Pivoto, D.; Waquil, P.D.; Talamini, E.; Finocchio, C.P.S.; Dalla Corte, V.F.; de Vargas Mores, G. Scientific development of smart farming technologies and their application in Brazil. *Inf. Process. Agric.* **2018**, *5*, 21–32. [CrossRef]
8. Shang, L.; Heckelei, T.; Gerullis, M.K.; Börner, J.; Rasch, S. Adoption and diffusion of digital farming technologies—Integrating farm-level evidence and system interaction. *Agric. Syst.* **2021**, *190*, 103074. [CrossRef]
9. Lezoche, M.; Hernandez, J.E.; Díaz, M.d.M.E.A.; Panetto, H.; Kacprzyk, J. Agri-food 4.0: A survey of the supply chains and technologies for the future agriculture. *Comput. Ind.* **2020**, *117*, 103187. [CrossRef]
10. Liu, Y.; Ma, X.; Shu, L.; Hancke, G.P.; Abu-Mahfouz, A.M. From Industry 4.0 to Agriculture 4.0: Current Status, Enabling Technologies, and Research Challenges. *IEEE Trans. Ind. Inform.* **2021**, *17*, 4322–4334. [CrossRef]
11. Bolfe, É.L.; Jorge, L.A.d.C.; Sanches, I.D.; Luchiari Júnior, A.; da Costa, C.C.; Victoria, D.d.C.; Inamasu, R.Y.; Grego, C.R.; Ferreira, V.R.; Ramirez, A.R. Precision and Digital Agriculture: Adoption of Technologies and Perception of Brazilian Farmers. *Agriculture* **2020**, *10*, 653. [CrossRef]
12. Chuang, J.-H.; Wang, J.-H.; Liang, C. Implementation of Internet of Things depends on intention: Young farmers' willingness to accept innovative technology. *Int. Food Agribus. Manag. Rev.* **2020**, *23*, 253–266. [CrossRef]
13. CREA: L'agro-Alimentare Italiano Settore Chiave Dell'economia Leader in Europa per Valore Aggiunto Agricolo. Available online: https://www.crea.gov.it/-/crea-l-agro-alimentare-italiano-settore-chiave-dell-economia-leader-in-europa-per-valore-aggiunto-agricolo-1 (accessed on 15 July 2022).
14. Lytos, A.; Lagkas, T.; Sarigiannidis, P.; Zervakis, M.; Livanos, G. Towards smart farming: Systems, frameworks and exploitation of multiple sources. *Comput. Netw.* **2020**, *172*, 107147. [CrossRef]
15. Nikkilä, R.; Seilonen, I.; Koskinen, K. Software architecture for farm management information systems in precision agriculture. *Comput. Electron. Agric.* **2010**, *70*, 328–336. [CrossRef]
16. Zhai, Z.; Martínez, J.F.; Beltran, V.; Martínez, N.L. Decision support systems for agriculture 4.0: Survey and challenges. *Comput. Electron. Agric.* **2020**, *170*, 105256. [CrossRef]
17. Khujamatov, K.E.; Toshtemirov, T.K. Wireless sensor networks based Agriculture 4.0: Challenges and apportions. In Proceedings of the 2020 International Conference on Information Science and Communications Technologies (ICISCT), Karachi, Pakistan, 8–9 February 2020; IEEE: Tashkent, Uzbekistan, 2020; pp. 1–5.
18. Raj, M.; Gupta, S.; Chamola, V.; Elhence, A.; Garg, T.; Atiquzzaman, M.; Niyato, D. A survey on the role of Internet of Things for adopting and promoting Agriculture 4.0. *J. Netw. Comput. Appl.* **2021**, *187*, 103107. [CrossRef]
19. Quy, V.K.; Hau, N.V.; Anh, D.V.; Quy, N.M.; Ban, N.T.; Lanza, S.; Randazzo, G.; Muzirafuti, A. IoT-Enabled Smart Agriculture: Architecture, Applications, and Challenges. *Appl. Sci.* **2022**, *12*, 3396. [CrossRef]
20. Radoglou-Grammatikis, P.; Sarigiannidis, P.; Lagkas, T.; Moscholios, I. A compilation of UAV applications for precision agriculture. *Comput. Netw.* **2020**, *172*, 107148. [CrossRef]
21. Araújo, S.O.; Peres, R.S.; Barata, J.; Lidon, F.; Ramalho, J.C. Characterising the Agriculture 4.0 Landscape—Emerging Trends, Challenges and Opportunities. *Agronomy* **2021**, *11*, 667. [CrossRef]

22. Ramin Shamshiri, R.; Weltzien, C.; Hameed, A.I.; Yule, J.I.; Grift, E.T.; Balasundram, S.K.; Pitonakova, L.; Ahmad, D.; Chowdhary, G. Research and development in agricultural robotics: A perspective of digital farming. *Int. J. Agric. Biol. Eng.* **2018**, *11*, 1–14. [CrossRef]
23. Agrimonti, C.; Lauro, M.; Visioli, G. Smart agriculture for food quality: Facing climate change in the 21st century. *Crit. Rev. Food Sci. Nutr.* **2021**, *61*, 971–981. [CrossRef]
24. Branca, G.; Arslan, A.; Paolantonio, A.; Grewer, U.; Cattaneo, A.; Cavatassi, R.; Lipper, L.; Hillier, J.; Vetter, S. Assessing the economic and mitigation benefits of climate-smart agriculture and its implications for political economy: A case study in Southern Africa. *J. Clean. Prod.* **2021**, *285*, 125161. [CrossRef]
25. Klerkx, L.; Jakku, E.; Labarthe, P. A review of social science on digital agriculture, smart farming and agriculture 4.0: New contributions and a future research agenda. *NJAS-Wagening. J. Life Sci.* **2019**, *90–91*, 100315. [CrossRef]
26. Schneider, P. Managerial challenges of Industry 4.0: An empirically backed research agenda for a nascent field. *Rev. Manag. Sci.* **2018**, *12*, 803–848. [CrossRef]
27. Zambon, I.; Cecchini, M.; Egidi, G.; Saporito, M.G.; Colantoni, A. Revolution 4.0: Industry vs. Agriculture in a Future Development for SMEs. *Processes* **2019**, *7*, 36. [CrossRef]
28. Norman, W.; MacDonald, C. Getting to the Bottom of "Triple Bottom Line". *Bus. Ethics Q.* **2004**, *14*, 243–262. [CrossRef]
29. Rose, D.C.; Wheeler, R.; Winter, M.; Lobley, M.; Chivers, C.-A. Agriculture 4.0: Making it work for people, production, and the planet. *Land Use Policy* **2021**, *100*, 104933. [CrossRef]
30. Porter, M.E. Technology and competitive advantage. *J. Bus. Strategy* **1985**, *5*, 60–78. [CrossRef]
31. Prathibha, S.R.; Hongal, A.; Jyothi, M.P. IOT Based Monitoring System in Smart Agriculture. In Proceedings of the 2017 International Conference on Recent Advances in Electronics and Communication Technology (ICRAECT), Bangalore, India, 16–17 March 2017; pp. 81–84.
32. Marinoudi, V.; Sørensen, C.G.; Pearson, S.; Bochtis, D. Robotics and labour in agriculture. A context consideration. *Biosyst. Eng.* **2019**, *184*, 111–121. [CrossRef]
33. Santiteerakul, S.; Sopadang, A.; Yaibuathet Tippayawong, K.; Tamvimol, K. The Role of Smart Technology in Sustainable Agriculture: A Case Study of Wangree Plant Factory. *Sustainability* **2020**, *12*, 4640. [CrossRef]
34. da Silveira, F.; Lermen, F.H.; Amaral, F.G. An overview of agriculture 4.0 development: Systematic review of descriptions, technologies, barriers, advantages, and disadvantages. *Comput. Electron. Agric.* **2021**, *189*, 106405. [CrossRef]
35. Petrillo, A.; Felice, F.D.; Cioffi, R.; Zomparelli, F. Fourth Industrial Revolution: Current Practices, Challenges, and Opportunities. In *Digital Transformation in Smart Manufacturing*; Petrillo, A., Cioffi, R., Felice, F.D., Eds.; IntechOpen: Rijeka, Croatia, 2018.
36. Rea, L.M.; Parker, R.A. *Designing and Conducting Survey Research*; Jossey-Boss: San Francisco, CA, USA, 1992.
37. Rossi, P.H.; Wright, J.D.; Anderson, A.B. *Handbook of Survey Research*; Academic Press: New York, NY, USA, 2013.
38. Malhotra, M.K.; Grover, V. An assessment of survey research in POM: From constructs to theory. *J. Oper. Manag.* **1998**, *16*, 407–425. [CrossRef]
39. Wacker, J.G. A Definition of Theory: Research Guidelines for Different Theory-Building Research Methods in Operations Management. *J. Oper. Manag.* **1998**, *16*, 361–385. [CrossRef]
40. McMurray, A. *Research: A Commonsense Approach*; Cengage Learning: Melbourne, Australia, 2004.
41. Oppenheim, A.N.; Oppenheim, A.N. *Questionnaire Design, Interviewing, and Attitude Measurement, New ed.*; Distributed exclusively in the USA and Canada by St. Martin's Press; Pinter Publishers: London, UK; New York, NY, USA, 1992; ISBN 978-1-85567-043-3.
42. Dillman, D.A.; Phelps, G.; Tortora, R.; Swift, K.; Kohrell, J.; Berck, J.; Messer, B.L. Response rate and measurement differences in mixed-mode surveys using mail, telephone, interactive voice response (IVR) and the Internet. *Soc. Sci. Res.* **2009**, *38*, 1–18. [CrossRef]
43. Yun, G.W.; Trumbo, C.W. Comparative Response to a Survey Executed by Post, E-mail, & Web Form. *J. Comput.-Mediat. Commun.* **2000**, *6*, JCMC613. [CrossRef]
44. Forza, C. Survey research in operations management: A process-based perspective. *Int. J. Oper. Prod. Manag.* **2002**, *22*, 152–194. [CrossRef]
45. Series Foreword: Critical Specialities in Treating Autism and Other Behavioral Challenges. In *A Practical Guide to Finding Treatments That Work for People with Autism*; Critical Specialties-Treating Autism & Behavioral Challenge; Wilczynski, S.M. (Ed.) Academic Press: San Diego, CA, USA, 2017; pp. vii–viii. ISBN 978-0-12-809480-8.
46. Chaney, M.A. So You Want to Write a Narrative Review Article? *J. Cardiothorac. Vasc. Anesth.* **2021**, *35*, 3045–3049. [CrossRef] [PubMed]
47. Struttura Delle Aziende Agricole. Available online: http://dati.istat.it/Index.aspx?DataSetCode=DCSP_SPA (accessed on 15 July 2022).

applied sciences

MDPI

Article

Bench Test and Analysis of Cleaning Parameter Optimization of 4 L-2.5 Wheat Combine Harvester

Peng Liu [1], Xiangyou Wang [1] and Chengqian Jin [1,2,*]

[1] School of Agricultural Engineering and Food Science, Shandong University of Technology, Zibo 255000, China
[2] Nanjing Institute of Agricultural Mechanization, Ministry of Agriculture and Rural Affairs, Nanjing 210014, China
* Correspondence: jinchengqian@caas.cn

Abstract: Inaccurate and untimely adjustments of cleaning parameters during the operation of wheat combine harvesters lead to high cleaning losses and impurity rates. For this reason, a self-made 4 L-2.5 threshing and cleaning experiment table was employed for cleaning parameter optimization experiments for wheat combine harvesters in this paper. The influence of the cleaning parameters on thecleaning loss and impurity rates was analyzed, and the optimum combination of cleaning parameters was predicted and verified. The contribution hierarchy of cleaning parameters to cleaning loss rate is as follows: crank speed of shale shaker > opening of chaffer > operation speed > fan speed > throttle opening. Meanwhile, the contribution hierarchy of cleaning parameters to impurity rate is as follows: operation speed > fan speed > throttle opening > crank speed of shale shaker > opening of chaffer. The predicted optimum combination of cleaning parameters, i.e., when the cleaning loss and impurity rates are both at a minimum and the feed quantity is at the maximum, is as follows: operating speed—2.2 m/s; opening of chaffer—26 mm, throttle opening—20°; fan speed—1100 r/min; and Crank speed of shale shaker—350 r/min. With these settings, the cleaning loss rate was 1.5% and the impurity rate was 1.9%. In the validation experiment, the average cleaning loss rate was found to be 1.47%, the average impurity rate was 1.96%, and the relative error of the predicted values was 0.03% and 0.06%, respectively. Compared with the cleaning index of combine harvesters with commonly used parameters, the cleaning loss rate was reduced by 0.12% and the impurity rate was reduced by 0.19%.

Keywords: wheat; 4 L-2.5 combine harvester; threshing and cleaning testbed; cleaning parameters optimization; response surface methodology

Citation: Liu, P.; Wang, X.; Jin, C. Bench Test and Analysis of Cleaning Parameter Optimization of 4 L-2.5 Wheat Combine Harvester. *Appl. Sci.* **2022**, *12*, 8932. https://doi.org/10.3390/app12188932

Academic Editors: José Miguel Molina Martínez, Ginés García-Mateos and Dolores Parras-Burgos

Received: 15 July 2022
Accepted: 29 August 2022
Published: 6 September 2022

1. Introduction

The cleaning device is one of the core structures of a wheat combine harvester; it is used to perform threshing mixture cleaning operations in order to complete the separation of wheat grains and impurities, as well to clean the wheat grain itself [1–5]. Therefore, the operative quality of the cleaning device is directly related to the overall operating quality of the combine harvester [6,7]. Cleaning parameters refer to adjustable settings that affect the operation quality of the cleaning device of the combined harvester, whereas the cleaning loss and impurity rates are evaluation indexes for the cleaning quality of the harvester [8–10].

Extensive research has been conducted on theoptimization of wheat harvester cleaning parameters and their influence on cleaning quality. Zhong et al. [11] undertook orthogonal experiments on thecleaning components of a 4lz-1.0q rice wheat combine harvester; the results of wheat field tests were analyzed using the fuzzy comprehensive evaluation method. The optimal parameter combinations of vibrating screen crank speed, screen surface structure, centrifugal fan speed, and vibrating screen amplitude in the cleaning process, as well as the primary and secondary order cleaning parameters affecting cleaning performance, were obtained, thereby improving the machine's adaptability and ensuring good harvesting quality. Geng et al. [12] attempted to optimize the cleaning parameters, such as cleaning screen amplitude, cleaning screen frequency, fan wind speed, and air flow direction angle, of a wheat combine harvester to investigate their effect on thecleaning loss and impurity content rates. In their study, the optimal combination of cleaning parameters was determined, providing a reference for the parameter selection and optimization. Jin et al. [13,14] undertook a bench optimization test on themain operating parameters (feed rate, air door opening, fan speed, and angle of upper and lower air guide plates) of a double outlet, multi-air duct cleaning device. Those authors studied the influence of these factors on thecleaning loss rate, impurity content rate, and secondary impurity content rate, determining the optimal parameter combination thereof. A response surface bench test was carried out with the mass fraction of grains, stems, and sundries in the total mass of the desorbed matter of a double desorber as the test index. Feed rate, the opening of the air door, fan speed and the angle of the upper and lower air guide plates were used as test factors. The influences of these factors on thetest index were analyzed, and the best matching parameters were obtained, serving as a reference for the study of the harvester's performance and the structural design of a multi air channel cleaning device. Li et al. [15] carried out a wheat bench test on themain influences (feed rate, air door opening and fan speed) on theoperation of a multi duct centrifugal fan with double outlets. The impact of each factor on thecleaning quality evaluation index (cleaning loss rate and impurity content) was investigated, and the optimal parameter combination was determined, providing a reference for research and design of multi duct centrifugal fans. Tong et al. [16] proposed improvements to the fan structure, i.e., longitudinal axial flow full feed double air channels and six outlets. Those authors then performed an orthogonal optimization simulation of fan speed, fan incidence angle, and fish scale screen angle using the improved cleaning device. The influence law of various factors on theairflow field was analyzed, and the optimal parameter combination was obtained to improve the performance of the cleaning device even further. Using the cleaning device of a small-scale semi-feeding wheat combine harvester in intercropping mode as the research object, Zhang et al. [17] took the cleaning and loss rates as evaluation indices, and the feeding speed, suction pressure, cylinder height, and angle of the lower cone as test factors in an orthogonal simulation of the cleaning process. In this way, the optimal cleaning parameter combination was identified. Furthermore, a field test validated the simulation optimization results in order to improve the cleaning performance of a wheat combine harvester in intercropping mode. Using the test-bed of a cyclone separation and cleaning system with double threshers and taking wheat as the test object, the rotation speed of the first-stage thresher, second-stage thresher, and suction fan as test factors, and cleaning rate as the test index, Shi et al. [18–20] performed orthogonal and regression tests to optimize the parameters, providing a basis for the design of a portable grain chopper cleaning system. Using the cyclone separation system of the double thresher within a micro grain combine harvester as the research object, the rotation speed of the two-stage thresher and the suction fan as test factors, and the grain cleaning rate, cleaning loss rate, and grain crushing rate as test indexes, a cleaning performance regression test was performed and the motion parameters for the cleaning system were optimized; this research serves as an experimental foundation for the parameter design of cleaning systems for micro-grain combine harvesters. As a result of that research, a composite device was developed for conveying, separating, and cleaning grain effluent; the optimal parameters of this device were obtained through indoor tests, and its cleaning capacity was improved. Using orthogonal tests, general rotation combination tests, and a novel design, Ni et al. [21] reported the optimum combination of structural parameters and motion parameters of each part of the cyclone separation and cleaning system without a guide vane in the separation cylinder. The results revealed improved cleaning performance of the micro wheat combine harvester. Liu et al. [22] selected the main parameters that affect the cleaning performance index using a self-made cleaning system test-bed with the upper cone angle of the separation cylinder, the height of the cylinder, the number of guide vanes, the speed of the suction fan, and the speed of the thresher as design variables. The optimal combination of parameters and performance were obtained through orthogonal tests, a second general rotation combination test, and a novel design, which serves as a further foundation for the design of cleaning systems for micro wheat combine harvesters.

In spite of the aforementioned studies, research on theoptimization of wheat harvester cleaning parameters and their impact on cleaning quality is still insufficient. The cleaning parameters studied are few and incomplete, resulting in inaccurate and untimely cleaning parameter adjustments during wheat combine harvester field operation. Wheat machine cleaning is often inefficient; it restricts the cleaning operation level of the combine harvester and limits the annual wheat crop output.

In light of these issues, this paper takes all five adjustable parameters (operation speed, chaffer opening, throttle opening, fan speed, and shale shaker crank speed) that affect the cleaning quality of wheat combine harvesters as cleaning parameters, takes cleaning loss rate and impurity rate as cleaning quality evaluation indicators, and uses Design Expert software to complete the response surface experiment design. A bench test of the optimized wheat harvester cleaning parameters was completed using a self-made 4 L-2.5 threshing and cleaning test bed. The contribution rate and response effect of the five cleaning parameters relative to the two cleaning quality evaluation indexes were analyzed using the contribution rate and response surface methods, and the hierarchy of wheat mechanical cleaning parameters was obtained. The present research is an expansion and improvement of previous research on theoptimization of existing wheat machine cleaning parameters and their influence on cleaning quality. The research findings can be used to guide the setting and adjustment of cleaning parameters during the field operation of wheat combine harvesters, as well as to provide a theoretical foundation for the research and development of future wheat combine harvester adaptive cleaning systems.

2. Materials and Methods

2.1. 4 L-2.5 Threshing and Cleaning Experiment Table and Cleaning Device Structure

The 4 L-2.5 threshing and cleaning experiment table was composed of a material conveying device, a material feeding device, a threshing device, a cleaning device, a control system, and a touch screen to electronically control of the operation speed, feed auger speed, conveyor chain harrow speed, threshing drum speed, and fan speed. In contrast, the crank speed of the shale shaker opening as well as the opening of the chaffer were adjusted manually. Figure 1 shows the structure and workflow of the 4 L-2.5 threshing and the cleaning experiment table.

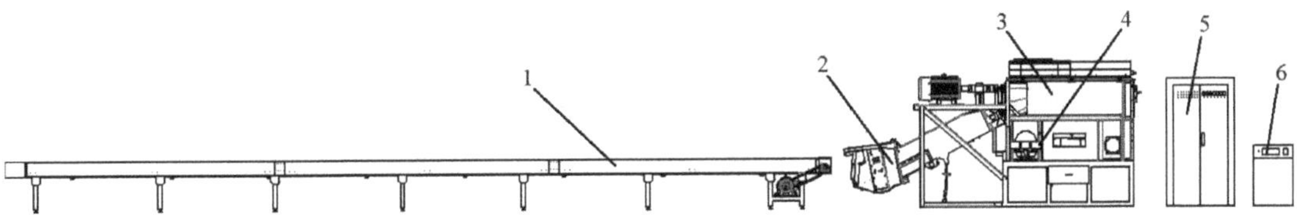

(**a**) Structure of 4 L-2.5 threshing and the cleaning experiment table

Figure 1. *Cont.*

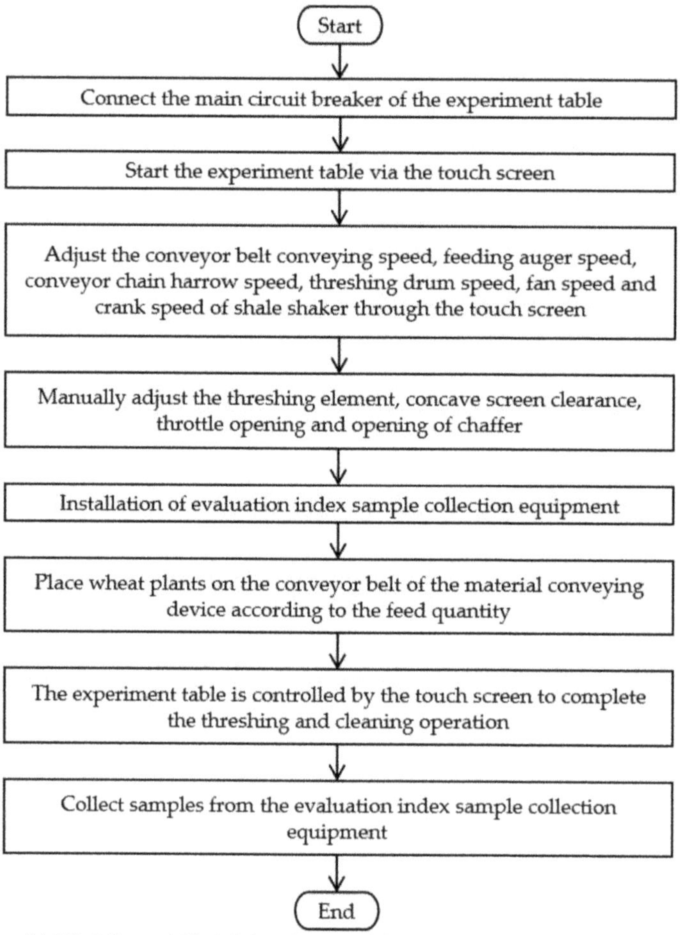

(**b**) Workflow of 4 L-2.5 threshing and the cleaning experiment table

Figure 1. Structure and workflow of the 4 L-2.5 threshing machine and the cleaning experiment table. 1. Material conveying device. 2. Material feeding device. 3. Threshing device. 4. Cleaning device. 5. Control system. 6. Touch screen.

The cleaning device consisted of a shale shaker, throttle opening adjusting plate, screen opening adjusting plate, sampling box, crank connecting rod mechanism, crank torque sensor, crank motor, fan motor, fan torque sensor, fan, and a frame and subframe. The structure is shown in Figure 2.

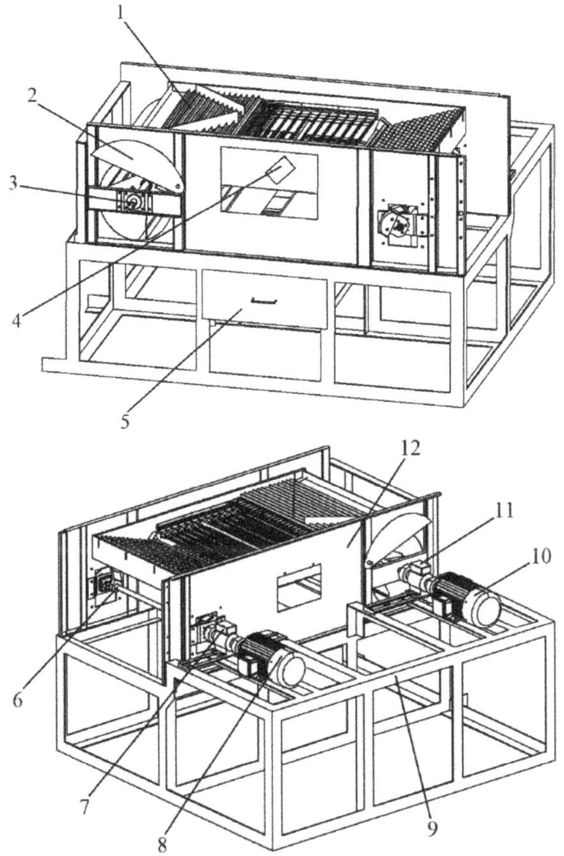

Figure 2. 4 L-2.5 Threshing and cleaning experiment table. 1. Shale shaker. 2. Throttle opening adjusting plate. 3. Fan. 4. Screen opening adjusting plate. 5. Sampling box. 6. Crank connecting rod mechanism. 7. Crank torque sensor. 8. Crank motor. 9. Subframe. 10. Fan motor. 11. Fan torque sensor. 12. Frame.

2.2. Control Method of Cleaning Parameters

The cleaning parameters affecting cleaning operations are operation speed, fan speed, throttle opening, crank speed of the shale shaker, and the opening of the chaffer. Therefore, this paper studied the optimization of these parameters. To improve the accuracy of the experiment, the five parameters needed to be adjusted with a high degree of precision. Operation speed, fan speed, crank speed of shale shaker, and the other devices were set via a touch screen interface, while throttle opening and the opening of chaffer were adjusted manually. The experiment table was controlled by a touch screen interface. The opening of chaffer X describes the vertical distance between the adjacent parallel screens, and the throttle opening is the relative change angle between the throttle opening regulator and the fan support plate. Figure 3 shows the manual adjustment mechanism of the cleaning device.

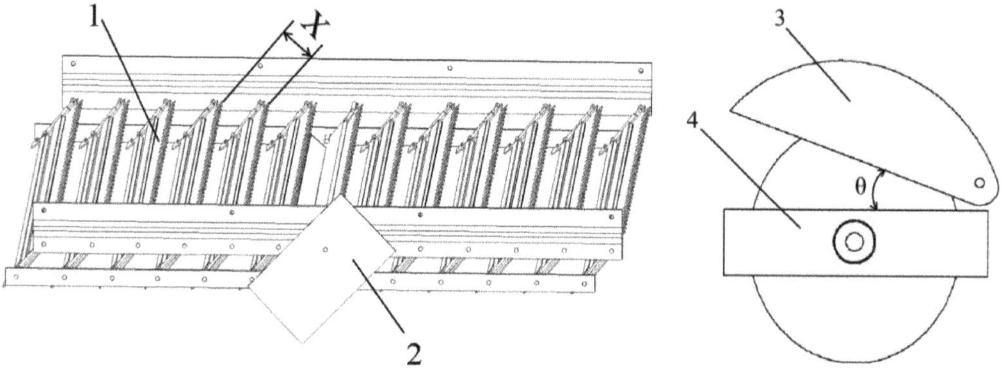

(a) Opening of chaffer control part (b) Throttle opening control part

Figure 3. Manual control of cleaning parameters. 1. Chaffer sieve. 2. Screen opening adjusting plate. 3. Throttle opening regulator. 4. Fan support plate. X. Opening of chaffer. θ. Throttle opening.

2.3. Experiment Parameters

The collection of 1 m^2 wheat plant samples and the measurements of parameters were completed at the Hedong wheat experimental base, Linyi City, Shandong Province, China. Statistics are provided in Table 1.

Table 1. Wheat characteristics.

Name	Parameter
Wheat varieties	Lin Mai 4
Growth period	235 days
1 m^2 wheat plant quality	1600 g
1 m^2 wheat yield	647 g
Wheat spike length	77.6 mm
Natural height of crop	649 mm
Grass Valley ratio	0.78
Moisture content of grain	13.34%
100 grain weight	3.76 g

The implemented operation parameters were determined in various ways, i.e., by taking measurements on theexperiment table, using previous field experiment research and relevant literature, and field wheat harvest practical experience, as shown in Table 2 [23–34].

Table 2. Implemented parameters.

Name	Parameter
Feeding auger speed	180 r/min
Conveyor chain harrow speed	440 r/min
Threshing drum speed	950 r/min
Cut width	1.8 m
Conveyor belt width	1.5 m

2.4. Experiment Data Calculation Method

We referred to relevant literature to set the calculation equation of the feed quantity, cleaning loss rate, and impurity rate [13,15].

The feed quantity was calculated using Equation (1) as follows:

$$Q = \frac{W}{t} = \frac{M \times B \times L}{t} = M \times B \times v \quad (1)$$

where Q is the Feed quantity in kg/s; W is wheat plant quantity in kg; T is operation time in seconds; M is 1 m^2 wheat plant quantity; B is cut width in meters; L is operation distance in meters, and V is operation speed in m/s.

The cleaning loss rate was calculated using Equation (2) as follows:

$$S = \frac{W_{sh}}{W_{ch}} = \frac{S_0 \ / \ (B \times L)}{W_{ch}} \times 100\ \% \tag{2}$$

where S is the percentage of cleaning loss; $S_0/(B \times L)$ is 1 m^2 loss quality of wheat cleaning in g/m^2; S_0 is loss quality of wheat cleaning; B × L is the Harvest area in m^2; W_{sh} is the wheat cleaning loss quality of 1 m^2 in g/m^2; W_{ch} is 1 m^2 wheat harvest quality in g/m^2; B is the cut width in meters; and L is operation distance in meters.

The impurity rate was calculated using Equation (3) as follows:

$$Z = \frac{W_z}{W_{zy}} \times 100\ \% \tag{3}$$

where Z is the percentage of impurity; W_z is the quality of impurities in samples containing impurities; and W_{zy} is the quality of the impurity samples.

2.5. Experiment Design and Data Statistics

The values and ranges of the five cleaning parameters during the routine operation of a wheat combine harvester were selected based on relevant literature and field experience [11–22]. During routine operation thevalues of the five parameters were set to the middle their respective ranges. The codes −1, 0, and 1 represent the low, middle, and high values of the cleaning parameter range, as determined by the conversion function in the Design Expert software. The following is the code-to-actual-value conversion formula, as well as the representative symbols for the five cleaning parameters:

$$\begin{cases} 1 = \max X \\ 0 = \mathrm{mid} X \\ -1 = \min X \\ X = A, B, C, D, E \end{cases} \tag{4}$$

where X represents five cleaning parameters; 1 represents the maximum value for each parameter; 0 represents the middle value; −1 represents the minimum value; A is operation speed in m/s; B is opening of chaffer in mm; C is throttle opening degree; D is the fan speed in r/min; and E is the crank speed of the shale shaker in r/min.

The Design Expert software was applied for the design of five factors and three horizontal response surface experiments. The calculation and analysis of the cleaning loss and the impurity rates were completed using the experiment data calculation method. Tables 3 and 4 show the statistics for the five cleaning parameter levels and the cleaning loss and impurity rates.

Table 3. Cleaning parameters.

Code	Operation Speed A (m/s)	Opening of Chaffer B (mm)	Throttle Opening C (°)	Fan Speed D (r/min)	Crank Speed of Shale Shaker E (r/min)
−1	1.5	22	0	900	350
0	2.0	24	10	1000	400
1	2.5	26	20	1100	450

Table 4. Experiment data for five factors and three horizontal response surfaces.

NO.	A (m/s)	B (mm)	C (°)	D (r/min)	E (r/min)	A (m/s)	B (mm)	C (°)	D (r/min)	E (r/min)	Cleaning Loss Rate S (%)	Impurity Rate Z (%)
	Code Value					Actual Value						
1	0	1	1	0	0	2.0	26	20	1000	400	1.26	1.90
2	−1	0	0	0	1	1.5	24	10	1000	450	0.63	0.63
3	1	0	0	−1	0	2.5	24	10	900	400	2.61	4.13
4	0	0	0	−1	1	2.0	24	10	900	450	2.02	2.52
5	0	1	0	1	0	2.0	26	10	1100	400	1.28	1.83
6	0	0	0	1	−1	2.0	24	10	1100	350	1.37	1.87
7	0	0	0	0	0	2.0	24	10	1000	400	1.65	2.17
8	0	−1	0	−1	0	2.0	22	10	900	400	2.33	2.54
9	0	−1	0	1	0	2.0	22	10	1100	400	2.37	1.82
10	0	0	1	0	−1	2.0	24	20	1000	350	1.36	1.91
11	0	0	0	1	1	2.0	24	10	1100	450	2.09	1.81
12	1	1	0	0	0	2.5	26	10	1000	400	2.62	2.83
13	−1	0	0	1	0	1.5	24	10	1100	400	0.55	0.57
14	0	0	0	0	0	2.0	24	10	1000	400	1.51	2.26
15	0	0	1	−1	0	2.0	24	20	900	400	1.83	2.45
16	−1	0	0	0	−1	1.5	24	10	1000	350	0.45	0.77
17	0	0	−1	0	1	2.0	24	0	1000	450	2.01	2.33
18	0	0	0	0	0	2.0	24	10	1000	400	1.77	2.22
19	0	1	0	0	1	2.0	26	10	1000	450	1.29	1.97
20	0	1	0	0	−1	2.0	26	10	1000	350	1.13	2.32
21	1	0	0	1	0	2.5	24	10	1100	400	2.65	2.62
22	0	0	0	0	0	2.0	24	10	1000	400	1.53	2.08
23	1	0	1	0	0	2.5	24	20	1000	400	2.68	2.75
24	0	0	1	1	0	2.0	24	20	1100	400	1.85	1.74
25	0	1	0	−1	0	2.0	26	10	900	400	1.22	2.58
26	−1	0	0	−1	0	1.5	24	10	900	400	0.51	0.88
27	0	−1	0	0	1	2.0	22	10	1000	450	2.41	1.96
28	−1	0	1	0	0	1.5	24	20	1000	400	0.53	0.52
29	0	0	0	0	0	2.0	24	10	1000	400	1.46	2.14
30	−1	1	0	0	0	1.5	26	10	1000	400	0.38	0.65
31	0	−1	0	0	−1	2.0	22	10	1000	450	2.31	2.31
32	0	−1	−1	0	0	2.0	22	0	1000	400	2.32	2.36
33	−1	−1	0	0	0	1.5	22	10	1000	400	0.66	0.64
34	0	0	0	−1	−1	2.0	24	10	900	350	1.35	2.57
35	0	−1	1	0	0	2.0	22	20	1000	400	2.34	1.86
36	0	0	−1	−1	0	2.0	24	0	900	400	1.41	2.68
37	−1	0	−1	0	0	1.5	24	0	1000	400	0.48	1.21
38	0	0	0	0	0	2.0	24	10	1000	400	1.62	2.03
39	1	−1	0	0	0	2.5	22	10	1000	400	3.33	2.81
40	1	0	−1	0	0	2.5	24	0	1000	400	2.62	3.50
41	0	0	−1	0	−1	2.0	24	0	1000	350	1.35	2.40
42	0	0	−1	1	0	2.0	24	0	1100	400	1.80	1.93
43	0	1	−1	0	0	2.0	26	0	1000	400	1.23	2.37
44	1	0	0	0	−1	2.5	24	10	1000	350	2.62	2.84
45	1	0	0	0	1	2.5	24	10	1000	450	3.76	2.85
46	0	0	1	0	1	2.0	24	20	1000	450	2.07	1.85

2.6. Experiment Process

Experiments were conducted from 1 July 2019, to 9 July 2019 at the Agricultural Machinery Research Office of the School of Agricultural Engineering and Food Science, Shandong University of Technology, China. Before the experiments, it was determined that the quality of the harvested wheat plants was to be 5.4 × 1600 g with to a cutting width of 1.8 m. 5.4 × 1600 g. Plants were evenly distributed at a distance of 2 m or more from the conveyor belt surface to the header, with a sharing area of 3 m × 1.8 m in order to ensure that the feed quantity was consistent.

A cleaning loss collection bag was fixed at the outlet of the cleaning room to collect the cleaning loss samples, while a sampling box was used to collect miscellaneous samples at the grain collection auger. The optimization bench experiment and sample processing of the cleaning parameters were performed according to the response surface experiment design table. Then, cleaning loss and impurity quantity and quality data were recorded. The experimental setup is shown in Figure 4.

Figure 4. Experiment table.

3. Results

3.1. Cleaning Loss Rate

3.1.1. Establishment of the Regression Model of Cleaning Loss Rate and Significance Test

According to the experiment results shown in Table 4, a variance analysis of the cleaning loss rate was undertaken; the results are shown in Table 5. The p-value is used to analyze the significance of objects; $p \leq 0.01$ implies that the response model is extremely significant, $p \leq 0.05$ means that it is relatively significant, and $p > 0.05$ means that it is insignificant. The regression equation of the cleaning loss rate was as follows:

$$S = 1.59 + 1.18A - 0.51B + 0.060C + 0.063D + 0.29E - 0.017AB + 0.017AC + 0.000AD + 0.19AE - (2.500E - 003BC) - (5.000E - 003BD) - (2.500E - 003BE) - 0.088CD + (2.500E - 003CE) + (5.000E - 003DE) + 0.065A^2 + 0.15B^2 + 0.031C^2 + 0.035D^2 + 0.11E^2 \quad (5)$$

Table 5. Analysis of variance of cleaning loss rate.

Source	Sum of Squares	df	F	p	
Model	28.48	20	42.35	<0.0001	significant
A	22.4	1	666.09	<0.0001	
B	4.19	1	124.68	<0.0001	
C	0.058	1	1.71	0.2025	
D	0.064	1	1.9	0.1807	
E	1.33	1	39.5	<0.0001	
AB	1.23×10^{-3}	1	0.036	0.8502	
AC	1.23×10^{-3}	1	0.036	0.8502	
AD	0	1	0	1.0000	
AE	0.15	1	4.41	0.0460	
BC	2.50×10^{-5}	1	7.44×10^{-4}	0.9785	
BD	1.00×10^{-4}	1	2.97×10^{-3}	0.9569	
BE	2.50×10^{-5}	1	7.44×10^{-4}	0.9785	
CD	0.031	1	0.91	0.3490	
CE	2.50×10^{-5}	1	7.44×10^{-4}	0.9785	
DE	1.00×10^{-4}	1	2.97×10^{-3}	0.9569	
A^2	0.037	1	1.1	0.3035	
B^2	0.21	1	6.12	0.0205	
C^2	8.41×10^{-3}	1	0.25	0.6214	
D^2	0.011	1	0.32	0.5756	
E^2	0.1	1	2.96	0.0975	
Residual	0.84	25			
Lack of Fit	0.78	20	3.04	0.1102	not significant
Pure Error	0.064	5			
Cor Total	29.32	45			
R^2	0.9713				

From Table 5, it can be seen that the *p*-value of the cleaning loss rate model was less than 0.01, indicating that the established model was extremely significant. The decision coefficient R^2 of the model was 0.9713, indicating that the model reflected a response value change of 97.13%. In the regression model, the *p*-values of A, B, and E were less than 0.01, while those of AE and B^2 were greater than 0.01 but less than 0.05. The *p*-values of the other items were greater than 0.05.

3.1.2. Contribution of Each Parameter to Cleaning Loss Rate

Contribution rate Δ_j reflects the degree of influence of a given parameter on theestablished regression model; the greater the value of Δ_j, the greater the degree of influence. Δ_j may be calculated as follows:

$$\delta = \begin{cases} 0 & F \leq 1 \\ 1 - \frac{1}{F} & F > 1 \end{cases} \tag{6}$$

$$\Delta_j = \delta_j + \frac{1}{2} \sum_{\substack{i=1 \\ i \neq j}}^{m} \delta_{ij} + \delta_{jj} \; j = 1, 2, \cdots, m \tag{7}$$

where F is the value of variance analysis; δ is the assessment value; Δ_j is the contribution rate of the first power term of the j-th factor; Δ_{jj} is the contribution rate of the second power term of the j factor; and Δ_{ij} is the contribution rate of interaction between the factor j and other factors.

According to Equations (6) and (7), the contribution of each parameter to the cleaning loss rate was calculated, as shown in Table 6. The hierarchy of the cleaning parameters relative to the cleaning loss rate was as follows: crank speed of shale shaker E > opening of chaffer B > operation speed A > fan speed D > throttle opening C.

Table 6. Contribution of each parameter to cleaning loss rate.

Name	Contribution Rate (%)
Operation speed A	1.4760
Opening of chaffer B	1.8286
Throttle opening C	0.4152
Fan speed D	0.4737
Crank speed of shale shaker E	2.0235

3.1.3. Analysis of Response of Each Parameter to Cleaning Loss Rate

From the data provided in Tables 5 and 6, it can be seen that operation speed (A), the opening of chaffer (B), and the crank speed of shale shaker (E) had a greater impact on thecleaning loss rate than throttle opening (C) and fan speed (D). From Figure 5a–d, it can be seen that operation speed (A) had a positive correlation with the cleaning loss rate, and that operation speed A and the crank speed of the shale shaker (E) interacted with the cleaning loss rate, as the former determines the feeding quantity, and a larger feeding quantity leads to an increase in the cleaning quantity of the threshing mixture over time. The cleaning device failed to separate the grains and impurities, increasing the quantity of such materials that was discharged out of the cleaning room. The crank speed of shale shaker (E) determines the rate of separation and the discharge speed of the threshing mixture, and the interaction of these variables increases the cleaning loss rate. Additionally, the opening of chaffer B had a negative correlation and a secondary effect on thecleaning loss rate due to the increase of opening b of the chaffer. As the distance between the adjacent parallel sieves increased, the number of wheat grains passing through the shaker increased proportionally, and the loss of wheat grains was reduced, along with the loss rate due to cleaning. Furthermore, throttle opening C had a positive correlation with the cleaning loss rate, since the former determines the air inlet area of the fan, and increasing this variable increased the air inlet area and volume of the air field in the cleaning room. As the wind force on thethreshing mixture increased, the number of grains discharged from the cleaning chamber by the threshing mixture and the cleaning loss rate also increased. Additionally, fan speed D had a positive correlation with the cleaning loss rate, since the former determines the wind speed in the cleaning room. The number of seeds blown out of the cleaning room by the threshing mixture over time increased, and the cleaning loss rate increased accordingly. Finally, the crank speed of the shale shaker (E) had a positive correlation with the cleaning loss rate, because the former determines the frequency of the shale shaker, and as such, with an increase in speed, the frequency of the threshing mixture pushed back and forth by the vibrating screen, the number of grains discharged from the cleaning chamber by the threshing mixture, and the cleaning loss rate increased.

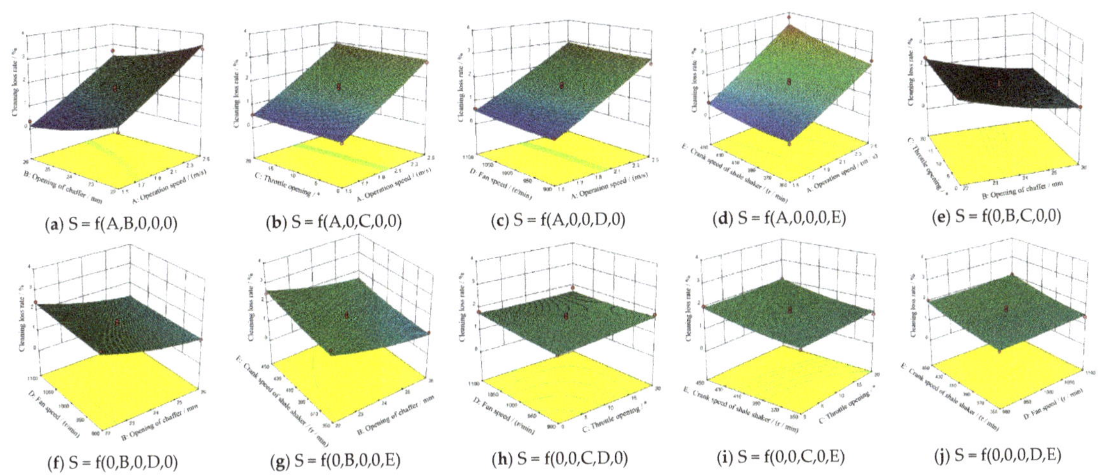

(**a**) S = f(A,B,0,0,0) (**b**) S = f(A,0,C,0,0) (**c**) S = f(A,0,0,D,0) (**d**) S = f(A,0,0,0,E) (**e**) S = f(0,B,C,0,0)
(**f**) S = f(0,B,0,D,0) (**g**) S = f(0,B,0,0,E) (**h**) S = f(0,0,C,D,0) (**i**) S = f(0,0,C,0,E) (**j**) S = f(0,0,0,D,E)

Figure 5. Response surface analysis of the effect of each parameter on cleaning loss rate.

3.2. Impurity Rate

3.2.1. Establishment of the Regression Model of Impurity Rate and Significance Test

According to the experiment results shown in Table 4, variance analysis was carried out for the impurity rate; the results are shown in Table 7. The regression equation for the impurity rate was as follows:

$$\begin{aligned} Z = {} & 2.15 + 1.15A + (9.375E - 003)B - 0.24C - 0.38D - 0.067E + (2.500E - 003)AB - 0.015AC - 0.30AD + 0.038AE + \\ & (7.500E - 003)BC - (7.500E - 003)BD + 0.000BE + (1.000E - 002)CD + (2.500E - 003)CE - (2.500E - 003)DE - \\ & 0.27A^2 - 0.056B^2 + 0.028C^2 + 0.093D^2 - 0.043E^2 \end{aligned} \tag{8}$$

From Table 7, it can be seen that the p-value of the impurity rate model was less than 0.01, indicating that the established regression model was extremely significant. Furthermore, the decision coefficient R^2 of the model was 0.9783, which implies that the model can reflect a response value change of 97.83%, indicating that the obtained linear regression equation fitting effect was very good. In the regression model, the p-values of A, C, D, AD, and A^2 were all less than 0.01, indicating that the influence on themodel was very significant. The p-values of the other items were greater than 0.05, indicating insignificant influence.

Table 7. Analysis of variance according to impurity rate.

Source	Sum of Squares	df	F	p	
Model	25.96	20	56.24	<0.0001	significant
A	21.3	1	922.99	<0.0001	
B	1.41×10^{-3}	1	0.061	0.807	
C	0.9	1	39.11	<0.0001	
D	2.37	1	102.78	<0.0001	
E	0.072	1	3.1	0.0905	
AB	2.50×10^{-5}	1	1.08×10^{-3}	0.974	
AC	9.00×10^{-4}	1	0.039	0.845	
AD	0.36	1	15.6	0.0006	
AE	5.63×10^{-3}	1	0.24	0.6258	
BC	2.25×10^{-4}	1	9.75×10^{-3}	0.9221	
BD	2.25×10^{-4}	1	9.75×10^{-3}	0.9221	
BE	3.55×10^{-15}	1	1.54×10^{-13}	1	
CD	4.00×10^{-4}	1	0.017	0.8963	
CE	2.50×10^{-5}	1	1.08×10^{-3}	0.974	
DE	2.50×10^{-5}	1	1.08×10^{-3}	0.974	
A^2	0.63	1	27.23	<0.0001	
B^2	0.027	1	1.18	0.2879	
C^2	7.01×10^{-3}	1	0.3	0.5865	
D^2	0.076	1	3.29	0.0815	
E^2	0.016	1	0.68	0.4163	
Residual	0.58	25			
Lack of Fit	0.54	20	3.67	0.0772	not significant
Pure Error	0.037	5			
Cor Total	26.53	45			
R^2	0.9783				

3.2.2. Contribution of Each Parameter to Impurity Rate

Equations (6) and (7) were used to quantify the contribution of each parameter to the impurity rate, as shown in Table 8. The hierarchy of these parameters was as follows: operation speed A > fan speed D > throttle opening C > crank speed of shale shaker E > opening of chaffer B.

Table 8. Hierarchy of parameters relative to impurity rate.

Name	Contribution Rate (%)
Operation speed A	2.4301
Opening of chaffer B	0.1525
Throttle opening C	0.9744
Fan speed D	2.1543
Crank speed of shale shaker E	0.6774

3.2.3. Analysis of the Response Effect of Each Parameter to the Impurity Rate

From the data provided in Tables 7 and 8, it can be seen that operation speed (A), throttle opening (C), and fan speed (D) had greater impacts on theimpurity rate than the chaffer (B) and crank speed of shale shaker E. From Figure 6, it can be seen that operation speed (A) had a positive correlation with the impurity rate. The operation speed of A and fan speed D influenced the impurity rate, with the former having a secondary effect, as it determined the feeding quantity, and an increase in feeding quantity leads to an increase in threshing mixture cleaning. The cleaning device was unable to separate the grains and impurities in a timely manner. As such, the number of grains and impurities passing through the vibrating screen increased, and the impurity rate increased accordingly. An increase of fan speed D increased the separation amount in the threshing mixture and the number of impurities blown out while reducing the impurity rate. As also shown in Figure 6, the opening of chaffer B had a positive correlation with the impurity rate, as with an increased distance between the adjacent parallel sieves of the chaffer, the sieve penetration of the threshing mixture increased, and the separation of the grain and the impurity decreased, thereby increasing the impurity rate. Additionally, throttle opening C had a negative correlation with the impurity rate, because this parameter determines the air inlet area of the fan; with a greater air inlet area and air volume of the air field in the cleaning room, the wind force on thethreshing mixture, the amount of separation between the threshing mixture and the impurities, and the number of impurities blown out all increase while the impurity rate decreases. Furthermore, fan speed D had a negative correlation with the impurity rate, as increasing the fan speed increased the wind field speed, thereby increasing the amount of separation of the threshing mixture grain and impurity, as well as the amount of impurity blown out, resulting in a decrease in the impurity rate. Finally, the crank speed of shale shaker E had a negative correlation with the impurity rate, as this factor determines the frequency of the shale shaker. With an increase in the crank speed, the frequency and screening rate of the threshing mixture pushed back and forth by the vibrating screen increased, as did the amount of separation of wheat threshing mixture grain and impurities and the number of impurities blown out, thereby reducing the impurity rate.

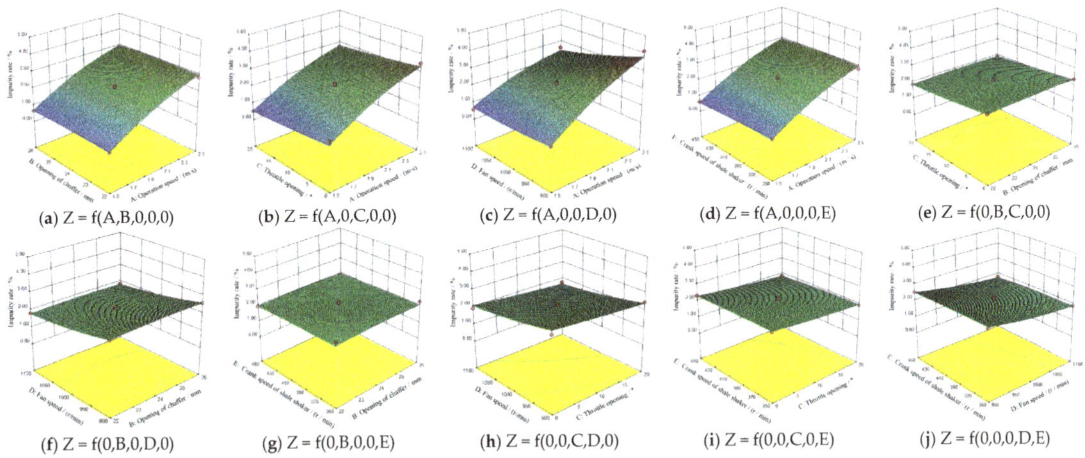

(**a**) Z = f(A,B,0,0,0) (**b**) Z = f(A,0,C,0,0) (**c**) Z = f(A,0,0,D,0) (**d**) Z = f(A,0,0,0,E) (**e**) Z = f(0,B,C,0,0)
(**f**) Z = f(0,B,0,D,0) (**g**) Z = f(0,B,0,0,E) (**h**) Z = f(0,0,C,D,0) (**i**) Z = f(0,0,C,0,E) (**j**) Z = f(0,0,0,D,E)

Figure 6. Response surface analysis of each parameter on impurity rate.

3.3. Optimization and Verification of Cleaning Parameters

3.3.1. Optimization of Cleaning Parameters

It is desirable that the cleaning loss and impurity rates are as low as possible and the operation speed is maximal to achieve the optimum cleaning operation level and the maximum feed quantity. According to the response effect analysis of each parameter relative to cleaning loss rate, it can be seen that to ensure the minimum cleaning loss rate, it is necessary to ensure the maximum opening of the chaffer and the minimum throttle opening, fan speed, operation speed, and crank speed of a shale shaker. To minimize the impurity rate, it is necessary use the minimum operation speed and opening of chaffer and the maximum throttle opening, fan speed and crank speed. To ensure the collection efficiency and undertake an effective cleaning operation of a wheat machine, the feed quantity and the operation speed are required to be at the maximum values. To predict the optimum combination of the cleaning parameters when the cleaning loss and impurity rates were the smallest and the feed quantity was the largest, it was necessary to perform an optimization analysis of five cleaning parameters. To this end, the following model was established:

$$\begin{cases} \max A \\ \min S \\ \min Z \\ -1 \leq A, B, C, D, E \leq 1 \end{cases} \tag{9}$$

By using the Design Expert software to optimize the cleaning parameters, it was concluded that the optimum combination of parameters was as follows: operation speed—2.2 m/s; opening of chaffer—26 mm; throttle opening—20°, fan speed—1099.8 r/min; and crank speed of shale shaker—350 r/min. By applying these settings, a cleaning loss rate of 1.5% and an impurity rate of 1.9% were obtained.

3.3.2. Verification of the Optimum Cleaning Parameters

From 1 July 2019, to 9 July 2019, the optimum combination of cleaning parameters was verified in the Agricultural Machinery Research Office of the School of Agricultural Engineering and Food Science, Shandong University of Technology, China. Table 1 shows the wheat characteristic parameters for the wheat plants used in the experiment. Since it was difficult to set the fan and crank speeds of the shale shaker to 1 decimal place, to facilitate the smooth operation of the experiment, the cleaning parameter combination was adjusted as follows: operation speed—2.2 m/s; opening of chaffer—26 mm; throttle opening—20°; fan speed—1100 r/min; and crank speed of shale shaker—350 r/min. Using this combination of parameters, three sets of bench verification experiments were carried out; see Table 2.

The cleaning loss and impurity rates of the three groups of validation experiments were calculated using Equations (2) and (3) and average values of 1.47% and 1.96%, respectively, were calculated, as shown in Table 9. The relative errors between the experimental data and the predicted values were 0.03% and 0.06%, respectively.

Table 9. Optimum cleaning parameter combination verification experiment data.

Experiment Serial Number	Cleaning Loss Rate (%)	Impurity Rate (%)
1	1.25	1.89
2	1.54	1.98
3	1.61	2.01
Average value	1.47	1.96

4. Discussion

Our bench test and analysis of cleaning parameter optimization of a 4 L-2.5 wheat combine harvester are an extension and improvement on previous research on theoptimization of cleaning parameters of wheat combine harvesters and their influence on cleaning quality. To complete the research in this paper, a bench test under closed conditions was used, i.e., excluding the influence of complex working conditions such as weather, environmental conditions, and the driver's level of experience. Such an approach provides more realistic results than a simulation test. The cleaning parameters investigated included all five adjustable parameters that influence the cleaning quality of wheat combine harvesters. Notably, we investigated how to effectively reduce the loss rate and impurity content of wheat harvester cleaning, as well as how to effectively improve the accuracy and timeliness of cleaning parameter setting and adjustment during field operation.

The cleaning parameters of the six groups of experiments (see Table 4) with test numbers 7, 14, 18, 22, 29, and 38 were the same. These six groups of experiments revealed average cleaning loss and impurity rates of 1.59% and 2.15%, respectively. The cleaning loss and impurity rates obtained using the optimum cleaning parameter combination were reduced by 0.12% and 0.19%, respectively, when compared to the experimental data of the optimum cleaning parameter combination.

The apparatus studied in this paper was an air screen cleaning device, similar to those discussed in the literature [11–16]. The cleaning parameters studied in the literature have included shale shaker crank speed and centrifugal fan speed (fan speed) [11], cleaning screen frequency (crank speed of shale shaker), fan wind speed (throttle opening) [12], feed quantity (operation speed), throttle opening, and fan speed [13], feed quantity (operation speed), throttle opening, and fan speed [14], feed quantity (operation speed), throttle opening and fan speed [15], and fan speed and the included angle of the fish scale screen (opening of chaffer) [16]. Given that in each of these papers, only two or three parameters were studied, it may be stated that the optimization of cleaning parameters in this paper and the study of their impact on cleaning quality are more comprehensive.

5. Conclusions

An optimization bench experiment of the cleaning parameters of a wheat combine harvester was completed by using the bench experiment method. The contribution rate method was used to analyze the experimental data, and the influences of operation speed, fan speed, throttle opening, the opening of chaffer, and crank speed of the shale shaker on thecleaning quality were determined. The contribution hierarchy of cleaning parameters relative to cleaning loss rate is as follows: crank speed of shale shaker > opening of chaffer > operation speed > fan speed > throttle opening. Meanwhile, the contribution hierarch of the cleaning parameters relative to impurity rate is as follows: operation speed > fan speed > throttle opening > crank speed of shale shaker > opening of chaffer. The response effects of five cleaning parameters on cleaning loss rate and impurity rate were obtained by analyzing the response surface graph. An optimization model was established, and a response surface data analysis was carried out using Design Expert software. The optimal combination of cleaning parameters (operation speed—2.2 m/s; opening of chaffer—26 mm; throttle opening—20°; fan speed—1100 r/min; and crank speed of shale shaker—350 r/min) was predicted when the cleaning loss and impurity rates were at the minima and the feed quantity was at the maximum. Subsequently, using a 4 L-2.5 threshing and cleaning experiment table, a bench experiment intended to determine the optimum cleaning parameters combination was carried out. The result revealed a cleaning loss rate of 1.47% and an impurity rate of 1.96%. Compared with the evaluation indexes of the cleaning quality of the combine harvester with routine cleaning parameters, the cleaning loss rate was reduced by 0.12% and the impurity rate by 0.19%. The results of this paper provide a theoretical basis for the research and development of a self-adaption cleaning system for wheat combine harvesters.

Author Contributions: Conceptualization, P.L.; methodology, P.L.; formal analysis, P.L.; investigation, P.L.; data curation, P.L.; writing—original draft, P.L.; visualization, P.L.; supervision, X.W.; resources, C.J. All authors have read and agreed to the published version of the manuscript.

Funding: This research was funded by the National Natural Science Foundation of China, grant number 32171911.

Institutional Review Board Statement: Not applicable.

Informed Consent Statement: Not applicable.

Data Availability Statement: Not applicable.

Conflicts of Interest: The authors declare no conflict of interest.

References

1. Xu, L.; Li, Y.; Li, Y.; Chai, X.; Qiu, X. Research progress on cleaning technology and device of grain combine harvester. *Trans. Chin. Soc. Agric. Mach.* **2019**, *50*, 1–16.
2. Guo, H.; Jin, C.; Yin, X.; Kang, Y.; Ning, X.; Liu, P. Research status of the cleaning system of the corn combine harvester. *J. Chin. Agric. Mech.* **2020**, *41*, 26–36.
3. He, X.; Chen, X.; Qu, Z.; Lan, M.; Wang, W. Review on application research and development of intelligent technology in the process of wheat harvest. *J. Henan Agric. Univ.* **2022**, *56*, 341–354.
4. Yang, M.; Hu, Z.; Zhang, Y.; Xu, H.; Gu, F.; Wu, F. Research status and prospect of pneumatic cleaning device for agricultural granular materials. *J. Chin. Agric. Mech.* **2020**, *41*, 121–127. [CrossRef]
5. Wu, P. Cleaning technology and adaptability analysis of grain combine harvester. *Agric. Mach. Us Main.* **2020**, 36–37.
6. Liu, D.; Lu, W.; Wang, X.; Li, X.; Xie, F. Research status and development trend of cyclone separating cleaning unit of combine harvesters. *Agric. Equipm. Veh. Eng.* **2016**, *54*, 8–13.
7. Su, T.; Han, Z.; Cui, J.; Wang, G.; Hao, X.; Hao, F.; Han, K. Research status and development trend of cleaning unit of cereal combine harvesters. *J. Agric. Mech. Res.* **2016**, *38*, 6–11.

8. Liu, P.; Jin, C.; Yin, X.; Ning, X.; Li, Q. Research progress of soybean combine harvester cleaning equipment and key technologies. *Acta Agric. Zhejiangensis* **2019**, *31*, 1758–1766.
9. Liu, P.; Jin, C.; Liu, Z.; Zhang, G.; Cai, Z.; Kang, Y.; Yin, X. Optimization of field cleaning parameters of soybean combine harvester. *Trans. Chin. Soc. Agric. Eng.* **2020**, *36*, 35–45.
10. Ning, X.; Jin, C.; Yin, X.; Liu, P.; Li, Q. Research status and development trend of air-and-screen cleaning device for cereal combine harvesters. *J. Chin. Agric. Mech.* **2018**, *39*, 5–10.
11. Zhong, T.; Hu, Z.; Gu, F.; Cao, M.; Chen, N.; Wang, H. Optimization and experiment for threshing and cleaning parts of 4LZ-1.0 Q cereal combine harvester. *Trans. Chin. Soc. Agric. Mach.* **2012**, *43*, 76–81.
12. Geng, D.; Mu, X.; Zhang, G.; Wang, Z.; Zhu, J.; Xu, H. Analysis and optimization of cleaning mechanism of wheat combine harvester. *J. Jilin Univ. (Eng. Technol. Ed.)* **2022**, *52*, 219–230.
13. Jin, C.; Li, Q.; Ni, Y.; Wang, T.; Yin, X. Experimental study on double air outlet multi-ducts cleaning device of wheat combine harvester. *Trans. Chin. Soc. Agric. Eng.* **2020**, *36*, 26–34.
14. Jin, C.; Li, Q. Test and parameter optimization of regaining and cleaning in multiple-pass cleaning device. *J. Jiangsu Univ. (Nat. Sci. Ed.)* **2021**, *42*, 331–338.
15. Li, Q.; Jin, C.; Teng, Y.; Ning, X.; Liu, P. Experimental optimization of main parameters of multi-channel centrifugal fan with double outlet. *J. Agric. Mech. Res.* **2021**, *43*, 190–196.
16. Tong, S.; Shen, Q.; Tang, N.; Jia, Y.; Cong, F.; Gu, W. Numerical simulation and optimization experiment of mixed flow field on longitudinal axial flow cleaning device. *Tran. Chin. Soc. Agric. Mach.* **2016**, *47*, 135–142.
17. Zhang, K.; Fan, H.; Sun, B.; Chai, Q. The CFD-DEM gas-solid coupling simulation and experimental verification of cleaning device of wheat combine harvester for intercropping system. *Agric. Res. Arid. Areas* **2019**, *37*, 268–274.
18. Shi, Q.; Ma, M.; Yan, W.; Zhou, H.; Yuan, H.; Li, Y. Two-stage winnower cyclone separating cleaning system performance testing and optimization. *Trans. Chin. Soc. Agric. Mach.* **2014**, *45*, 124–128.
19. Shi, Q.; Zhang, X.; Geng, L.; Wang, S.; Lu, X. Experiment of cleaning system of micro grain combine harvester. *J. Henan Univ. Sci. Technol. (Nat. Sci.)* **2015**, *36*, 9, 82–86.
20. Shi, Q.; Xu, Z.; Geng, L.; Wang, S.; Luo, H. Composite device for new grain transportation separation and cleaning. *J. Henan Univ. Sci. Technol. (Nat. Sci.)* **2017**, *38*, 7, 61–66.
21. Ni, C.; Zhang, L.; Liu, S.; Shi, Q.; Gao, C.; Geng, L. Experimental analysis on cyclone separating cleaning system of no-guide vanes. *Trans. Chin. Soc. Agric. Eng.* **2008**, *24*, 135–138.
22. Liu, S.; Zhang, L.; Shi, Q.; Ni, C.; Wang, G.; Geng, L. Experimental research on cyclone separating cleaning system used on micro-combine harvester of wheat. *Trans. Chin. Soc. Agric. Mach.* **2006**, *37*, 45–48.
23. Teng, Y.; Jin, C.; Chen, Y.; Liu, P.; Yin, X.; Wang, T.; Yu, K. Design and optimization of segmented threshing device of combine harvester for rice and wheat. *Trans. Chin. Soc. Agric. Eng.* **2020**, *36*, 1–12.
24. Wang, J.; Wang, D.; Shang, S.; Wang, Y.; Kuai, J. Development and experiment on 4LZZ-1.0 type plot grain combine. *Trans. Chin. Soc. Agric. Eng.* **2016**, *32*, 19–25.
25. Fu, J.; Zhang, Y.; Cheng, C.; Chen, Z.; Tang, X.; Ren, L. Design and experiment of bow tooth of rigid flexible coupling for wheat threshing. *J. Jilin Univ. (Eng. Technol. Ed.)* **2020**, *50*, 730–738.
26. Li, X.; Han, Z.; Dai, F.; Gao, A.; Wei, L. Simulation research on working parameters of threshing device on plot-bred wheat based on EDEM. *Agric. Res. Arid. Areas* **2016**, *34*, 292–298.
27. Zhou, H.; Shi, Q.; Geng, L.; Ma, M.; Yuan, H.; Li, Y. Experiment of wheat supported threshing apparatus based on free-cutting harvest. *Trans. Chin. Soc. Agric. Mach.* **2013**, *44*, 31, 52–56.
28. Li, X.; Lei, Y.; Shi, Q.; Geng, L.; Wang, S. Experiment of double row wheat threshing device without cutting. *J. Henan Univ. Sci. Technol. (Nat. Sci.)* **2016**, *37*, 72–77.
29. Tang, Z.; Li, Y.; Xu, L.; Li, H.; Pang, J. Experiment and evaluating indicators of wheat threshing and separating on test-bed of longitudinal axial-threshing unit. *Trans. Chin. Soc. Agric. Eng.* **2012**, *28*, 14–19.
30. Tang, Z.; Li, Y.; Xu, L.; Pang, J.; Li, H. Experimental study on wheat feed rate of tangential-axial combine harvester. *Trans. Chin. Soc. Agric. Eng.* **2012**, *28*, 26–31.
31. Wei, L.; Dai, F.; Han, Z.; Li, X.; Gao, A. Experiment on plot wheat breeding combine harvester. *Acta Agric. Zhejiangensis* **2016**, *28*, 1082–1088.
32. Zhai, Y. Electrical control circuit optimization on theself-propelled wheat combine harvester. *J. Agric. Mech. Res.* **2019**, *41*, 233–237.
33. Meng, F.; Jiang, M.; Geng, D.; Lin, L.; Xu, H. The design of longitudinal-axial cylinder for the combine. *J. Agric. Mech. Res.* **2019**, *41*, 90–94.
34. Dai, F.; Han, Z.; Zhao, W.; Zhang, F.; Gao, A.; Wei, L. Design and experiment on threshing and separating unit of plot breeding wheat with double longitudinal axial cylinder. *Res. Agric. Mod.* **2016**, *37*, 1015–1020.

applied sciences

MDPI

Article

Impact of Combine Harvester Technological Operations on Global Warming Potential

Dainius Savickas *, Dainius Steponavičius, Liudvikas Špokas, Lina Saldukaitė and Michail Semenišin

Institute of Agricultural Engineering and Safety, Agriculture Academy, Vytautas Magnus University, LT-53362 Kaunas, Lithuania; dainius.steponavicius@vdu.lt (D.S.); liudvikas.spokas@vdu.lt (L.Š.); lina.saldukaite@vdu.lt (L.S.); michailsem@gmail.com (M.S.)
* Correspondence: dainius.savickas@vdu.lt

Abstract: The agricultural machinery is making a considerable negative contribution to the acceleration of global warming. In this study, we analyzed the impact of combine harvesters (CHs) on the global warming potential (GWP) by evaluating the telematics data from 67 CHs operating in Lithuania and Latvia between 2016 and 2020. This study examined the use of their technological operations and the associated impacts on ambient air and performed field tests using the same CH model to determine the composition of exhaust gases and the impact of different technological operations on GWP. The data confirmed the release of significant GWP during indirect operation, and it was estimated that considerable lengths of time were spent in idle (~20%) and transport (~13%) modes. During these operations, over 13% of the total GWP (~27.4 t year^{-1} per CH), affected by emissions, was released. It was calculated that a GWP reduction exceeding 1 t year^{-1} per machine can be achieved by optimizing the idling and transport operations. The dual telematics/field test data approach facilitates a comprehensive assessment of both the impact of CH exhaust gases on GWP and the methods for reducing the negative impact on the environment.

Keywords: air pollution; exhaust gas; telematics system; nitric oxide; carbon dioxide

Citation: Savickas, D.; Steponavičius, D.; Špokas, L.; Saldukaitė, L.; Semenišin, M. Impact of Combine Harvester Technological Operations on Global Warming Potential. *Appl. Sci.* **2021**, *11*, 8662. https://doi.org/10.3390/app11188662

Academic Editors: José Miguel Molina Martínez, Ginés García-Mateos and Dolores Parras-Burgos

Received: 27 August 2021
Accepted: 16 September 2021
Published: 17 September 2021

1. Introduction

One of today's critical challenges is to meet the needs of a growing human population with limited resources, while prioritizing both global food security and sustainability (based on increasing crop yields) within the context of reducing environmental impacts [1]. Simultaneously, the potential for the efficient use of soil by future generations must be increased [2]. The urgent threat of climate change requires specific action, especially by the main emitters of greenhouse gases (GHGs), like agriculture, which is closely associated with environmental pollution. Agricultural activities are responsible for around 24% of GHGs [3], which often lead to water quality deterioration, wastage of water resources, and loss of biodiversity [4]. According to the latest Food and Agriculture Organization reports on agricultural GHG emissions, the world has almost doubled GHG in the last 50 years, and this amount could increase by another 30% by 2050 [5]. In terms of agricultural technologies, diesel fuels and fertilizers account for most of the energy consumed [6,7]; furthermore, fertilizers and pesticides have been identified among the most important secondary sources of CO_2 emissions [8]. It is estimated that harvesting accounts for up to 30% of the total cost of using agricultural machinery. Reductions in fuel consumption, air pollution, and the associated costs incurred by farmers [9] can be realized by the optimization of work and the correct operation of combine harvesters (CHs). In 2017, the global agricultural sector produced CO_{2eq} emissions of 11.1 gt year^{-1} [10], with Lithuania alone responsible for 4387.0 kt year^{-1} [11]. As an EU member state, Lithuania participates in the global climate change mitigation process, contributes to EU commitments, prepares national documents, participates in the formation of climate change policy, and is one of the 195 countries that have ratified the United Nations Framework Convention on Climate Change.

Rapidly increasing concentrations of atmospheric GHGs, including CO_2, CH_4, and N_2O, are contributing to unprecedented changes in the earth's atmosphere. Collectively, these three GHGs account for over 90% of anthropogenic global warming [12]. GWP is a widely used, convenient, and quantifiable measure of environmental impacts. Thus, this paper uses GWP as a clear indicator of the unified concept. It should be noted that the impact of CH exhaust gases on GWP is primarily influenced by the direct use of diesel fuel and its conversion to CO_2 within internal combustion engines. Other emissions, including N_2O and CH_4, can be reduced by different engine technologies, (e.g., clean-up catalysts, NO control by selective catalytic reduction [13], and CO/HC emissions control by diesel oxidation catalysts), and a direct correlation has been noted between CO_2 emissions and fuel consumption. In order to reduce engine emissions, the following engine improvement systems can also be useful: cylinder shut-off [14,15], start–stop [16,17], and speed reduction [18,19]. Emissions can also be reduced by switching to alternative fuels, such as compressed natural gas (LNG) and liquefied petroleum gas (LPG) [20], or by installing electric hybrid engines into the agricultural machinery [21–23]. However, overall, emissions can only be reduced if less fuel is used [24–26]. Other researchers are also looking at ways to reduce fuel consumption and air pollution when harvesting with CHs [27,28]. In their 2019 research, Špokas et al. revealed that leaving longer-length stubble in the field when harvesting is an acceptable way of reducing fuel consumption (and thus emissions) while preserving the driving speed. They achieved a reduction in hourly fuel consumption of 6.2 l h^{-1}, which in turn resulted in 16.3 kg h^{-1} less CO_2, by increasing the stubble height of oilseed rape from 0.2 to 0.4 m (equating to 7 t h^{-1} less stem mass being fed into the CH). However, research on the environmental impacts (in terms of GHGs or the GWP) of CHs is still lacking.

Thus, this study aimed to evaluate the GWP of the exhaust gases of CHs by assessing different technological operations and to identify methods for its reduction. We substantiate the benefits of the telematics data and field test data analysis method for a comprehensive economic and environmental assessment. We also propose an improved model of a continuous process to achieve economic and environmental goals, presented in a previously published article [29].

2. Data and Methods

2.1. Telematics Data Collection and Analysis

This study analyzed telematics data from CHs (one of the most popular models in the region with a tangential threshing apparatus) operating in Lithuania and Latvia between 2016 and 2020. Access to the relevant databases required a username and password. Such access was obtained from the manufacturer's representative. In order to be able to objectively compare the telematics data with the data obtained during the field tests, a set of telematics data was selected only for CHs of the same model. In total, our filtering conditions were met by 67 different machines (from an array of 239 different machines). The information was then organized into a unified database to statistically process the collected data. In this study, the data for each individual machine for each year were downloaded to a personal computer; these were then imported into a data array for further processing. Only the information needed for the study was selected from the large telematics database, including working time and fuel consumption in different operational modes. A machine's working time (h $year^{-1}$) can be categorized as follows: idle with a partially full grain tank (T1), idle with a full grain tank (T2), unloading and not harvesting (T3), harvesting and unloading (T4), harvesting (T5), headland turn separator engaged (T6), transportation below 16 km h^{-1} (T7), and transportation above 16 km h^{-1} (T8). The categorizations for fuel consumption (l $year^{-1}$) were as follows: idle with a partially full grain tank (F1), idle with full grain tank (F2), unloading and not harvesting (F3), harvesting and unloading (F4), harvesting (F5), headland turn separator engaged (F6), transportation below 16 km h^{-1} (F7), and transportation above 16 km h^{-1} (F8).

In the analysis, the T1 and T2 times were obtained during the CH's idle mode, T7 and T8 were collected during the transport mode, and T3, T4, T5, and T6 were acquired during the CH's working mode. Fuel consumption was calculated: F1 and F2 represent the fuel consumed at idling, F7 and F8 are the quantities of fuel consumed during the transport mode, and F3, F4, F5, and F6 represent the fuel consumed during direct work.

2.2. Calculation Methodology for Greenhouse Gases and Global Warming Potential

For the GHG and GWP calculations, this study converted the fuel consumption data from the telematics database from l $year^{-1}$ and l h^{-1} to kg $year^{-1}$ and kg h^{-1}, respectively. A diesel fuel volume-to-mass conversion factor of 0.832 kg l^{-1} [30–32] was then employed.

The GHG emissions in the telematics data analysis were not measured directly but were assessed using the methodology outlined in Chapter 1.A.4 of the EMEP/EEA's Air Pollutant Emission Inventory Guidebook [33]:

$$E_{pollutant} = FC_{fuel\ type} \times EF_{pollutant}, \quad (1)$$

where $E_{pollutant}$ denotes the emission of a specified GHG (CO_2, N_2O, and CH_4), $FC_{fuel\ type}$ denotes the fuel consumption, and $EF_{pollutant}$ denotes the emission factor of the pollutant (g t^{-1} or kg t^{-1}) of the consumed diesel fuel (Table 1).

Table 1. Greenhouse gas (GHG) emission factors in agricultural transport using diesel fuel.

GHG Emission Factors		
CO_2, kg t^{-1}	N_2O, g t^{-1}	CH_4, g t^{-1}
3160	139	13

The GHG impact on GWP was estimated using Equation (2) [12,34–37]

$$GWP = E_{CO2} + 25 \times E_{CH4} + 298 \times E_{N2O}. \quad (2)$$

The GWP in CO_2 equivalents is such that 1 kg of CO_2 equates to 1 kg of CO_{2eq}, 1 kg of CH_4 equals 25 kg of CO_{2eq}, and 1 kg of N_2O corresponds to 298 kg of CO_{2eq}.

2.3. Field Test Measurement Device for CO_2, O_2, and NO

Concentrations of CO_2, O_2, and NO in the exhaust gases of the CHs during harvesting were analyzed using a hand-held AUTOplus 5-2 (Kane, Welwyn Garden, UK) gas analyzer, which has been used in previous research to analyze the emissions from mobile vehicles [38–41]. Table 2 provides the measurement range and resolution of the analyzer.

Table 2. Measurement range of the individual components of exhaust gases.

Component	Resolution	Accuracy	Measurement Range
CO_2	0.1%	±5% of reading ±0.5% volume	0–16% Over range: 25%
O_2	0.01%	±5% of reading ±0.1% volume	0–21% Over range: 25%
NO	1 ppm	±5% of reading ±25 ppm of volume	0–5000 ppm Over range: 5000 ppm

At the time of the measurements, the analyzer had a valid calibration passport. Calibration was performed using a reference calibration gas. In the field tests, the measuring device was connected to the exhaust pipe via a flexible hose. After the speed (engine load factor) of the CH was modified, the exhaust gas composition was allowed to stabilize for 1 min. Then, 20 measurements were recorded, which were stored in the measuring device at 10 s intervals. The received data were then transferred to a personal computer via KANE LIVE software for further processing.

2.4. Combine Harvester and Test Site Specifications

In 2019 and 2020, field trials were conducted on the land of an agricultural company engaged in crop production. The Wideband Global SATCOM system field coordinates of the study site in 2019 and 2020 were 54.7970, 23.0079. The 2019 study used the winter wheat variety *Ada* for harvesting, whereas *Balitus* was used in 2020. Five samples were collected from 0.25 m^2 areas in five randomly selected field locations to determine the weight of 1 m^2 of the crop. The *Ada* variety weighed 1702.9 ± 209.1 g m^{-2} (grain moisture: 12.2% ± 1.1%), while the weight of the *Balitus* variety was 1370.2 ± 154.6 g m^{-2} (grain moisture: 13.5% ± 1.3%).

The wheat flow mass delivered to the CH was calculated using the following formula:

$$q = CS \times HW \times WM, \quad (3)$$

where q denotes the wheat mass fed to the CH (feed rate) in kg s^{-1}, CS denotes the CH speed during harvesting in m s^{-1}, HW denotes the CH header width in m, and *WM* denotes the mass of the wheat in 1 m^2.

The same CH was used for the 2019 and 2020 field studies. There are already newer and higher environmental standards (i.e., Tier5) CHs. However, we chose to examine one of the most popular machines in the region at the moment. It should also be emphasized that the life cycle of a CH in the field is long, and this type of CH will work for many years to come. Table 3 presents the precise measurement specifications.

Table 3. Combine harvester characteristics.

Indicator	Measurement Unit	Value
Manufacturer	-	John Deere
Model	-	T670i
Year of manufacture	Year	2016
Emission level	-	Final Tier 4
AdBlue©	-	No
Header (cutter bar) width	M	9.14
Threshing cylinder diameter	M	0.66
Threshing channel width	M	1.67
Concave area	m^2	1.35
Cleaning shoe separation area	m^2	6.3
Grain tank capacity	l	11,000
Engine power	kW	292
Engine operational hours during the 2019 study	h	620
Engine operational hours during the 2020 study	h	828

In the field tests for both years, fuel consumption and engine load of the CHs were recorded using data from the machines' on-board computers. For each technological operation or change in harvesting speed, 10 on-board computers display values were recorded at equal 10 s intervals.

2.5. Statistical Analysis

The data were analyzed using Statistica 10.0 (TIBCO Software, Palo Alto, CA, USA) statistical software with a level of 0.05 as the significance criterion.

3. Results and Discussion

3.1. Results of the Telematics Data Analysis

Table 4 provides the summarized results of the telematics data analysis.

Table 4. Combine harvester operating time structure and global warming potential (GWP) in various engine modes.

Year	[1] Machine Utilization, h Year^{-1}	[1] GWP, kg h^{-1}	[1] GWP, kg Year^{-1}
		Idle mode	
2016	20.1 ± 9.0	23.9 ± 1.6	493.0 ± 243.8
2017	40.1 ± 17.7	25.1 ± 1.2	1039.1 ± 522.1
2018	39.7 ± 30.8	22.7 ± 0.9	863.1 ± 384.5
2019	69.2 ± 22.5	22.1 ± 1.1	1354.2 ± 374.0
2020	61.8 ± 13.8	22.2 ± 1.0	1233.5 ± 213.2
[2] 2016–2020	57.2 ± 9.3	22.5 ± 0.5	1166.2 ± 154.0
		Transport mode	
2016	22.3 ± 14.6	69.5 ± 9.4	1489.7 ± 952.9
2017	29.4 ± 8.6	64.6 ± 6.1	1849.0 ± 439.2
2018	26.8 ± 6.30	60.2 ± 4.5	1609.7 ± 323.5
2019	41.1 ± 8.83	63.4 ± 2.9	2654.4 ± 596.9
2020	38.0 ± 5.38	64.4 ± 2.7	2488.6 ± 376.2
[2] 2016–2020	35.9 ± 3.71	63.8 ± 1.6	2319.0 ± 253.2
		Working mode	
2016	97.5 ± 60.0	138.3 ± 7.4	13,618.9 ± 8442.6
2017	154.8 ± 51.7	134.9 ± 7.7	20,301.4 ± 5474.0
2018	147.5 ± 38.1	121.7 ± 5.7	17,833.1 ± 4032.2
2019	202.2 ± 47.7	125.1 ± 3.6	25,131.5 ± 5832.7
2020	209.6 ± 28.7	128.8 ± 2.9	26,828.7 ± 3660.7
[2] 2016–2020	189.0 ± 20.3	127.7 ± 2.0	23,932.0 ± 2516.6
		All modes	
2016	140.0 ± 82.0	107.7 ± 9.6	15,601.6 ± 9586.4
2017	224.3 ± 74.5	105.9 ± 8.2	23,189.5 ± 6269.6
2018	214.0 ± 54.9	95.5 ± 6.0	20,306.0 ± 4618.2
2019	312.5 ± 77.3	96.2 ± 3.7	29,140.1 ± 6704.7
2020	309.5 ± 47.0	101.2 ± 2.7	30,550.8 ± 4189.1
[2] 2016–2020	282.2 ± 32.5	99.7 ± 2.0	27,417.2 ± 2879.2

[1] The arithmetical means of working time and global warming potential were obtained from the telematics database containing data from 67 CHs. [2] 2016–2020: averages of parameters over 5 years.

The study revealed the following key results:

- On average, each CH spent 68.2% ± 1.1% of its time in direct harvesting work between 2016 and 2020. Accordingly, in this mode, a GWP of 87.2% ± 0.7% was released into the environment via the exhaust gases. In the other modes (idle/transport), the emission of air pollutants occurred while the CH was engaged in indirect work. A similar value was calculated in previous studies and for other CH models [42].
- The maximum recorded engine operation time of a single CH per year (in all modes) was 1036.9 h year^{-1} with a minimum of 18.1 h year^{-1}.
- The highest total annual GWP (in all modes) of a single CH was 94,262.7 kg year^{-1}, while the lowest was 1974.7 kg year^{-1}.
- Despite significant differences in absolute values between individual machines, the overall proportions of time spent engaged in direct work and in idle/transport modes were deemed comparable. The calculated correlation coefficient between the work (TW) and the idle (TI) modes was R^2 = 0.88 ($TW = 2.03TI + 72.87$), whereas the coefficient between the work and the transport (TT) modes was R^2 = 0.75 ($TW = 4.73TT + 18.88$). The calculated correlation coefficient between the GWP emitted in the work (FW)

and idle (*FI*) modes was $R^2 = 0.81$ ($FW = 14.72FI + 6761.8$), while the coefficient between the GWP emitted in the work and transport (*FT*) modes was $R^2 = 0.75$ ($FW = 8.61FT + 3961.1$). These strong correlations indicate a stable relationship between the different modes (idle, transport, and work); however, they can also suggest that implementing changes to reduce both indirect working time and GWP emissions can be problematic.

The data analysis revealed the potential to reduce both the duration of indirect work and the associated negative environmental impacts. However, previous research has identified the influence of poorly organized work, inappropriate operator practices, and lack of knowledge and skills in the inefficient use of agricultural machinery [43,44].

3.2. *Influence of Technological Operations on the Global Warming Potential in Idle, Transport, and Harvesting Modes*

It can be concluded from the telematics data that during idling, the length of time spent in this mode (20.3%) and the GWP released (4.3%) account for fairly significant shares of the totals. By comparison, the idle time of other non-road machines can be between 20% and 70% [45–47]. The 2019 field tests explored the impact of fuel consumption on the GWP and the use of various technological operations at the idle speed of the CH engine. Figure 1 presents the obtained results.

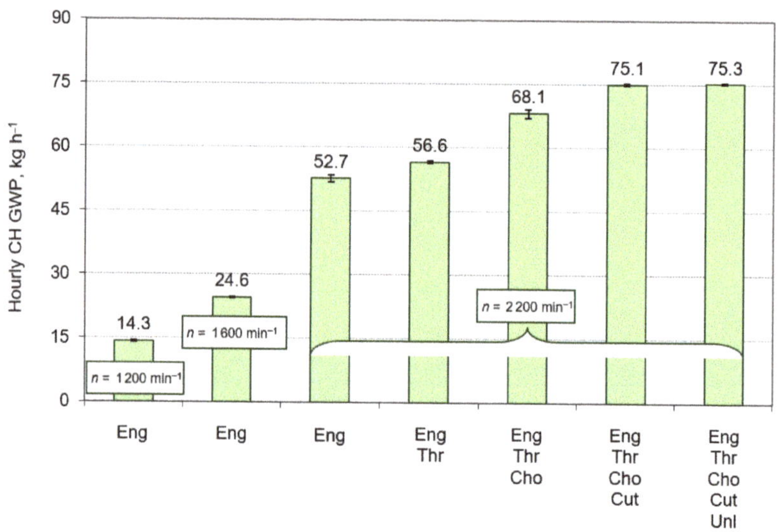

Figure 1. Hourly global warming potential (GWP) of a combine harvester (CH) in idle mode engaging different operational units (Eng = engine, Thr = threshing unit, Cho = straw chopper unit, Cut = cutter bar unit, and Unl = grain unloading unit, where n = engine crankshaft speed) in the 2019 tests. The value of each column is derived from an average of 20 replicates.

These results confirm that engine speed had the most significant impact on the hourly GWP in idle mode. An increase in engine speed from 1200 to 2200 min^{-1} represented a very significant 2.7-fold increase in GWP emissions. However, other operations produced lower increases in GWP emissions, as follows:

- Cereal threshing unit: 7.5%
- Straw chopper unit: 20.3%
- Cutter bar unit: 10.2%
- Grain unloading unit: 0.3%

Generally, maximum reductions should be made in the duration of idling; however, where this is not possible, engine speed should be reduced, and unused technological operations should thus be terminated.

During the transport mode, the length of time and the GWP accounted for 13.2% and 8.8% of the totals, respectively. In 2019, the same field tests explored the relationships between engine speed, the use of technological operations, driving speed, and GWP. Figure 2 provides the obtained results.

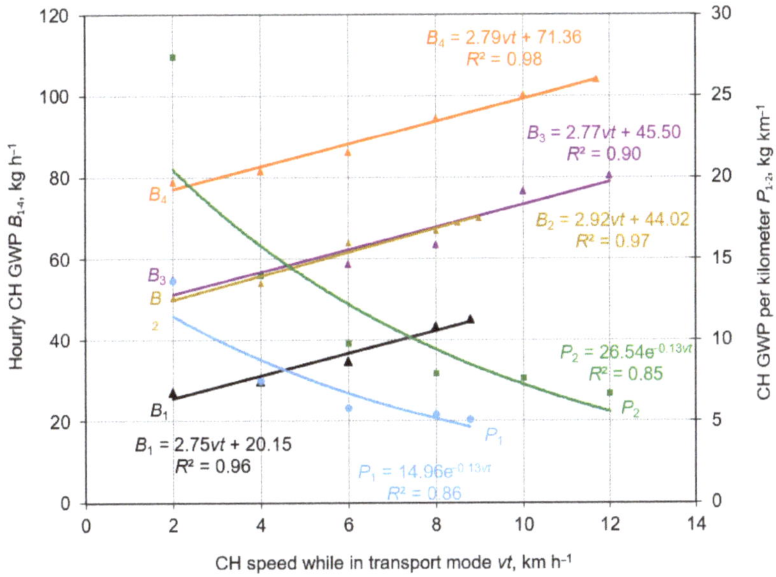

Figure 2. Relations between combine harvester (CH) speed in transport mode and global warming potential (GWP) in the 2019 tests (stubble height, 15 cm; soil moisture, 16.2%): B_1—All gears are switched off. Engine crankshaft speed n = 1690 min^{-1} B_2—Gears engaged: header, threshing unit, and straw chopper. n = 1690 min^{-1} B_3—All gears are switched off. n = 2340 min^{-1} B_4—Gears are engaged: header, threshing unit, and straw chopper. n = 2200 min^{-1} P_1—All gears are switched off. n = 1690 min^{-1} P_2—All gears are switched off. n = 2340 min^{-1} The value of each point was obtained from an average of 20 replicates.

According to the outcomes of the 2019 field test study, the absolute numerical value of the GWP increased commensurately with driving speed, while the GWP per km was reduced. The graphs in Figure 2 indicate that when the CH moves between fields, it is preferable to terminate any technological operations, reduce the engine speed, and increase the driving speed.

The telematics data collected over a 5-year period from 67 CHs of the same model were compared with data from a specific CH obtained during field research. From the results given in Table 4 and Figure 1, the hourly GWP in idle mode can be potentially lower and equal to the 14.29 kg h^{-1} achieved during the tests, although an analysis of the 5-year data revealed the value to be 22.54 kg h^{-1}. Considering that each CH spends an estimated average of 57.18 h year^{-1} in idle mode, a potential GWP reduction of 471.7 kg year^{-1} per machine is possible. This figure is even higher for the transport mode, for which the 5-year data provide an average hourly GWP of 63.84 kg h^{-1}. Switching off all technological gears, reducing the engine speed to 1690 min^{-1}, and driving at a speed of 8 km h^{-1} can achieve a GWP of 44.91 kg h^{-1} (Figure 2). The average time spent in the transport mode is 35.94 h year^{-1}, which allows for a potential GWP reduction of 680.3 kg year^{-1}. By optimizing the operation of CHs during their idling and transport modes, a potential GWP reduction of 1.15 t year^{-1} per machine is achievable. It should be stressed that reductions in GWP will not only produce positive environmental impacts but also have direct financial benefits for farmers.

The GWP per ton of wheat processed (Figure 3) was barely affected by changes to the speed of the threshing cylinder during threshing, particularly when the CH reached its optimal working speed (approx. 4–5 km h^{-1}). At a harvesting speed of 1 km h^{-1} with threshing cylinder rotational speeds of 850 and 950 min^{-1}, a 12.42% difference was noted in GWP. However, after reaching a maximum (and typical) harvesting speed of 5 km h^{-1}, the GWP difference was 3.88%. A decrease in GWP per ton of wheat processed was realized by significantly increasing the running speed of the CH during harvesting; a comparison between running speeds of 1 and 5 km h^{-1} at a threshing cylinder speed of 950 min^{-1} resulted in a significant decrease in GWP by 60.08%. In fact, the driving speed during harvesting (or, more precisely, the supplied wheat crop flow in the threshing apparatus and the engine load) is the most important economic and environmental factor. If the selected driving speed is too low, the GWP per ton of raw material processed will be very high. Conversely, if the speed is too high (at the same time overloading the engine), we will lose part of the crop due to grain damage and grain separation losses. The latest CH models have built-in technologies that automatically select the driving speed (at the same time, the supplied crop in the threshing apparatus and the engine load) in order to achieve optimal fuel consumption and reduce the environmental impact. In our study, assessing these factors and making decisions depend on the CH driver. That is why it is important to research and publicize research results and properly educate CH drivers.

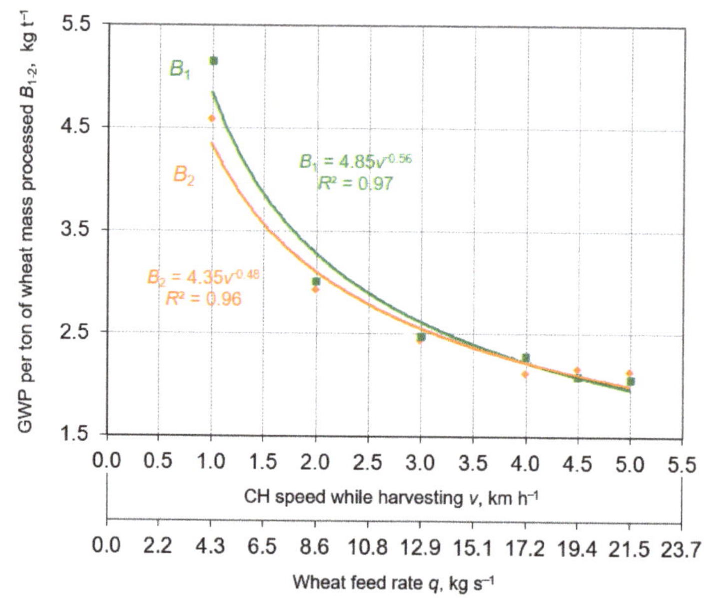

Figure 3. Relation between combine harvester (CH) speed in harvesting mode and global warming potential (GWP) per ton of wheat mass processed in the 2019 tests (engine crankshaft speed n = 2200 min^{-1}): B_1—threshing cylinder rotational speed nt = 950 min^{-1} B_2—nt = 850 min^{-1} The value of each point was obtained from an average of 20 replicates.

3.3. Link between Global Warming Potential in Threshing Mode and Exhaust Gas Concentrations and Engine Load

Figure 4 presents the results of the integrated field research conducted in 2020. The data show a connection between the GWP per ton of wheat processed, the exhaust gas content/concentration, the driving speed during harvesting, the feed rate, and the engine load factor. The most obvious result is that both NO content and CO_2 concentration increased commensurately with harvesting speed (and thus engine load), while the concentration of O_2 decreased. When comparing the values obtained at the slowest and fastest harvesting speeds (1 and 4.5 km h^{-1}, respectively), the following differences are noted:

- O_2 concentration decreased by 23.4%
- CO_2 concentration increased by 26.7%
- NO concentration increased by 118.9%

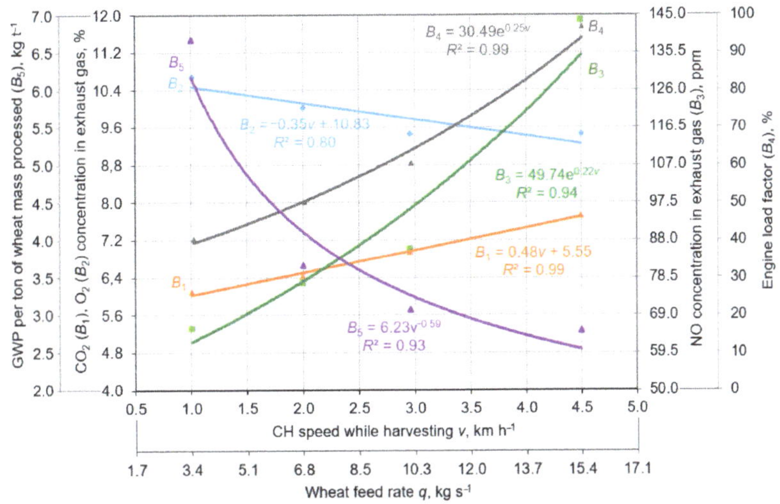

Figure 4. Relationships between combine harvester (CH) speed, wheat feed rate, engine load factor, global warming potential (GWP) per ton of wheat mass processed, and exhaust gas concentrations during threshing in the 2020 tests. The value of each point was obtained from an average of 20 replicates.

These variations are attributed to the fact that, as the speed of the CH increases, the amount of wheat flow per unit of time also increases. Simultaneously, there is an increase in engine load and corresponding rises in both gas compression and temperature inside the internal combustion engine's cylinder. Increases also occur in the interactions between N, O, and C in both ambient air and diesel fuel. The consequences of heat and pressure cause O_2 in the exhaust gases to decrease and CO_2 and NO levels to rise [48].

Despite the research estimates indicating that the lowest GWP per ton of wheat processed during harvesting is achieved at top operational speeds (Figures 3 and 4), there is no definitive conclusion or solution. Furthermore, when a CH reaches a certain critical speed during harvesting, there is an increase in both grain damage and grain separation losses [32]; therefore, other researchers indicate the need to include evaluations of both grain damage and grain loss [49]. Future research should investigate the determination of optimal harvesting speeds by comprehensively estimating GWP, grain damage, and grain losses.

After analyzing the accumulated multi-year telematics data and conducting field tests on a specific CH, we clearly see that there is potential for a more efficient use of the machine. We can propose a process that would allow continuous improvement to achieve economic and environmental goals (Figure 5). The proposed process has several scenarios and may include both machine data analysis with and without field testing to determine maximum measures to reduce fuel consumption, make optimal use of expensive agricultural machinery, and reduce GWP. The process offers the possibility of data analysis (only the data of a specific machine can be analyzed, or the whole array of machines of the same model can be compared), which shows the real situation in time and fuel consumption. The data obtained during the field tests are compared with the values recorded in the telematics system, and the possibilities of optimal work are analyzed. The analysis shall be followed by specific measures affecting economic and environmental factors, including, but not limited to, the training of CH operators, the regulation of machinery, and a better organization of ancillary transport.

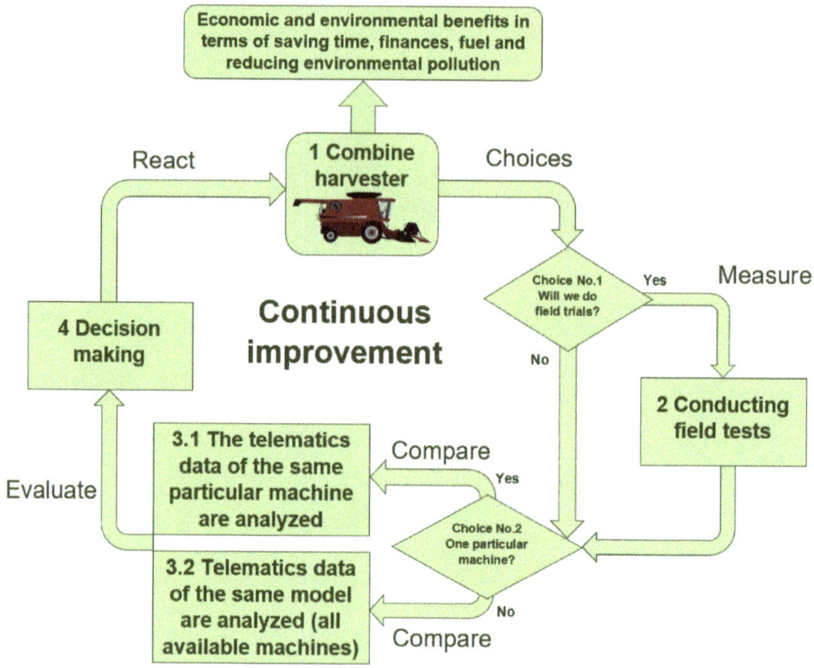

Figure 5. Dual telematics/field test data analysis method process to reduce fuel consumption and environmental pollution.

4. Conclusions

During a CH's relatively short annual operation (~282 h year^{-1}), substantial quantities of GWP (~27 t year^{-1}) are emitted. This study estimated the GWP released during the idle and transport modes to be significant (~1.2 and ~2.3 t year^{-1}, respectively) and confirmed that GWP can be reduced by 32.9% during idling and transportation by decreasing engine speeds and disabling unnecessary process operations. Additionally, it was confirmed that increasing the driving speed during the transport mode produces a reduction in GWP per kilometer travelled; an increase in speed from 2 to 12 km h^{-1} can cause the GWP to decrease from 27.4 to 6.7 kg km^{-1}. Furthermore, a fuel consumption–exhaust emission analysis revealed that the GWP per unit of wheat mass processed decreases significantly at higher harvesting speeds, and an estimated speed of $\geq$3 km h^{-1} with a wheat feed rate of at least 10 kg s^{-1} was proposed. It should be emphasized that driving speed, feed rate, and engine load during harvesting are the most important economic and environmental factors. Changes in the composition of exhaust gases are related directly to the driving speed during harvesting; the latter is also closely correlated with the load factor of the engine. This study also practically investigated the dependencies of changes in CO_2 (6.1–7.7%) and NO concentrations (65–143 ppm) on engine load (which ranged between 40% and 96% during harvesting). As a result of the research, the authors propose a wider application of the dual telematics/field test data analysis method. It makes it possible to pursue both economic and environmental goals.

Author Contributions: D.S. (Dainius Savickas), D.S. (Dainius Steponavičius), L.Š., L.S., and M.S.; Conceptualization, D.S. (Dainius Savickas), D.S. (Dainius Steponavičius), L.Š, L.S., and M.S.; methodology, D.S. (Dainius Savickas) and D.S. (Dainius Steponavičius); software, D.S. (Dainius Savickas); validation, D.S. (Dainius Savickas) and D.S. (Dainius Steponavičius); formal analysis, D.S. (Dainius Savickas), D.S. (Dainius Steponavičius), L.S. and M.S.; investigation, D.S. (Dainius Savickas), D.S. (Dainius Steponavičius), and L.Š.; resources, D.S. (Dainius Savickas) and D.S. (Dainius Steponavičius); data curation, D.S. (Dainius Savickas), D.S. (Dainius Steponavičius), and L.Š.; writing—original draft preparation, D.S. (Dainius Savickas); writing—review and editing, D.S. (Dainius Savickas), D.S. (Dainius Steponavičius); visualization, D.S. (Dainius Savickas); supervision, D.S. (Dainius Steponavičius). All authors have read and agreed to the published version of the manuscript.

Funding: This research received no external funding.

Institutional Review Board Statement: Not applicable.

Informed Consent Statement: Not applicable.

Data Availability Statement: The data that support the findings of this study are available from the corresponding author, D. Savickas, upon reasonable request.

Conflicts of Interest: The authors declare no conflict of interest.

References

1. O'Donoghue, C.; Chyzheuskaya, A.; Grealis, E.; Kilcline, K.; Finnegan, W.; Goggins, J.; Hynes, S.; Ryan, M. Measuring GHG Emissions across the Agri-Food Sector Value Chain: The Development of a Bioeconomy Input-Output Model. *Int. J. Food Syst. Dyn.* **2019**, *10*, 55–85. [CrossRef]
2. Davis, K.F.; Gephart, J.A.; Emery, K.A.; Leach, A.M.; Galloway, J.N.; D'Odorico, P. Meeting Future Food Demand with Current Agricultural Resources. *Glob. Environ. Chang.* **2016**, *39*, 125–132. [CrossRef]
3. Sykes, A.; Topp, C.; Rees, R. Modelling nutrient cycles in agriculture and their environmental impacts. In *Assessing the Environmental Impact of Agriculture*; Weidema, B., Ed.; Burleigh Dodds Science Publishing Limited: Series in Agricultural Science: Cambridge, UK, 2019; Volume 1, p. 69, ISBN 13: 9781786762283.
4. Garnett, T. Where Are the Best Opportunities for Reducing Greenhouse Gas Emissions in the Food System (Including the Food Chain)? *Food Policy* **2011**, *36*, S23–S32. [CrossRef]
5. Cordes, H.; Iriarte, A.; Villalobos, P. Evaluating the Carbon Footprint of Chilean Organic Blueberry Production. *Int. J. Life Cycle Assess.* **2016**, *21*, 281–292. [CrossRef]
6. Šarauskis, E.; Masilionytė, L.; Juknevičius, D.; Buragienė, S.; Kriaučiūnienė, Z. Energy Use Efficiency, GHG Emissions, and Cost-Effectiveness of Organic and Sustainable Fertilisation. *Energy* **2019**, *172*, 1151–1160. [CrossRef]
7. Hoffman, E.; Cavigelli, M.A.; Camargo, G.; Ryan, M.; Ackroyd, V.J.; Richard, T.L.; Mirsky, S. Energy Use and Greenhouse Gas Emissions in Organic and Conventional Grain Crop Production: Accounting for Nutrient Inflows. *Agric. Syst.* **2018**, *162*, 89–96. [CrossRef]
8. Lal, R. Carbon Emission from Farm Operations. *Environ. Int.* **2004**, *30*, 981–990. [CrossRef]
9. Sørensen, C.G. Workability and Machinery Sizing for Combine Harvesting. *Agric. Eng. Int. CIGR J. Sci. Res. Dev.* **2003**, *5*. PM 03.003.
10. FAO. The Contribution of Agriculture to Greenhouse Gas Emissions. Available online: http://www.fao.org/resources/infographics/infographics-details/en/c/218650/ (accessed on 25 August 2021).
11. *Lithuania's Greenhouse Gas Inventory Report 2020 Greenhouse Gas Emissions 1990–2018*; Environmental Protection Agency: Vilnius, Lithuania, 2020; p. 567.
12. Tian, H.; Chen, G.; Lu, C.; Xu, X.; Hayes, D.J.; Ren, W.; Pan, S.; Huntzinger, D.N.; Wofsy, S.C. North American Terrestrial CO_2 Uptake Largely Offset by CH_4 and N_2O Emissions: Toward a Full Accounting of the Greenhouse Gas Budget. *Clim. Chang.* **2015**, *129*, 413–426. [CrossRef]
13. Resitoglu, I.A.; Keskin, A.; Özarslan, H.; Bulut, H. Selective Catalytic Reduction of NOx Emissions by Hydrocarbons over Ag–Pt/Al_2O_3 Catalyst in Diesel Engine. *Int. J. Environ. Sci. Technol.* **2019**, *16*, 6959–6966. [CrossRef]
14. Gutarevych, Y.; Mateichyk, V.; Matijošius, J.; Rimkus, A.; Gritsuk, I.; Syrota, O.; Shuba, Y. Improving Fuel Economy of Spark Ignition Engines Applying the Combined Method of Power Regulation. *Energies* **2020**, *13*, 1076. [CrossRef]
15. Vopa, J.; Fridrichov, K.; Dlugo, J. Overview of the Potential and Limitations of Cylinder Deactivation. *Renew. Sustain. Energy Rev.* **2021**, *146*. [CrossRef]
16. Lijewski, P.; Szymlet, N.; Fuć, P.; Domowicz, A.; Rymaniak, Ł. The Effect of Start-Stop Systems on Scooter Exhaust Emissions. *Transp. Res. Part D Transp. Environ.* **2021**, *91*. [CrossRef]
17. Mueller, N.; Strauss, S.; Tumback, S.; Goh, G.-C.; Christ, A. Next Generation Engine Start/Stop Systems: "Free-Wheeling". *SAE Int. J. Engines* **2011**, *4*, 874–887. [CrossRef]

18. Warguła, Ł.; Krawiec, P.; Walus, K.J.; Kukla, M. Fuel Consumption Test Results for a Self-Adaptive, Maintenance-Freewood Chipper Drive Control System. *Appl. Sci.* **2020**, *10*, 2727. [CrossRef]
19. Warguła, Ł.; Kukla, M.; Lijewski, P.; Dobrzynski, M.; Markiewicz, F. Influence of Innovative Woodchipper Speed Control Systems on Exhaust Gas Emissions and Fuel Consumption in Urban Areas. *Energies* **2020**, *13*, 3330. [CrossRef]
20. Warguła, Ł.; Kukla, M.; Lijewski, P.; Dobrzyński, M.; Markiewicz, F. Impact of Compressed Natural Gas (CNG) Fuel Systems in Small Engine Wood Chippers on Exhaust Emissions and Fuel Consumption. *Energies* **2020**, *13*, 6709. [CrossRef]
21. Qin, Z.; Luo, Y.; Li, K.; Peng, H. Optimal Design of a Novel Hybrid Electric Powertrain for Tracked Vehicles. *Energies* **2017**, *10*, 2141. [CrossRef]
22. Troncon, D.; Alberti, L. Case of Study of the Electrification of a Tractor: Electric Motor Performance Requirements and Design. *Energies* **2020**, *13*, 2197. [CrossRef]
23. Moreda, G.P.; Muñoz-García, M.A.; Barreiro, P. High Voltage Electrification of Tractor and Agricultural Machinery—A Review. *Energy Convers. Manag.* **2016**, *115*, 117–131. [CrossRef]
24. Juostas, A.; Janulevičius, A. Tractor's Engine Efficiency and Exhaust Emissions' Research in Drilling Work. *J. Environ. Eng. Landsc. Manag.* **2014**, *22*, 141–150. [CrossRef]
25. Fontaras, G.; Zacharof, N.G.; Ciuffo, B. Fuel Consumption and CO_2 Emissions from Passenger Cars in Europe—Laboratory versus Real-World Emissions. *Prog. Energy Combust. Sci.* **2017**, *60*, 97–131. [CrossRef]
26. Helmers, E.; Leitão, J.; Tietge, U.; Butler, T. CO_2-Equivalent Emissions from European Passenger Vehicles in the Years 1995–2015 Based on Real-World Use: Assessing the Climate Benefit of the European "Diesel Boom". *Atmos. Environ.* **2019**, *198*, 122–132. [CrossRef]
27. Špokas, L.; Steponavičius, D. Fuel Consumption during Cereal and Rape Harvesting and Methods of Its Reduction. *J. Food Agric. Environ.* **2011**, *9*, 257–263.
28. Milan, F.; Antonín, D.; Ivo, C.; Martin, F.; Petr, B. Harvest of Cereals and Oilseed Rape by Combine Harvesters New Holand CX 8090 and New Holand CR 9080. *Poljopr. Teh.* **2017**, *42*, 19–24.
29. Savickas, D.; Steponavičius, D.; Domeika, R. Analysis of Telematics Data of Combine Harvesters and Evaluation of Potential to Reduce Environmental Pollution. *Atmosphere* **2021**, *12*, 674. [CrossRef]
30. Pandian, A.K.; Ramakrishnan, R.B.B.; Devarajan, Y. Emission Analysis on the Effect of Nanoparticles on Neat Biodiesel in Unmodified Diesel Engine. *Environ. Sci. Pollut. Res.* **2017**, *24*, 23273–23278. [CrossRef]
31. Chinnasamy, C.; Tamilselvam, P.; Ranjith, R. Influence of Aluminum Oxide Nanoparticle with Different Particle Sizes on the Working Attributes of Diesel Engine Fueled with Blends of Diesel and Waste Plastic Oil. *Environ. Sci. Pollut. Res.* **2019**, *26*, 29962–29977. [CrossRef]
32. Špokas, L.; Steponavičius, D.; Žebrauskas, G.; Čiplienė, A.; Bauša, L. Reduction in Adverse Environmental Impacts Associated with the Operation of Combine Harvesters during the Harvesting of Winter Oilseed Rape. *J. Environ. Eng. Landsc. Manag.* **2019**, *27*, 72–81. [CrossRef]
33. EMEP/EEA. *Air Pollutant Emission Inventory Guidebook*; European Environment Agency: Copenhagen, Denmark, 2019. Available online: https://www.eea.europa.eu/publications/emep-eea-guidebook-2019 (accessed on 23 August 2021).
34. Sainju, U.M.; Barsotti, J.L.; Wang, J. Net Global Warming Potential and Greenhouse Gas Intensity Affected by Cropping Sequence and Nitrogen Fertilization. *Soil Sci. Soc. Am. J.* **2014**, *78*, 248–261. [CrossRef]
35. Kliopova, I.; Staniškis, J.K.; Stunženas, E.; Jurovickaja, E. Bio-Nutrient Recycling with a Novel Integrated Biodegradable Waste Management System for Catering Companies. *J. Clean. Prod.* **2019**, *209*, 116–125. [CrossRef]
36. Huang, R.; Liu, J.; He, X.; Xie, D.; Ni, J.; Xu, C.; Zhang, Y.; Ci, E.; Wang, Z.; Gao, M. Reduced Mineral Fertilization Coupled with Straw Return in Field Mesocosm Vegetable Cultivation Helps to Coordinate Greenhouse Gas Emissions and Vegetable Production. *J. Soils Sediments* **2020**, *20*, 1834–1845. [CrossRef]
37. Azapagic, A.; Bore, J.; Cheserek, B.; Kamunya, S.; Elbehri, A. The Global Warming Potential of Production and Consumption of Kenyan Tea. *J. Clean. Prod.* **2016**, *112*, 4031–4040. [CrossRef]
38. Rosdia, S.M.; Mamata, R.; Azri, A.; Sudhakar, K.; Yusri, I.M. Evaluation of Properties on Performance and Emission to Turbocharged SI Engine Using Fusel Oil Blend with Gasoline. *IOP Conf. Ser. Mater. Sci. Eng.* **2019**, *469*. [CrossRef]
39. Sadhik Basha, J.; Al Balushi, M.; al Shibli, K.; Al Fazari, M.; Al Rumhi, S.; Al Mansouri, S.; Al Balushi, R. An Emission Control Strategy in a Low Capacity Single Cylinder Compression Ignition Engine Powered with Dee Blended Fuels. *Mater. Sci. Energy Technol.* **2020**, *3*, 770–779. [CrossRef]
40. Daniyan, I.A.; Daniyan, O.L.; Adeodu, A.O.; Uchegbu, I.D.; Mpofu, K. Performance Evaluation of a Smart Multi Feedstock Biodiesel Plant. *Procedia Manuf.* **2019**, *35*, 1117–1122. [CrossRef]
41. Awad, O.I.; Mamat, R.; Noor, M.M.; Yusop, F.; Yusri, I.M. The Impacts of Moisture Content on Performance and Emissions of a Four-Cylinder SI Engine Running with Fusel Oil—Gasoline Blends. *WSEAS Trans. Environ. Dev.* **2017**, *13*, 120–128.
42. Savickas, D.; Steponavičius, D.; Kliopova, I.; Saldukaitė, L. Combine Harvester Fuel Consumption and Air Pollution Reduction. *Water Air Soil Pollut.* **2020**, *231*, 1–11. [CrossRef]
43. Gasso, V.; Sørensen, C.A.G.; Oudshoorn, F.W.; Green, O. Controlled Traffic Farming: A Review of the Environmental Impacts. *Eur. J. Agron.* **2013**, *48*, 66–73. [CrossRef]
44. Tihanov, G. Study on the Duration of Operations Perfomed by Transport Vehicles during Unloading of the Grain Harvester Hopper. *Appl. Res. Tech. Technol. Educ.* **2018**, *6*, 232–236. [CrossRef]

45. McEwan, A.; Brink, M.; Spinelli, R. Efficiency of Different Machine Layouts for Chain Flail Delimbing, Debarking and Chipping. *Forests* **2019**, *10*, 126. [CrossRef]
46. Wargula, L.; Kukla, M.; Krawiec, P.; Wieczorek, B. Impact of Number of Operators and Distance to Branch Piles on Woodchipper Operation. *Forests* **2020**, *11*, 598. [CrossRef]
47. Spinelli, R.; Visser, R.J.M. Analyzing and Estimating Delays in Wood Chipping Operations. *Biomass Bioenergy* **2009**, *33*, 429–433. [CrossRef]
48. Sindhu, R.; Amba Prasad Rao, G.; Madhu Murthy, K. Effective Reduction of NOx Emissions from Diesel Engine Using Split Injections. *Alex. Eng. J.* **2018**, *57*, 1379–1392. [CrossRef]
49. Nik, M.A.E.; Khademolhosseini, N.; Abbaspour-Fard, M.H.; Mahdinia, A.; Alami-Saied, K. Optimum Utilisation of Low-Capacity Combine Harvesters in High-Yielding Wheat Farms Using Multi-Criteria Decision Making. *Biosyst. Eng.* **2009**, *103*, 382–388. [CrossRef]

applied sciences

MDPI

Article

Design of a Distributed Wireless Sensor Platform for Monitoring and Real-Time Communication of the Environmental Variables during the Supply Chain of Perishable Commodities

Roque Torres-Sanchez [1,*], María Teresa Martínez Zafra [1], Fulgencio Soto-Valles [1], Manuel Jiménez-Buendía [1], Ana Toledo-Moreo [1] and Francisco Artés-Hernández [2]

[1] Systems and Electronics Division Group, ETSII. Universidad Politécnica de Cartagena, 30202 Cartagena, Spain; mtmz87@gmail.com (M.T.M.Z.); pencho.soto@upct.es (F.S.-V.); manuel.jimenez@upct.es (M.J.-B.); ana.toledo@upct.es (A.T.-M.)

[2] Postharvest and Refrigeration Group, ETSIA and Institute of Vegetal Biotechnology, Universidad Politécnica de Cartagena, 30202 Cartagena, Spain; fr.artes-hdez@upct.es

* Correspondence: roque.torres@upct.es

Abstract: Monitoring the main environmental conditions during storage and transportation of perishable foods is necessary to predict quality losses throughout shelf life. By far, temperature is the main factor affecting quality and shelf life, but there are other variables that would greatly affect quality losses such us relative humidity, O_2, CO_2, ethylene, etc. Thus, the real-time knowledge of the evolution of these parameters during the whole supply chain allows suppliers to prevent for food losses. This paper deeply describes the design of a flexible monitoring system with real-time communication to be used in the supply chain of perishable commodities, using Wi-Fi wireless communication as collaborative networks between different measurement points. Aspects such as consumption, performance and feasibility of the system are described in detail to check the adaptability of its use.

Keywords: shelf life; temperature monitoring; wireless sensor networks; commodities quality

Citation: Torres-Sanchez, R.; Zafra, M.T.M.; Soto-Valles, F.; Jiménez-Buendía, M.; Toledo-Moreo, A.; Artés-Hernández, F. Design of a Distributed Wireless Sensor Platform for Monitoring and Real-Time Communication of the Environmental Variables during the Supply Chain of Perishable Commodities. *Appl. Sci.* **2021**, *11*, 6183. https://doi.org/10.3390/app11136183

Academic Editors: Paolo Visconti and Piera Centobelli

Received: 30 May 2021
Accepted: 30 June 2021
Published: 3 July 2021

1. Introduction

Most of the perishable food to be exported requires several transportation days to reach the destination. To preserve quality of the transported products, recommended environmental conditions are required [1]. According to FAO (2019) [2], 1.3 billion tons are annually wasted, corresponding to more than 30% of all the food produced, from which approximately 50% are fruit and vegetables. These data are unsustainable in our society and require an effort to minimize these extremely high food losses. New technologies such as Internet of Things (IoT) and artificial intelligence (AI) used during the postharvest supply chain would allow us to know more precisely where these losses occur and propose actions to minimize them [3–5].

Temperature is the most important environmental factor to be preserved between certain recommended values during the postharvest stages [6]. Many works have related the evolution of the temperature of the perishable commodities with the quality losses, obtaining representative models that allow for predicting the shelf life [7–11]. Shelf life is the time in which a food product remains at high-quality standards while being microbiologically safe [12]. These models allow for estimating commodities' shelf life based on the temperature data recorded during the postharvest stages [13]. Several suppliers provide devices able to record temperature data for a long time (months or years). Small card-shaped devices with USB connection are one of the most frequently used as temperature loggers during transport or other postharvest periods [14,15]. Other devices such as time temperature integrators (TTI) are also used to show temperature preservation during a certain period [16]. Radio frequency identification (RFID) and IoT technologies, such as Bluetooth or NFC, are recently being used to distribute sensors inside pallets or cardboard boxes, making it possible to determine the temperature variation between different areas of the same load [17], providing an either efficient or not pallet distribution [18,19] The time-temperature data registered by these portable and autonomous devices are downloaded at the destination for temperature abuse detection. A brief review is detailed in Section 2.

Quality standards and consumer requirements, as well as the necessary guarantee by the transporter to ensure the correct logistical handling of the commodities, means that this sector demands precise, autonomous and flexible devices that are capable of data recording and real-time reporting those variables that influence the dynamic shelf life of a perishable product, as defined by postharvest technology. Even though the temperature is the main factor, other environmental variables such as relative humidity, luminosity, vibrations and physiological active gases such as oxygen (O_2), carbon dioxide (CO_2) and C_2H_4 (ethylene) influence the quality degradation processes, most of them resulted to maturation, so that their detection and control throughout the postharvest supply chain determines optimal preservation conditions [20].

Wireless sensor network (WSN) platforms have been widely used in placements where a high density and an intelligent allocating of sensors is required for precision measurement [21]. Areas such as precision agriculture [22,23], oceanography [24] and home automation [25–27] use IoT technology and cloud solutions [28,29] to deploy collaborative sensor networks. This WSN collects the information of the several nodes and routes the data to the Internet by using some of the protocols available such as GSM/GPRS, NoB-IoT, SigFox, LoRa, etc., which provide real-time access to the data.

In summary, the data-logger devices currently used for postharvest commodity monitoring are divided into: (i) passive communication devices with low power requirements but without real-time communication, i.e., they are not capable of transmitting the variables during the commercial life of the commodity; thus, it is possible only when the device is connected or approached to a reader in any of the phases of the product's commercial cycle (RFID, indicator strips, USB loggers, I-button, etc.); (ii) IoT devices, which use some wireless communication bands to send the data to cloud servers on the Internet without operator handling (Bluetooth, Wi-Fi or ZigBee).

IoT devices can be divided between those that use narrowband networks such as SigFox, (SigFox, Labège (France), NB-IoT, LTE-M, or LoRA with low power consumption that have data-sending limitations and, currently, limited coverage and other devices with wideband networks. These devices that use wide bands, such as GSM, Iridium and CDMA, require devices with high power consumption that depend on sampling, the data frame protocol and frequency of data sending but have worldwide coverage. The use of these devices for postharvest commodity monitoring makes it necessary to manage power consumption optimally to achieve the same autonomy that the theoretical commodity shelf life.

The needs demanded by the land transportation sector in issues, such as integration, flexibility, autonomy and multi-point variable measurement, have motivated the design of the system presented in this article, which aims to overcome the limitations observed in the market equipment, highlighting the following:

- Flexibility for any type of sensor measurement using analogic signals (voltage or current) or digital signals (1-wire, RS-232, SDI-12 [30]).
- Compatibility with cloud servers. Standard J-Son type frames are used for real-time data allocation.
- Use of non-proprietary networks and as standard as possible to allow for worldwide use.
- Management of collaborative networks for master-slave multi-point measurement.
- Energy management for a minimum autonomy of 1 month with 15 min of data sampling.
- Reduced size for flexible placement.

- Low-cost design for assuming its loss or breakage during all stages of the commodity handling life.

The designed and described in this paper is a flexible multi-point measurement tool, able to communicate in real time the measured variables with a power consumption capable of providing a minimum autonomy of 1 month. The system is developed as a sensor nodes with two roles, the 'slave' node and 'gateway' node. The 'slave' node uses the Wi-Fi network for communication. The 'gateway' node uses GSM/GPRS for real-time cloud data allocating through the Internet but, in addition, it generates a Wi-Fi infrastructure for receiving multi-point measurement from the slave nodes. Slave nodes are designed for independent working. It means that they are able to connect and send data through any Wi-Fi network available during the postharvest chain and registered in the setup configuration of the device. The gateway node can be used as a mono-point measurement device and as a harvesting node in a WSN with the slave nodes.

The objective of this article is to demonstrate that a wireless network system based on Wi-Fi communication is able to operate correctly under the specifications of flexibility, robustness to signal loss and power autonomy enough to be used in the cold supply chain of perishable commodities.

This paper is divided into 8 sections. After the first section with the introduction, Section 2 reviews the commercial devices used for commodities controlling transportation. Section 3 describes the system architecture and the different working configurations. Section 4 describes the hardware design and the different components used for the nodes. Software design for communication, time synchronization and power management are described in Section 5. Section 6 explains the methodology followed for conducting the tests, followed by the presentation of the results in Section 7 and the conclusions and future work in Section 8.

2. Review of State-of-Art of Commercial Sensors for Shelf-Life Prediction Used in Transportation

A market study was carried out to review the technologies offered by different companies worldwide. It focuses on how the equipment controls the commodities during transportation, irrespective of kind of goods, industries, features controlled or transport.

An analysis of the dataloggers used yielded the following results: the simplest passive communication one is based on time-temperature indicators (TTI) and just controls temperature [31], but it is not the most frequently used because RFID is the most extensive. The Faubel (Faubel & Co. Nachf. GmbH, Melsungen, Germany) company offers a system based on RFID and NFC, but it just controls temperature [32]. NFC communication is used by Marathon Products (Marathon Products, Inc., San Leandro, CA, USA) [33]. The disadvantage of these systems is the dependence on other devices for the manual or low-distance reading process, which avoids the possibility of network sensor reading and the real-time data transferring to proprietary servers for remote access. To avoid these inconveniences, some companies deploy gateway devices able to manage the measuring nodes, allowing for remote sensing and collecting data. This is the case of Secure System (Secure System GmbH, Munich, Germany), which focuses on security and personalized container access control, and recently also controls temperature and humidity [34]; Sensitech (Sensitech, Beverly, MA, USA), which uses a router that receives the information transferred by a RFID sensors network [15]; and OnAsset Intelligence (OnAsset Intelligence, Inc., Irving, TX, USA), which connects its sensor tags by Bluetooth Low Energy (BLE) [35] to a harvesting data-system. With this evolution, these companies offer a flexible tool that allows for wireless measurement at several points simultaneously. In the last few years, other companies have promoted the use of devices connected via LoRa as an improvement. However, although they allow for wireless communications between measurement points, they need routers or gateways to send the data to the Internet [36–39]. ORBCOMM [40] (ORBCOMM Networks LLC, Rochelle Park, NJ, USA) and Aeris (Aeris Communications, Inc., San Jose, CA, USA) [41] implement their connected devices directly into the microprocessor installed into the refrigerator container to register the temperature readings to use their cloud connectivity.

A synergy between the connection by BLE or LoRa and the communication using GSM is made by Roambee (Roambee Corporation, Santa Clara, CA, USA) [42], Swiftsensor (Swift Sensors, Inc., Austin, TX, USA) [43] or JRI (JRI, Fesches le Châtel, France) [44], whose systems send the registered data to a user interface. Laird Connectivity (Laird Connectivity, Akron, OH, USA) [45] has replaced BLE with NB and LTE. Meanwhile, HyperTech (Hyper-Tech Systems, Petach Tikva, Israel) [46] offers sensors based on Zigbee wireless devices, in the same way as hIOTron (hIOTron Gateway and application, Ashoka Nagar, Maharashtra) [47], which includes with in-built GNSS module to track the vehicle position. Other companies including SenseAware from FedEx (FedEx Corporate Services, Inc., Memphis, TN, USA) [48] or TagBox (TagBox Solutions Private Limited, Bengaluru, India) [49] prefer to work with wireless devices that use Wi-Fi and GSM technologies for communication facilities.

As an evolution, some companies design devices that work simultaneously as a sensor and router collecting and send the data using just GSM [50–53].

3. System Architecture

During the postharvest handling stages, the commodity is located in several places where Wi-Fi access points belonging to logistic companies that commercialize the product are available; thus, the system could be programmed with the lists of SSIDs and passwords to which it would connect in each stage. To solve the problems of lack of connection, the sensor nodes have storage capacity. As mentioned before, the difference between the gateway and the slave sensor nodes lies in the communication tasks they can perform. The gateway node acts as sink and coordinator, managing the information measured by itself and the information sent by the slave nodes. The combined information is transmitted to the servers through the Internet using commercial data communication networks (GPRS, SigFox or LTE). In addition, unlike a traditional wireless sensor network (WSN), the slave nodes can send the measured and recorded information without being connected to the gateway since they comply with the 802.11 (Wi-Fi) standard and this allows them, if properly configured and if Wi-Fi infrastructure is available, to send the information to the servers without a gateway node. In this way, the slave nodes can be located among the commodities, inside the card boxes or pallets, monitoring the conservation variables during the whole supply chain. The two working scenarios of the system are shown in Figure 1. Both scenarios manage multi-point measurement: (a) configuration is used during land transport, where Wi-Fi connection is not available; (b) configuration is used in a logistic warehouse, where Wi-Fi infrastructure is available. The combination of both types of nodes allows for the use of the system in the different situations of commodities handling.

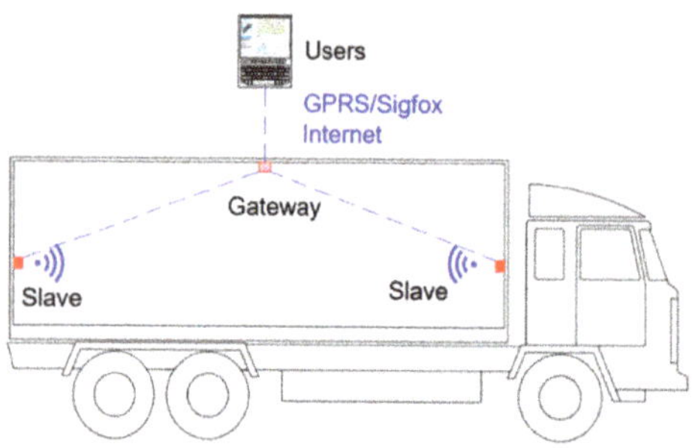

(a)

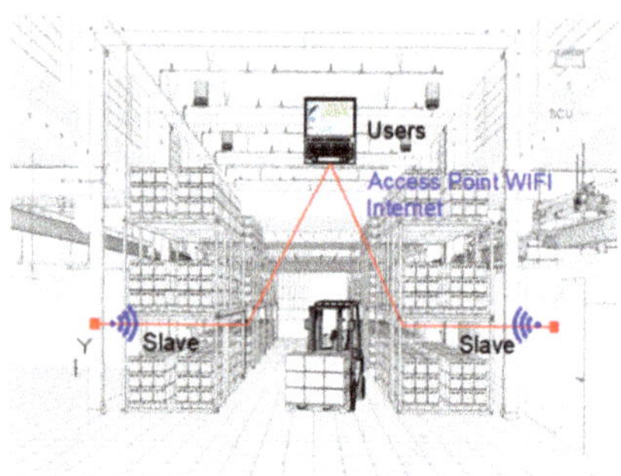

(b)

Figure 1. (**a**) Two slave nodes are communicating with one gateway node, which provides the cloud access; (**b**) two slave nodes are sending the measured information to the cloud servers through an existing Wi-Fi infrastructure.

4. Hardware Description

The sensor nodes were developed under hardware platforms based on ESP32 (ESPRESSIF SYSTEMS, Co., Ltd., Shanghai, China) that provide embedded Wi-Fi and BLE interfaces besides several input/output pins for reading and control of the sensors. ESP32 is capable of functioning reliably in industrial environments, with an operating temperature ranging from –40 °C to +125 °C. It is engineered for mobile devices, wearable electronics and IoT applications, ESP32 achieves ultra-low power consumption with a combination of several types of proprietary software. ESP32 also includes state-of-the-art features, such as fine-grained clock gating, various power modes and dynamic power scaling. It includes built-in antenna switches, RF balun, power amplifier, low-noise receiver amplifier, filters and power management modules [54]. WiPy (Pycom Ltd., Eindhoven, The Netherlands) is a microPython-enabled ESP32-based development board with a full suite of free services [55]. WiPy 3.0 is an updated enterprise grade IoT development platform. The tiny MicroPython enables Wi-Fi and a Bluetooth IoT development platform with a ESP32 chipset and dual processor [56] (see Figure 2). Pycom provides software libraries to control the device and the interfaces that can be used.

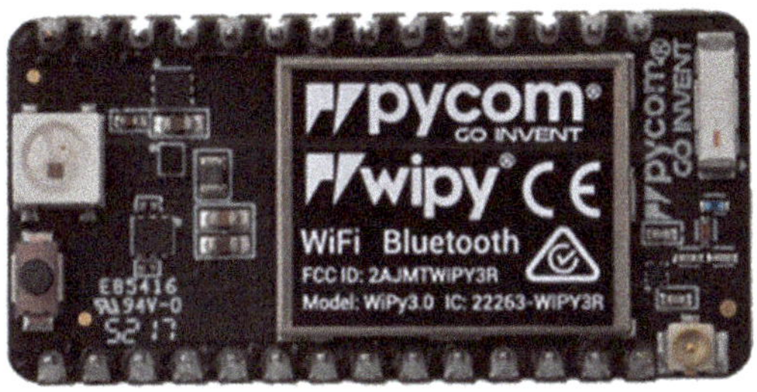

Figure 2. WiPy 3.0 development board based on ESP32 chip.

Both the sensor and gateway nodes are designed using this low-cost hardware module.

4.1. Sensor Node Design

The diagram of the sensor node can be seen in Figure 3.

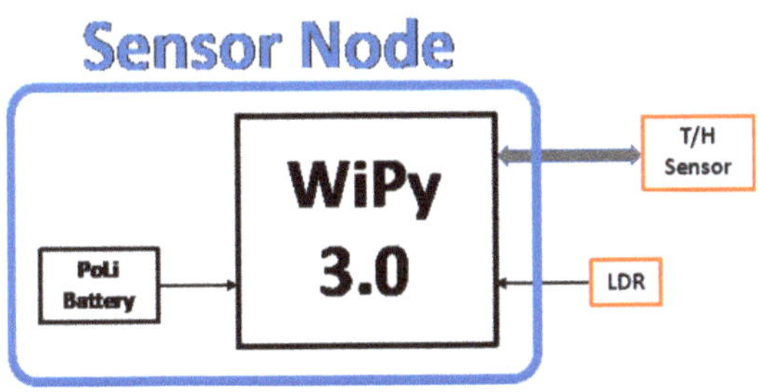

Figure 3. Sensor node architecture diagram.

The sensor node is powered by a 3.7 V 3500 mAh PoLi battery, which directly supplies the WiPy module (its power range is 3.5 V to 5.5 V) by connecting it to the Vin pin. The WiPy has a voltage regulator that provides a stable voltage of 3.3 V and 550 mA, externally connected to pin 3V3 of the module. This voltage is used to provide power to the sensors of the node through a PNP transistor that allows, controlling its state (cut off/saturation), for the power supply activation of the sensors. The activation of this transistor is controlled by pin P9 of the module, configured as a digital output. As the 3V3 connection to the sensor power supply can be controlled by software, maximum energy optimization is achieved as the sensors are only powered when measurements are needed.

The sensor nodes are able to measure temperature, relative humidity and light conditions through the use of these components:

LDR NSL06S53 (Advanced Photonics, Camarillo, CA, USA). This sensor is used for light detection in undesired truck door opening events. LDR is connected to a fixed 113 kΩ resistor, making a resistive divider that is powered by 3.3 V sourced by the DC/DC converter of the WiPy. This resistive divider provides a voltage output proportional to the LDR value and connected to P19 (ADC1_4) of the WiPy module.

DHT22 module (Aosong Electronics Co., Ltd. Guangzhou, China). This device is based on AM2302 that includes one NTC temperature and a humidity-sensing component formed by two electrodes with a moisture-holding substrate between them. The temperature range is −40 to 125 °C (accuracy ±0.5 °C and resolution 0.1 °C) and the relative humidity range is 0% to 100% (accuracy ±2% and resolution 0.1%). A one-wire proprietary bus is used for communication between digital I/O P11 of WiPy module and DHT22. The sensor data frames are composed of 40 bits [57] (see Equation (1)).

$$\text{DATA} = 16 \text{ bits}_{\text{RH\_data}} + 16 \text{ bits}_{\text{Temperature\_data}} + 8 \text{ bits}_{\text{check-sum}} \quad (1)$$

In addition, the device can measure the battery voltage using another resistive divider connected to analogic P16 (ADC1_3 of WiPy module).

The size of the Electronic PCB is 73 × 28 mm. A 3D view of the PCB is depicted in Figure 4, where the battery and programming connectors can be observed, and the plugs for LDR and DHT22 (JP4) are presented. Two push-buttons are added for reset and programming sequence purposes.

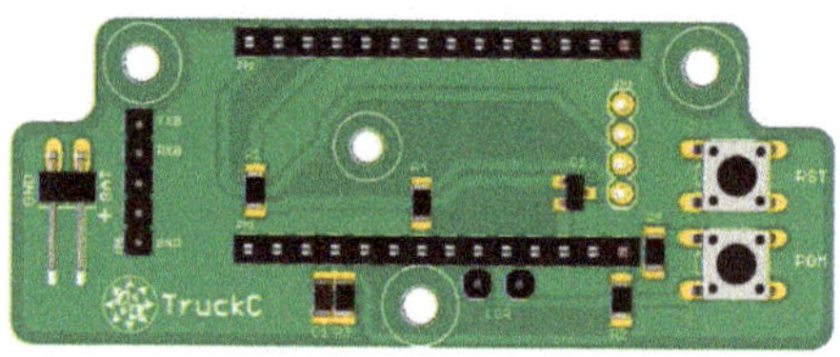

Figure 4. Sensor node 3D view of the PCB.

4.2. Gateway Node Design

The diagram of the gateway node can be seen in Figure 5.

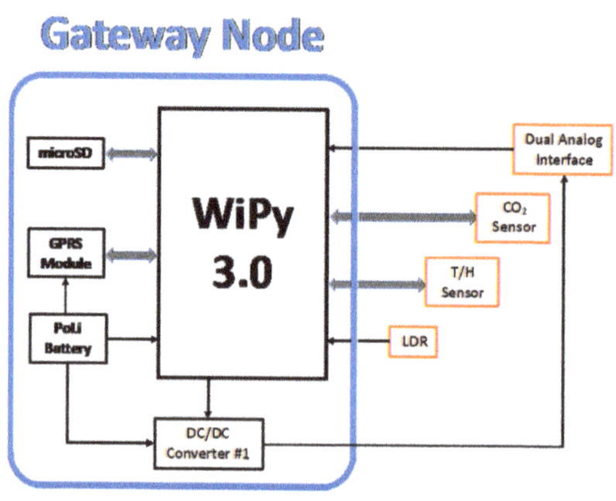

Figure 5. Gateway node architecture diagram.

The gateway node is also based on WiPy 3.0. In addition to sensors managing, the gateway node has extended functions such as:

- A MicroSD slot card for data storage. Three pins are used for SD managing. DAT0, SCLK and CMD are driven by P8, P4 and P20 using a UART1.
- A GPRS communication module (SIM800, SimComm, Shanghai, China) powered directly from the battery and controlled by AT commands through the UART of the Pycom module. The WiPy module has three UARTs, which can be easily redirected to several pins for a flexible design. The Tx and Rx signals of the SIM800 are connected to P5 and P6 pins and the reset line is fed through P7 of the WiPy. UART1 is dynamically assigned to drive the AT commands because the UART1 is shared with SD storing.
- A CO_2 sensor (COZIR-GC16 CO2Meter, Ormond Beach, FL, USA) for commodities' breathe monitoring. This sensor has a 0–20000 ppm range with an accuracy of 70 ppm ±5%. The sensor is serial controlled through UART2 directed to P20 (Tx) and P19 (Rx) pins of WiPy module. LDR light sensor is connected to P18 pin in the gateway node.
- A DC/DC boost switching regulator (MIC2288, Microchip Technology Inc. Chandler, Arizona, USA) that provides 12 V power supply to additional sensors.
- Two analogic inputs are either voltage (0–3 V) or current (4–20 mA) for additional sensor measurements, P14 and P15.

The power supply is provided by the WiPy module (3.3 V) to the microSD and the CO_2 sensor. It is switched using a transistor for optimum energy saving as explained in Section 3.

The size of the gateway node board is 82 × 40 mm. The 3D view of the PCB is shown in Figure 6.

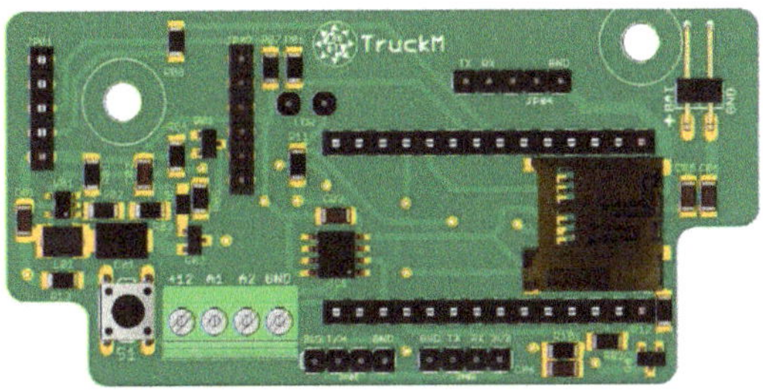

Figure 6. Gateway node 3D view of the PCB.

The GPRS SIM800 module is placed on the board using the JP01 and JP02 plugs. A push-button S1 is installed to run the setup sequence.

5. Software Architecture

The code programmed in the nodes was developed in microPython. Besides the standard microPython modules, Pycom company supplies specific libraries to control the GPIO, UARTs, DC-DC and other peripheral components included in the WiPy 3.0 module. The next flowcharts show the general operation of the gateway node (Figure 7) and sensor node (Figure 8).

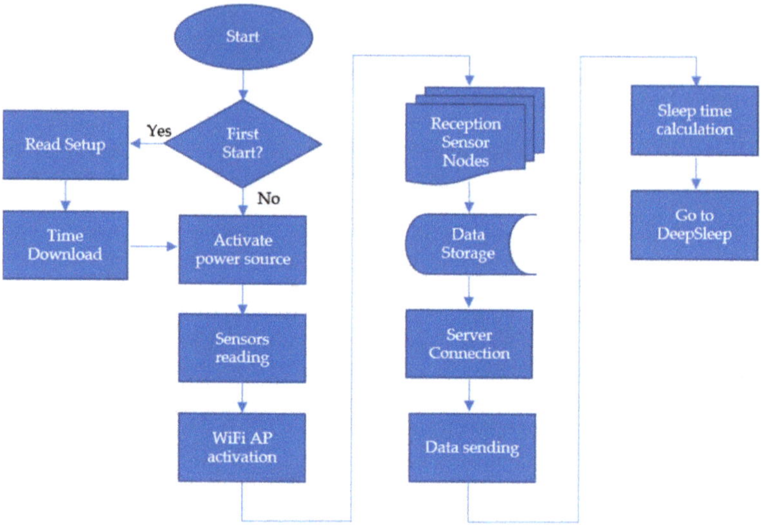

Figure 7. The general operation of Gateway node.

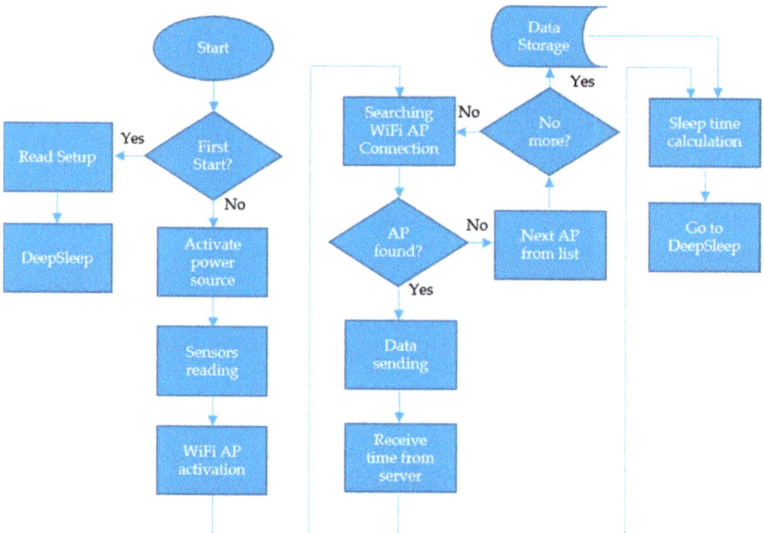

Figure 8. The general operation of the sensor node.

Both nodes were developed around four operational groups:

- Instrumentation, sampling and information storage.
- Energy management.
- Management of the communications of the nodes.
- Remote setup.

5.1. Instrumentation, Sampling and Information Storage

The instrumentation procedure is executed according to the programmed sampling time, which defines the frequency that the system wakes up, measures the sensors, records the information and (if necessary) sends the data. In the designed nodes, the instrumentation time was defined as an integer value of the number of samples desired per hour, with a maximum of 12. This process makes the system always instrument in the same minute every hour. The equation to calculate the remaining seconds that the system has to fall into the deep sleep before the next sample is calculated with the following equations.

Equation (2) calculates the exact minute of the hour when the next sampling will take place.

$$minute_{sampling} = sampling * int\left(\frac{minute_{current}}{sampling}\right) + 1 \quad (2)$$

where sampling is the minutes between measurements programmed in the setup file.

Equation (3) calculates the remaining seconds to reach the sampling time and, therefore, the time that the node must be in deep-sleep ultra-low power mode before it wakes up.

$$seconds_{wakeup} = minute_{sampling} * 60 - minute_{current} * 60 - second_{current} \quad (3)$$

Regarding the deviation in the time measurement between sampling periods, a test was performed to measure the time differences when the node was awakened from deep sleep. The average deviation for a period of 900 s (15 min) was 2.7 s, which implies an accuracy of 0.3%. This low accuracy is because the ESP32 does not use the RTC to wake up the device, but an ultra-low power timer. However, the device compares at each wakeup RTC and resets the second's wakeup value for the next sampling, so the timer accuracy value is not too important.

Analogic sensors are connected to ADC pins integrated into the module. In addition, P16 is used for battery level measurement. UARTs and digital pins are used for digital sensors, as previously written.

To improve the stability of the sensor readings, a biased average of 20 sensor readings was performed every 25 ms.

A Python dictionary with seven keys in the format [ID_nodo; Temp; Hum; %CO_2; %Lum; Vbatt; ppm C_2H_4] is returned after the measurement procedure is executed.

These data are temporarily stored in a partition of the device's flash memory. If the communication procedure is not possible due to coverage lacks or other problems, the data are stored together with the time stamp assigned to that dictionary in the SD card.

5.2. Power Management

Power management is a priority function of the system since the operation of the nodes is necessary for several weeks. Most of the time, the nodes are in an inactive mode, so the associated deep-sleep (time-out) function is used, which requires only 25 μA [56]. This function can be "awakened" through a timeout specified as a parameter of this function. This parameter is calculated at the end of the current process using Equations (2) and (3).

Using Wi-Fi communication for information exchange is very convenient due to the high compatibility and the large amount of Wi-Fi equipment available. However, the high consumption of this type of communication makes it necessary to incorporate energy managers to optimize battery life. The best way to reduce the power consumption of Wi-Fi transceivers is to reduce the time they must be online to send and receive information.

Regarding this issue, a time-synchronization algorithm was developed, so that the sensor nodes and the gateway nodes are synchronized with each other minimizing the connection time between them. The time pattern is always handled by the gateway node since it must "wait" for the sensor nodes to connect and send the information. The flowchart of the algorithm is presented in Figure 9.

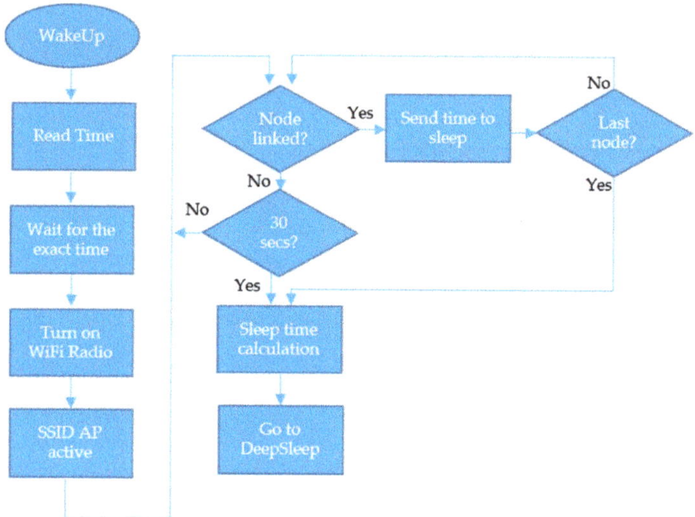

Figure 9. Execution diagram of the time-synchronizing procedure.

5.3. Communications

To optimize the time that the process is involved in the communication procedure, sockets over TCP are used for sending/receiving information to/from the server.

As mentioned in Section 3, the gateway node can be used as a stand-alone sensor device that performs socket communication with the server through GPRS/SigFox/LTE networks or, in addition, used as an access point Wi-Fi server for reception data messages from sensor nodes.

The information frames generated from the sensor nodes are sent directly to the database server if there is a recorded SSID available. Otherwise, the information is sent through the gateway. In this situation, the information registered by the nodes is carried out through an internal IP enabled by the gateway (192.168.4.1) common to all the slave nodes that communicate with the outside through it. In this case, the gateway node receives and coordinates the information from the slave nodes, synchronizing in each communication the "time-synchronization" slot and creating the necessary frames for the socket of the server that hosts the data.

Table 1 shows the format of the data frames sent by the gateway node to the server once the information from four Slave nodes distributed and connected to the generated access point was received.

Table 1. Format of the data frames sent by the gateway node to the server.

Node	Data Frame
Gateway	[GTW_1; Temp; Hum; %CO_2; %Lum; Vbatt; ppm C_2H_4]
Slave 1	[SLV_1; Temp; Hum; %Lum; Vbatt]
Slave 2	[SLV_2; Temp; Hum; %Lum; Vbatt]
Slave 3	[SLV_3; Temp; Hum; %Lum; Vbatt]
Slave 4	[SLV_4; Temp; Hum; %Lum; Vbatt]

After the data frames are sent and ACK from the server is received, the gateway node sends the timestamp in Unix epoch format and the geolocation coordinates using the engineering mode of the SIM800 to get detailed network information [53].

5.4. Remote Setup

Remote setup is a critical procedure for easy operational management of the devices. Bluetooth BLE was used to carry out the programming of the setup parameters since ESP32 incorporates BLE and personal devices, such as smartphones or tablets, also feature this technology. Consequently, BLE can easily be used by the operator as a tool for setting up the nodes. Configuration parameters include operating values such as the sampling period, cloud data storage and the SSIDs and passwords of the Wi-Fi networks through which the nodes are connected in the different locations of the transported product. These data can also be inserted using QR codes.

The node activates the BLE reception after a push-button is pressed that enables the transfer of data from an APP that uses the Bluetooth services of the device on which it is installed (see Figure 10).

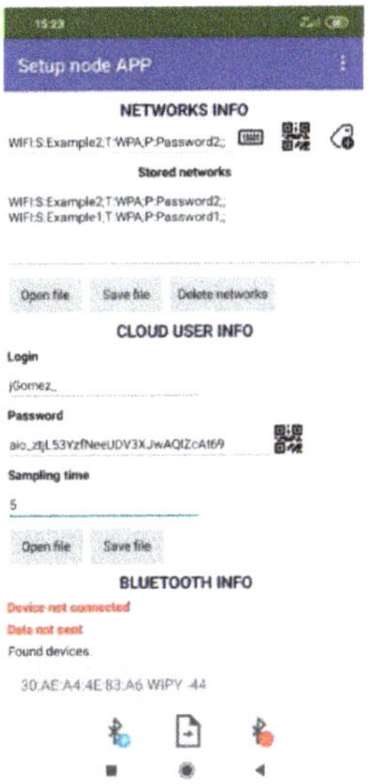

Figure 10. APP designed for setup procedure of the nodes.

6. Test Methodology

The performance of the system cannot only be exclusively described by the correct measurement of the parameters, it has to be subjected to different temperature and humidity conditions of the storage and transport rooms, which affect the transmission performance and energy autonomy.

Tests were carried out with the designed devices installed with different master-slave or slave-only configurations in three stages of the distribution process of the commodities: (i) center for sorting and packaging after harvest, (ii) cold storage and (iii) transportation and distribution stages (see Figure 11).

The center for sorting and packaging where the devices were installed was Fruca Marketing (Ctra. Fuente Álamo, km 6, 30,332 Balsapintada, Murcia, Spain). The devices were configurated with Wi-Fi network provided by the company.

Cold storage tests were carried out in the Intituto de Biotecnologia Vegetal facilities (IBV–Campus Muralla del Mar 30202, Cartagena, Murcia).

Figure 11. Stages of the postharvest supply chain used for real test of the system.

Finally, the land transport trials were carried out on several trips where the devices were installed at strategic points in the refrigerated cabin. The transports to different locations in Europe were carried out by the companies Transportes Mesa (Calle Los Carriones, 20, 30,594 Cartagena, Murcia) and Transportes el Segura (Poligono Industrial de Lorqui, S/N, Lorqui, Murcia). The setpoint temperatures were 10, 5 and 2.5 °C to check the operation and thermal distribution in the different locations of the devices.

7. Results

7.1. Performance of the Communication System

Six trials were conducted for communication testing purposes. Lettuces were used as a model of perishable food to be monitored in the trials, which lasted 10–12 days approximately. Trials 1 to 4 used two slave nodes for the three stages and one gateway node to provide internet access during the transportation and distribution stage. Trials 5 and 6 used three slave nodes during the three stages connected to several Wi-Fi networks previously configured in the setup file of the nodes. The devices were deployed at different places inside the commodities boxes. To check the communication performance in the different tests, the device records in the SD card packets, which were successfully sent at their corresponding sampling time (received packet). When the communication is unsuccessful, the device records the frame to send it later, and in that case, the attempt is marked as lost packet in order to evaluate the success rate in the communication tests of the devices in the different trials. Table 2 shows the total amount of data monitored and sent successfully to the server.

The ratio between received and lost data packets shows the communication performance. Although lost data were recovered as soon as the nodes restored Internet access, the valid data ratio demonstrates that Wi-Fi can be used as a communication carrier in these processes.

Wireless transmission performance at 2.4 GHz frequencies, such as those used in Wi-Fi, is influenced by temperature and humidity conditions [58,59].

These disturbances were reflected in the tests by some inter-sensor sending rates. However, 14 of the 18 tests that were carried out in temperature conditions between 10 and 2 °C had performance rates above 90%, which supports the use of this wireless transmission technique for the specific conditions of transport and storage of perishable products. However, it should be noted that non-communicated data are stored internally by the nodes and sent later when conditions are suitable.

Table 2. Data communication results for performance analysis.

Trial 1 (2018) DOY 81–93	Total Data Measured	Received Packet	Lost Packet	% Valid Data
Gateway	1181	1180	1	99.9%
Slave node 1	1181	1103	78	93.4%
Slave node 2	1181	1171	10	99.1%
Trial 2 (2018) DOY 100–120	**Total Data Measured**	**Received Packet**	**Lost Packet**	**% Valid Data**
Gateway	1691	1691	0	100%
Slave node 1	1691	1682	9	99.5%
Slave node 2	1691	1663	28	98.3%
Trial 3 (2019) DOY 91–100	**Total Data Measured**	**Received Packet**	**Lost Packet**	**% Valid Data**
Gateway	920	920	0	100%
Slave node 1	920	431	489	46.8%
Slave node 2	920	897	23	97.5%
Trial 4 (2019) DOY 102–112	**Total Data Measured**	**Received Packet**	**Lost Packet**	**% Valid Data**
Gateway	947	947	0	99.9%
Slave node 1	947	716	231	75.6%
Slave node 2	947	645	302	68.1%
Trial 5 (2019) DOY 245–256	**Total Packets Measured**	**Received Packet**	**Lost Packet**	**% Valid Data**
Slave node 1	959	910	49	94.9%
Slave node 2	959	959	0	100%
Slave node 3	959	878	81	91.5%
Trial 6 (2019) DOY 258–271	**Total Packets Measured**	**Received Packet**	**Lost Packet**	**% Valid Data**
Slave node 1	1281	1205	75	94.1%
Slave node 2	1281	1214	66	94.8%
Slave node 3	1281	1078	202	84.2%

7.2. Energy Management

Energy management is also a key factor in highly temperature-dependent and humidity-dependent conditions. The operational duration of the nodes depends on the capacity of the battery and the sampling period, so the analysis of the energy behavior of the node in an operational cycle is representative of the energy expenditure of the node in operational conditions. A digital power meter Yokowaga WT210 (Musashino, Tokio, Japan) was used for energy monitoring of the nodes.

Wi-Fi SSID searching and data sending to the cloud server are the processes with the greatest consumption in the sensor node. In order to calculate the energy consumption in the worst scenario, the sensor node was programmed with five SSID, but only the last one was available. The different modes and routines are described in Figure 12.

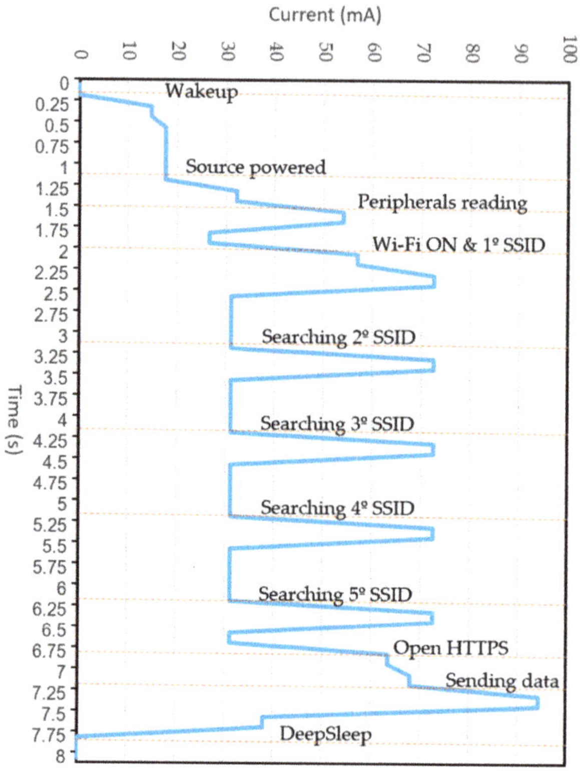

Figure 12. Power required by the different processes used in a complete execution cycle of a sensor node.

A large part of the consumption in each sampling cycle depends on the attempts to connect to the available communication network (Figure 12). For this reason, proper management of the networks allows for a reduction in energy consumption.

During the tests carried out at land transport, different conditions of communication losses with the gateway were observed, as shown in Table 2. The environmental conditions and the location of each node within the transport cabin influence the communication quality and the power consumption. Figures 13 and 14 show the results of the evolution of the voltage during trials 5 and 6, where the different slopes and voltage variations can be seen in accordance with the communication packets lost due to not being able to connect to the first network and trying to connect to the five access points available in the list.

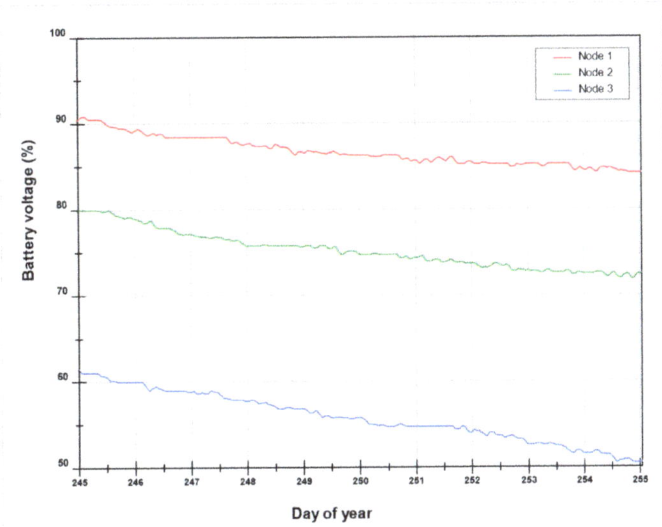

Figure 13. Trial 5: voltage evolution of the three nodes distributed in several places inside of the refrigerated truck container.

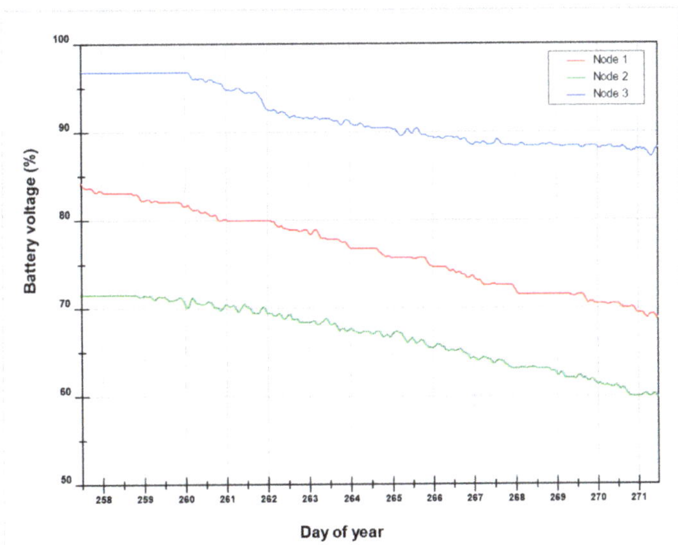

Figure 14. Trial 6: voltage evolution of the three nodes distributed in several places inside of the refrigerated truck container.

7.3. Temperature Measuring Performance

The system was tested in real conditions for one month. The commodities were placed in a cold storage room and a land transport container. Three different setpoint temperatures, at 10, 5 and 2 °C, were selected for system performance purposes. Two slave nodes placed in different locations inside the container were used. Position 1 was placed just below the evaporator of the freezer, and position 2 was placed at the back of the container on the opening doors. A gateway node was used to communicate the data (see Figure 15).

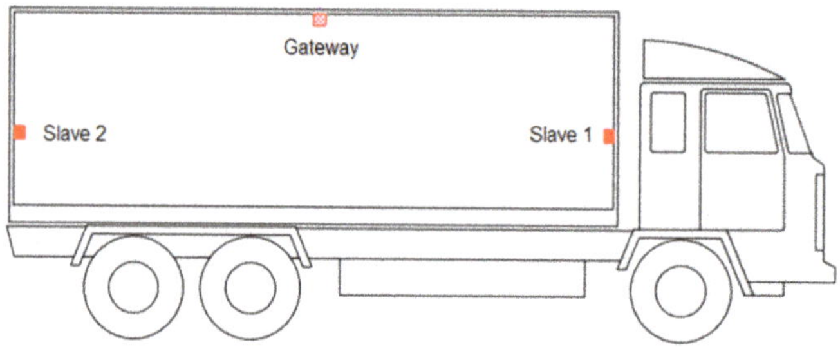

Figure 15. Distribution of the nodes during land transportation.

Charts in Figure 16 show the performance of the system. A temperature difference can be observed between the node positions. This difference is due to the different storage and transport conditions of the pallets containing the commodities.

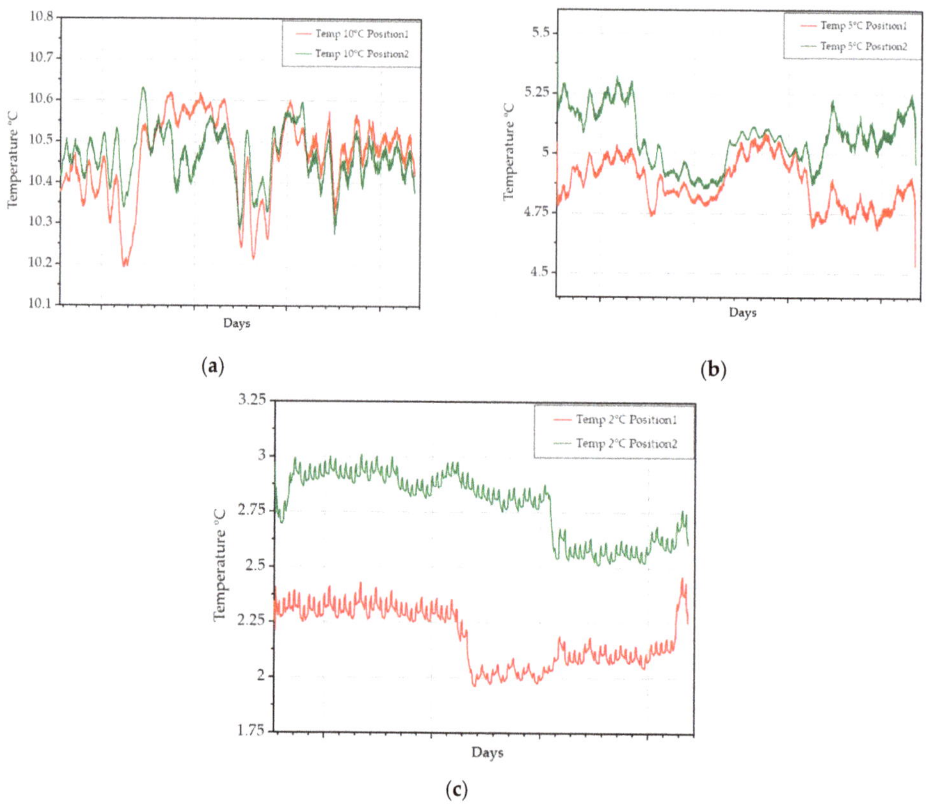

Figure 16. Evolution of the temperature between the two node locations at different temperatures: (**a**) 10 °C; (**b**) 5 °C; (**c**) 2 °C.

8. Conclusions

During storage and transportation of perishable commodities, several environmental factors should be kept within recommended thresholds to preserve quality throughout the supply chain, maximizing their shelf life. Temperature is the most important environmental factor affecting quality; however, most of the temperature control systems used in transportation media do not have multi-point measurement and temperature-recording systems, which are based on temperature cards or tags, and also do not have real-time communication protocols to know the conditions in which the commodities are being stored.

Due to the thermal differences observed in the tests, the real-time knowledge of such conditions during storage and transportation can be used not only as a warning mechanism, but also as a tool for predicting shelf life.

In this sense, a portable, multi-configurable system with real-time communication is presented in this paper.

The main advantages of the designed system over other similar devices are that it combines the measurement of multiple parameters with the flexibility of sending the measured information through Wi-Fi networks, whose infrastructure is widespread and easy to install. However, the Wi-Fi infrastructure is a broadband network, and therefore, its energy consumption is higher than other wireless technologies such as SigFox, LoRa or ZigBee. Therefore, to be able to use this technology taking advantage of the aforementioned advantages with sufficient autonomy, specific time synchronization algorithms were developed to allow for device usage ranges greater than 28 days.

On the other hand, the temperature and humidity conditions of the different stages of the supply chain affect the quality of data transmission, with tests showing data loss rates of less than 10% even in unfavorable conditions. Temporary storage of unsent data allows data to be sent when transmission conditions are favorable.

The main disadvantage is that the devices cannot be used for maritime transport as designed. The metal containers typically used in this type of transport affect the data transmission capacity. This could be solved by installing external antennas, although this would affect the ease of installation of the nodes. On the other hand, in the ocean there is no coverage for networks such as Sigfox, LoRa or NB-IoT, but ships usually have Wi-Fi communication (with satellite networks such as Iridium), so the nodes described in this article could be valid for this type of transport.

Regarding the use of the system by the agents involved in the supply chain, the main drawback observed was the programming of the nodes with the different networks (SSIDs and passwords) that the commodities were going to travel through during the journey from the harvest to the warehouse. This problem was solved by means of the APP, which allows the nodes to be programmed in an intuitive way with the different configurations. However, the input of the information is manual, and errors can be made that can cause the system to malfunction. For this reason, in future work, we proposed to improve the configuration system by using QR coding associated with the product and the route it will follow to the destination, making it unlikely that errors will be made when entering the information of the different Wi-Fi networks. On the other hand, the QR coding will allow for access to the traceability of the product conditions during the whole process. As future research, further experiments will be conducted to compare our proposal with other systems using different communication technologies.

9. Patents

The system has been patented in Spain (Registration number: ES201730301A). 'Dispositivo, sistema y método de monitorización en tiempo real de las variables físicas y ambientales durante el transporte de mercancías perecederas'.

Author Contributions: Conceptualization, R.T.-S. and F.A.-H.; methodology, R.T.-S. and F.S.-V.; software, R.T.-S.; validation, M.T.M.Z. and F.A.-H.; formal analysis, M.T.M.Z.; investigation, F.A.-H.; resources, R.T.-S.; writing—original draft preparation, M.J.-B. and A.T.-M.; writing—review and editing, M.J.-B. and A.T.-M.; visualization, F.S.-V. and M.J.-B.; supervision, F.A.-H.; project administration, R.T.-S.; funding acquisition, R.T.-S. and F.A.-H. All authors have read and agreed to the published version of the manuscript.

Funding: This research was funded by Fundación Séneca, Agencia de Ciencia y Tecnología de la Región de Murcia under the 'Excelence Group Program 19895/GERM/15'. The authors are grateful to RTI2018-099139-B-C21 project, funded by FEDER-EU/Ministry of Science and Innovation—National Research Agency.

Acknowledgments: The authors are grateful to Fruca Marketing S.L. for providing the lettuce used in this research, Transportes Directos el Segura SL and Transportes Mesa SL for the logistic support.

Conflicts of Interest: The authors declare no conflict of interest.

References

1. Domínguez, M.; García, C.; Arias, J.M. *Recomendaciones Para La Conservación y Transporte de Alimentos Perecederos*; Grupo Dominguez, 2009. Available online: https://digital.csic.es/bitstream/10261/15514/1/RECOMENDACIONES%20PARA%20LA%20CONSERVACI%C3%93N%20Y%20TRANSPORTE%20DE%20ALIMENTOS%20PERECEDEROS.pdf (accessed on 15 May 2021).
2. FAO. *Moving Forward on Food Loss and Waste Reduction*; FAO: Rome, Italy, 2019; ISBN 9789251317891.
3. Badia-Melis, R.; Ruiz-Garcia, L. Real-Time Tracking and Remote Monitoring in Food Traceability. In *Advances in Food Traceability Techniques and Technologies: Improving Quality Throughout the Food Chain*; Elsevier Inc.: Amsterdam, The Netherlands, 2016; pp. 209–224. ISBN 9780081003213.
4. Torres-Sánchez, R.; Martínez-Zafra, M.T.; Castillejo, N.; Guillamón-Frutos, A.; Artés-Hernández, F. Real-Time Monitoring System for Shelf Life Estimation of Fruit and Vegetables. *Sensors* **2020**, *20*, 1860. [CrossRef] [PubMed]
5. Calatayud, A.; Mangan, J.; Christopher, M. The Self-Thinking Supply Chain. *Supply Chain Manag.* **2019**, *24*, 22–38. [CrossRef]
6. De Venuto, D.; Mezzina, G. Spatio-Temporal Optimization of Perishable Goods' Shelf Life by a pro-Active WSN-Based Architecture. *Sensors* **2018**, *18*, 2126. [CrossRef] [PubMed]
7. Derossi, A.; Mastrandrea, L.; Amodio, M.L.; de Chiara, M.L.V.; Colelli, G. Application of Multivariate Accelerated Test for the Shelf Life Estimation of Fresh-Cut Lettuce. *J. Food Eng.* **2016**, *169*, 122–130. [CrossRef]
8. Achour, M. A New Method to Assess the Quality Degradation of Food Products during Storage. *J. Food Eng.* **2006**, *75*, 560–564. [CrossRef]
9. Ansorena, M.R.; Goñi, M.G.; Aguëro, M.V.; Roura, S.I.; Di Scala, K.C. Application of the General Stability Index Method to Assess the Quality of Butter Lettuce during Postharvest Storage Using a Multi-Quality Indices Analysis. *J. Food Eng.* **2009**, *92*, 317–323. [CrossRef]
10. Tsironi, T.; Dermesonlouoglou, E.; Giannoglou, M.; Gogou, E.; Katsaros, G.; Taoukis, P. Shelf-Life Prediction Models for Ready-to-Eat Fresh Cut Salads: Testing in Real Cold Chain. *Int. J. Food Microbiol.* **2017**, *240*, 131–140. [CrossRef] [PubMed]
11. Giménez, A.; Ares, F.; Ares, G. Sensory Shelf-Life Estimation: A Review of Current Methodological Approaches. *Food Res. Int.* **2012**, *49*, 311–325. [CrossRef]
12. *Institute of Food Science and Technology (UK) Shelf Life of Foods: Guidelines for Its Determination and Prediction*; Institute of Food Science & Technology: London, UK, 1993; ISBN 9780905367118.
13. Konovalenko, I.; Ludwig, A.; Leopold, H. Real-Time Temperature Prediction in a Cold Supply Chain Based on Newton's Law of Cooling. *Decis. Support Syst.* **2021**, *141*, 113451. [CrossRef]
14. TempDot®Plus Time-Temperature Indicator Labels, Modelo 51050-51062-DeltaTrak Mexico. Available online: https://deltatrakmexico.com/tempdot-plus-labels/51050-tempdot-plus-time-temperature-indicator-label (accessed on 10 February 2020).
15. Temperature Monitors. Sensitech. Available online: https://www.sensitech.com/en/products/monitors/ (accessed on 10 February 2020).
16. Koutsoumanis, K.P.; Gougouli, M. Use of Time Temperature Integrators in Food Safety Management. *Trends Food Sci. Technol.* **2015**, *43*, 236–244. [CrossRef]
17. Lorite, G.S.; Selkälä, T.; Sipola, T.; Palenzuela, J.; Jubete, E.; Viñuales, A.; Cabañero, G.; Grande, H.J.; Tuominen, J.; Uusitalo, S.; et al. Novel, Smart and RFID Assisted Critical Temperature Indicator for Supply Chain Monitoring. *J. Food Eng.* **2017**, *193*, 20–28. [CrossRef]
18. Mercier, S.; Villeneuve, S.; Mondor, M.; Uysal, I. Time–Temperature Management Along the Food Cold Chain: A Review of Recent Developments. *Compr. Rev. Food Sci. Food Saf.* **2017**, *16*, 647–667. [CrossRef] [PubMed]
19. Chen, J.; Huang, Y.; Xia, P.; Zhang, Y.; Zhong, Y. Design and implementation of real-time traceability monitoring system for agricultural products supply chain under Internet of Things architecture. *Concurr. Comput. Pract. Exp.* **2019**, *31*, e4766. [CrossRef]
20. Cantwel, M.; Kasmire, R.E. Postharvest handling systems: Flower, leafy and stem vegetables. In *Postharvest Technology of Horticultural Crops*; Kader, A.A., Ed.; UC Davis: Davis, CA, USA, 2002; Volume 3311, pp. 423–434.
21. Tkach, I.; Edan, Y. Distributed Heterogeneous Multi Sensor Task Allocation Systems. In *Automation, Collaboration, & E-Services*; Springer International Publishing: Cham, Switzerland, 2020; Volume 7, ISBN 978-3-030-34734-5.
22. Jawad, H.M.; Nordin, R.; Gharghan, S.K.; Jawad, A.M.; Ismail, M. Energy-Efficient Wireless Sensor Networks for Precision Agriculture: A Review. *Sensors* **2017**, *17*, 1781. [CrossRef] [PubMed]
23. Navarro-Hellín, H.; Torres-Sánchez, R.; Soto-Valles, F.; Albaladejo-Pérez, C.; López-Riquelme, J.A.; Domingo-Miguel, R. A Wireless Sensors Architecture for Efficient Irrigation Water Management. *Agric. Water Manag.* **2015**, *151*, 64–74. [CrossRef]
24. Albaladejo, C.; Sánchez, P.; Iborra, A.; Soto, F.; López, J.A.; Torres, R. Wireless Sensor Networks for Oceanographic Monitoring: A Systematic Review. *Sensors* **2010**, *10*, 6948–6968. [CrossRef] [PubMed]
25. Yang, L.; Ji, M.; Gao, Z.; Zhang, W.; Guo, T. Design of Home Automation System Based on ZigBee Wireless Sensor Network. In Proceedings of the 2009 1st International Conference on Information Science and Engineering, Nanjing, China, 26–28 December 2009; pp. 2610–2613.

26. Agalya, M.; Nancy, S.; Selvarasu, R. Home Automation System Using Wireless Sensor Networks. *Int. J. Emerg. Technol. Comput. Sci. Electron.* **2015**, *13*, 230–234.
27. Vikram, N.; Harish, K.S.; Nihaal, M.S.; Umesh, R.; Shetty, A.; Kumar, A. A Low Cost Home Automation System Using Wi-Fi Based Wireless Sensor Network Incorporating Internet of Things (IoT). In Proceedings of the 7th IEEE International Advanced Computing Conference, Hyderabad, India, 12 July 2017; pp. 174–178.
28. Pavón-Pulido, N.; López-Riquelme, J.A.; Torres, R.; Morais, R.; Pastor, J.A. New Trends in Precision Agriculture: A Novel Cloud-Based System for Enabling Data Storage and Agricultural Task Planning and Automation. *Precis. Agric.* **2017**, *18*, 1038–1068. [CrossRef]
29. López-Riquelme, J.A.; Pavón-Pulido, N.; Navarro-Hellín, H.; Soto-Valles, F.; Torres-Sánchez, R. A Software Architecture Based on FIWARE Cloud for Precision Agriculture. *Agric. Water Manag.* **2017**, *183*, 123–135. [CrossRef]
30. SDI-12 Support Group. Available online: https://sdi-12.org/ (accessed on 30 April 2021).
31. Irreversible Temperature Indicators-JRI. Available online: https://www.jri-corp.com/products/temperature-indicators/irreversible-temperature-indicators (accessed on 11 May 2021).
32. RFID Labels-Faubel. Available online: https://www.faubel.de/en/pharma-services/pharma-chemistry/smart-label.html (accessed on 13 April 2020).
33. Temperature Data Collection Device Supplier & Manufacturer-Marathon Products, Inc. Available online: https://marathonproducts.com/ (accessed on 10 May 2021).
34. SecureSystem Smart Shipping Container Technology. Available online: https://securesystem.net/ (accessed on 2 April 2021).
35. OnAsset Intelligence. Available online: https://onasset.com/ (accessed on 15 March 2021).
36. TempSen-Cold Chain Monitoring Solutions. Available online: https://www.tempsen.com/index.html (accessed on 2 April 2021).
37. Globe Tracker-GT Sense Reefer. Available online: https://www.globetracker.com/ (accessed on 24 March 2021).
38. InTemp Monitoring Solutions-Onset InTemp Data Loggers. Available online: https://www.onsetcomp.com/intemp/ (accessed on 10 March 2020).
39. Wireless Temperature Monitoring-OCEASOFT. Available online: https://www.oceasoft.com/ (accessed on 3 April 2020).
40. ORBCOMM. La Transformación de Las Cadenas de Suministro de Contenedores Con Conectividad M2M e IoT; 2017. Available online: http://www2.orbcomm.com/remote-container-management-es (accessed on 2 April 2020).
41. Cellular IoT Connectivity and Connected Vehicle Platform-Aeris. Available online: https://www.aeris.com/ (accessed on 13 March 2020).
42. Real-Time Shipment Tracking and Asset Monitoring-Roambee. Available online: https://www.roambee.com/ (accessed on 2 April 2021).
43. Wireless Monitor System for Improved Productivity-Swift Sensors. Available online: https://www.swiftsensors.com/ (accessed on 13 April 2020).
44. Wireless Temperature & Humidity Monitoring Systems-JRI. Available online: https://www.jri-corp.com/ (accessed on 13 April 2020).
45. Wireless Module, Antenna & IoT Manufacturer-Laird. Available online: https://www.lairdconnect.com/ (accessed on 15 March 2021).
46. Hyper-Tech Systems. Available online: https://www.hypertech.co.il/ (accessed on 2 April 2021).
47. IoT Solutions & Service Provider-HIOTron. Available online: https://www.hiotron.com/ (accessed on 10 March 2021).
48. SenseAware. Available online: https://www.senseaware.com/ (accessed on 15 March 2020).
49. Tagbox. Supply Chain Monitoring at Your Fingertips. Available online: https://www.tagbox.co/ (accessed on 20 April 2020).
50. Tempmate.®-Data Logger & Supply Chain Monitoring. Available online: https://www.tempmate.com/ (accessed on 10 April 2021).
51. 7PSolutions. Available online: https://www.7pgps.com/ (accessed on 23 April 2020).
52. SmartSense. Available online: https://www.smartsense.co/ (accessed on 15 April 2021).
53. Controlant. Available online: https://controlant.com/ (accessed on 20 March 2021).
54. ESP32 Overview. Espressif Systems. Available online: https://www.espressif.com/en/products/hardware/esp32/overview (accessed on 25 August 2019).
55. Pycom. Available online: https://docs.pycom.io/ (accessed on 25 August 2019).
56. WiPy 3.0-Pycom. Available online: https://pycom.io/product/wipy-3-0/ (accessed on 25 August 2019).
57. Aosong Electronics Co. DHT22 Temperature and Humidity Sensor Datasheet. 2003. Available online: http://www.aosong.com (accessed on 10 May 2020).
58. Luomala, J.; Hakala, I. Effects of Temperature and Humidity on Radio Signal Strength in Outdoor Wireless Sensor Networks. In Proceedings of the 2015 Federated Conference on Computer Science and Information Systems, Lodz, Poland, 13–16 September 2015; pp. 1247–1255.
59. Boano, C.A.; Brown, J.; He, Z.; Roedig, U.; Voigt, T. Low-Power Radio Communication in Industrial Outdoor Deployments: The Impact of Weather Conditions and ATEX-Compliance. In *Lecture Notes of the Institute for Computer Sciences, Social-Informatics and Telecommunications Engineering*; Springer: Berlin, Germany, 2009; Volume 29, pp. 159–176.

applied sciences

MDPI

Article

Towards In Vivo Monitoring of Ions Accumulation in Trees: Response of an in Planta Organic Electrochemical Transistor Based Sensor to Water Flux Density, Light and Vapor Pressure Deficit Variation

Davide Amato [1], Giuseppe Montanaro [1,*], Filippo Vurro [2], Nicola Coppedé [2], Nunzio Briglia [1], Angelo Petrozza [3], Michela Janni [2,4], Andrea Zappettini [2], Francesco Cellini [3] and Vitale Nuzzo [1]

1 Università degli Studi della Basilicata, Via Nazario Sauro, 85, 85100 Potenza, Italy; davide.amato@unibas.it (D.A.); nunzio.briglia@gmail.com (N.B.); vitale.nuzzo@unibas.it (V.N.)
2 Istituto dei Materiali per l'Elettronica e il Magnetismo (IMEM-CNR)–Parco Area delle Scienze, 43124 Parma, Italy; filippo.vurro@imem.cnr.it (F.V.); nicola.coppede@imem.cnr.it (N.C.); michela.janni@ibbr.cnr.it (M.J.); andrea.zappettini@imem.cnr.it (A.Z.)
3 ALSIA Centro Ricerche Metapontum Agrobios, s.s. Jonica 106, km 448, 2, 75010 Metaponto, Italy; angelo.petrozza@alsia.it (A.P.); francesco.cellini@alsia.it (F.C.)
4 Istituto di Bioscienze e Biorisorse (IBBR-CNR), Via Amendola 165/A, 70126 Bari, Italy
* Correspondence: giuseppe.montanaro@unibas.it

Citation: Amato, D.; Montanaro, G.; Vurro, F.; Coppedé, N.; Briglia, N.; Petrozza, A.; Janni, M.; Zappettini, A.; Cellini, F.; Nuzzo, V. Towards In Vivo Monitoring of Ions Accumulation in Trees: Response of an in Planta Organic Electrochemical Transistor Based Sensor to Water Flux Density, Light and Vapor Pressure Deficit Variation. *Appl. Sci.* **2021**, *11*, 4729. https://doi.org/10.3390/app11114729

Academic Editors: Dolores Parras-Burgos and Simone Morais

Received: 24 March 2021
Accepted: 19 May 2021
Published: 21 May 2021

Abstract: Research on organic electrochemical transistor (OECT) based sensors to monitor in vivo plant traits such as xylem sap concentration is attracting attention for their potential application in precision agriculture. Fabrication and electronic aspects of OECT have been the subject of extensive research while its characterization within the plant water relation context deserves further efforts. This study tested the hypothesis that the response (R) of an OECT (bioristor) implanted in the trunk of olive trees is inversely proportional to the water flux density flowing through the plant (Jw). This study also examined the influence on R of vapor pressure deficit (*VPD*) as coupled/uncoupled with light. R was hourly recorded in potted olive trees for a 10-day period concomitantly with Jw (weight loss method). A subgroup of trees was bagged in order to reduce *VPD* and in turn Jw, and other trees were located in a walk-in chamber where *VPD* and light were independently managed. R was tightly sensitive to diurnal oscillation of Jw and at negligible values of Jw (late afternoon and night) R increased. The bioristor was not sensitive to the *VPD per se* unless a light source was coupled to trigger Jw. This study preliminarily examined the suitability of bioristor to estimate the mean daily nutrients accumulation rate (Ca, K) in leaves comparing chemical and sensor-based procedures showing a good agreement between them opening new perspective towards the application of OECT sensor in precision agricultural cropping systems.

Keywords: bioristor; mineral nutrition; OECT; precision agriculture; PEDOT; sap concentration

1. Introduction

The management of mineral nutrition in agricultural systems contributes to the adequate nutrients availability for biochemical and structural functions in planta. Matching plant nutrient demand with nutrient soil supply (and availability) at short interval time (days or even hours) might help to avoid excessive (supply>>demand) or deficient (supply<<demand) nutrition and in turn contributing to healthy plant and environment. A number of in-vivo sensors and non-touch image-based technologies are increasingly proposed within innovative agriculture to monitor plant traits including water status, diseases and minerals content [1–4]. Real time monitoring of nutrient availability in xylem sap is pivotal for that purpose and to minimize agriculture dependency on mineral fertilization and/or face nutrition stress.

A relatively new set of organic electrochemical transistors (OECT) devices is emerging for the determination of physical and chemical characteristics of liquid samples and biological system including the ionic content [5–7]. Through mathematical models the change of the OECT device bulk conductivity allows the detection of the concentration of ions in solution making them increasingly used for bio-interfacing, bio-sensing, and electrophysiological recording [7–9]. Due to their ions' sensitivity and selectivity (e.g., K, Mg, Na) [10,11], OECT-based biosensors are promising tool within a smart plant nutrition management context. For example, current open-source application includes leaf nutrient content (sourced by foliar analysis) for the implementation of the monthly fertigation plan in olive [12]. Having a real time monitoring of ions carried in the sap as input data would strength that kind of app.

An OECT sensor (hereafter named bioristor) has been proposed for the in vivo real time monitoring of some physiological traits (including concentration of ions in the xylem sap) under variable environmental conditions and drought stress in herbaceous crops (e.g., tomato) [13–15]. However, due to the differences in the vascular system between herbaceous and trees [16], testing similar sensors for perennials is highly desirable.

The sensitivity of bioristor to positively charged ions (e.g., Ca, K) [15] led to a close relationship between the molar concentration of aqueous solutions and the sensor response (R). Moreover, previous work reported R and *VPD* and stomatal conductance highly correlated ($R^2 = 0.82$) in tomato [14,15] under a *VPD* ranging within 0–0.8 kPa. The *VPD* is a well-known driver of transpiration (and in turn of sap flow) which often lags behind (or occurs in advance) of *VPD* particularly under high *VPD* [17,18]. In addition, values of *VPD* in cultivated areas (e.g., Mediterranean-type) usually peak at 3–4 KPa [19], hence testing an OECT sensor in a perennial crop under Mediterranean growing conditions and in relation to plant transpiration would integrate current knowledge on OECT sensor.

Concentration of xylem sap (Ci, mol m^{-3}) at any instant is satisfactorily approximated by the ratio of the flux density of solutes (Js, mol m^{-2} s^{-1}) and that of (volumetric) water (Jw, m^3 m^{-2} s^{-1}) entering the xylem [20]. Values of Js depend on several factors including nutrients uptake by roots, xylem tissues loading-unloading, nutrients delivery and utilization along the transportation pathway [21]. In addition, roots might passively (diffusion) and/or actively uptake ions [22]. A mass flow of ions into root also occurs because they are dragged in by Jw (activated by transpiration flux) and/or because Jw induces a continuous removal of the ions at membrane interface after they have been released into the xylem lumen [23,24]. Concerning the flow of water in the xylem (Jw), by recalling the Poiseuille's Law and based on the cohesion theory, it depends on plant conductance and is driven by the hydrostatic soil-plant-atmosphere pressure gradient (ΔP) which is generated by diffusion of water vapor from leaf to the atmosphere in response to the leaf-to-air *VPD* [20]. Hence, Jw is influential on Ci because it contributes to the load of ions (proportionally to the soil solution concentration and the reflection coefficient) and because it operates as the solvent of xylem sap [20].

It appears that while fabrication and electronic aspects of OECT have been the subject of roughly extensive research, its characterization within a plant water relation and ion transport mechanisms context deserve further efforts. Therefore, this study tested the hypothesis that the response of bioristor (R) which is candidate to measure Ci would be inversely proportional to water flux density (Jw).

Within a mechanistic approach *VPD* drives the water flux (*VPD* $\rightarrow$ Jw). Water flux is also triggered by light availability due to its influence on stomatal behavior and in turn on transpiration [25]. Under outdoor conditions changes of *VPD* occur simultaneously with that of other meteorological variables (e.g., light) making their effect(s) on transpiration (and on R) difficult to separate. To infer new information on the influence of *VPD per se* on R, this study examined whether the response of the bioristor to changing environmental conditions is dominated by *VPD per se* or by water flux. For this purpose, olive trees growing in a walk-in chamber were subjected to various combinations of light dark and *VPD* cycles.

In order to progress towards the application of OECT sensor for precision agriculture, this study also examined the suitability of the bioristor to estimate ion accumulation rate in leaves. Considering that accumulation of ions in a certain time period would be proportional to the flux density of water and sap concentration in the same time period (i.e., Jw × Ci), that accumulation was calculated in parallel using Ci inferred from the OECT sensor and from actual analytically determined tissue concentration.

2. Materials and Methods

2.1. Plant Material and Experimental Design

The experiment has been conducted in a greenhouse at the ALSIA-CRMA research center, Southern Italy (40°23′31.4″ N, 16°47′10.9″ E) using 12 olive trees (Olea europaea L., var. Picholine). Trees were 2 years old and grown in a 6.2 L PVC pot and were fertilized at 15 days interval before the experiment (total of 4 applications) using 3 g/pot of NPK fertilizer 14.7.14 (Slowenne 212, Valagro Spa, Atessa, Italy). Each pot was covered with a plastic film and aluminum foil to minimize warming and direct evaporation of water from the soil. bioristors were installed on the 2nd of April (hereafter 0 day after sensor installation, DASI). At 6 DASI (7 p.m.) the whole plant canopy of 6 trees was enclosed in a plastic bag (130 L) to minimize the *VPD* of the air surrounding the canopy. Bag closure was ensured by wrapping the bag around the trunk with tight sealing film (Parafilm-M, Sigma Aldrich, St. Louis, MI, USA). After 9 days (15 DASI, 7 p.m.), bags were removed.

2.2. Determination of *Jw*

Each potted tree was positioned on a digital scale (5 g resolution) (FieldScale, Phenospex, Heerlen, Netherlands) programmed to measure and record the weight every 15 min interval. Irrigation was supplied daily to fully restore water consumption using tap water in order to maintain soil moisture close to 85% of field capacity. The daily water consumption (g d^{-1}) was calculated as the difference of weigh recorded at 7 p.m. of two consecutive days. Plant water consumption (W) (mol H_2O h^{-1}) was calculated hourly throughout the 24 h interval from 00:00 to 23:00 h solar time as the difference between two consecutive weights measured at 1 h interval. Plant transpiration (E) (mol H_2O m^{-2} h^{-1}) was determined as W/LA where LA was the plant leaf area (m^2).

The initial LA was determined by counting the total number of leaves per plant and considering a mean area (see below) of 3.53 cm^2 per leaf. The number of leaves newly developed during the experiment was determined by counting at the end of the trial the leaves standing on the new part of the shoots. For mean leaf area determination, 3 bulk samples (total 200 leaves) were collected from 3 trees not included in the trial, leaves were then pictured, and the surface area of each single bulk was determined via image analysis (ImageJ, https://imagej.nih.gov/ij/). Then a linear correlation between the number of leaves of each bulk and the corresponding leaf area was employed to satisfactorily (R^2 = 0.99) estimate the mean area of a single leaf as the slope value. Leaf area was assumed to linearly develop during the experimental period.

Values of water flux density (Jw, mol m^{-2} h^{-1}) was calculated as W/TA where TA was the trunk cross sectional area (m^2) estimated by measuring the trunk diameter close to the point where the bioristor was installed.

2.3. OECT Installation

Bioristor (i.e., biological resistor) sensors were installed across trunk at approx. 40 cm from the ground following the procedure reported in [15]. Briefly, bioristor is composed by two PEDOT:PSS (poly-3,4-ethylenedioxythiophene polystyrene sulfonate) functionalized threads: a main channel of the transistor and a gate of the transistor that, subjected to a positive voltage, generates the electric field that pushes the cations present in the sap in the polymer deposited on the main wire, changing its conductivity [13]. Sensors were prepared and inserted in two trunk holes (⌀ = 0.8 mm, 5 mm distance) drilled with a dremel allowing them to fully cross the trunk. The constant voltage on the source-drain channel was of $-0.1V$ while the voltage applied at the gate was 1V. Figure 1 shows the setup of the bioristor.

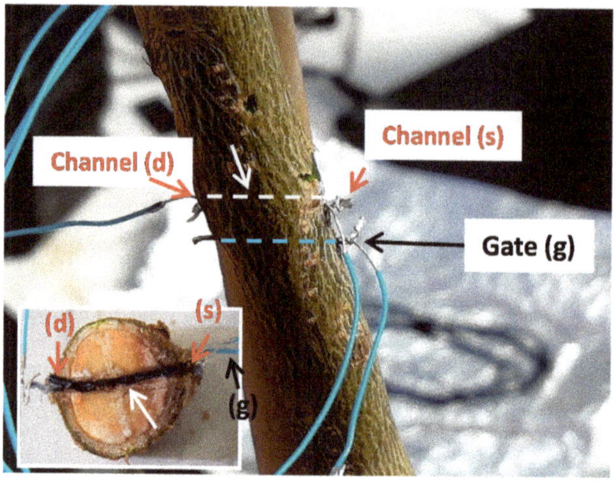

Figure 1. Side view of the olive tree trunk showing the setup of the bioristor with channel and gate wires which were connected to the multifunction I/O device for supply, record and store of electrical currents. In the inset: illustrative trunk cross sectional view (at the channel insertion point) showing the textile fiber functionalized with Pedot:PSS (white arrow).

All bioristors were then connected to a multichannel digital analogic converter I/O (USB-6343, National Instruments, Austin, TX, USA) to supply voltages controlled by a home-made software which, measure and store in the PC the values of electrical currents. Values of the OECT sensor response (R) was determined according to Coppedè et al. [13] as $R = (I_{ds} - I_{ds0})/I_{ds0}$, where I_{ds0} is the current at gate voltage equal to zero, and I_{ds} was the resulting current at the voltage equal 1. For data analysis covering successive days, data of R before plotting were scaled (min–max normalization, (0, 1) range) to ensure comparisons between days of experiment.

2.4. Walk-in Chamber Experiment

To examine the effect of light and *VPD* on R, 3 additional trees were located in a $2 \times 3 \times 2.5$ m walk-in chamber (KW Apparecchi Scientifici, Monteriggioni–Siena, Italy). The chamber was equipped with 6 lamps (mod. APO 6 OrtoLED42-6, INDOORLINE s.r.l., Garzigliana, TO, Italy). Each lamp had 90 led ensuring an illumination equivalent to a light PAR intensity of approx. 1100 μmol m^{-2} s^{-1}. Inside the chamber the regulation of parameter (air temperature, humidity) was operated once it deviated $\pm 10\%$ of the target value. Trees inside the chamber were separately positioned on 3 additional digital scales and their W, E, and Jw determined as reported above.

2.5. Ions Accumulation Rate

2.5.1. Bioristor Based Estimates

The mean daily accumulation rate (mg day^{-1}) of Ca and K in newly developing leaves was calculated as follow:

$$\text{accumulation rate} = N\_Jw \times [R] \times N\_LA/n \text{ (mg day}^{-1})$$

where N_Jw was the hourly water flux density suppling the new leaf area reported in volumetric values (m h-1) measured over n (17) days of experiment; [R] was the hourly nutrient concentration estimted from R values converted in mol m^{-3} of K and Ca [15];

N_LA, was the leaf area (m^2) of the new developing leaves. The N_Jw and [R] where reported to daily values by summing the diurnal 24 records collected over the 0:23 h. The flux Jw (mol m^{-2} h^{-1}) was partitioned between new developing (N_Jw) and old leaves according to their leaf area as destructively determined at the end of the experiment under the assumption they had similar E.

2.5.2. Analytical Determination

Leaves collected for final leaf area determination from 6 plants were dried in a ventilated oven (65 °C, minimum 48 h till constant weight) to determine the total dry matter per plant (DM). An aliquot of dry matter was used for mineral element (me) (i.e., Ca e K) concentration ([me]) determination according to Celano et al. [26]. The mean daily accumulation rate was determined considering the number of days of the experiment (n) by means of the formula DM × [me]/n.

2.6. Meteorological Data

Air temperature (°C) and relative humidity (%RH) were monitored through digital probes (mod. CS215, Campbell Scientific Inc., Logan, UT, USA) positioned close to the canopy (×1 a tree). Probes were connected to a datalogger (CR10X, Campbell Scientific Inc., UT, Logan, USA) programmed to record a 60s interval and to compute the mean value every 15 min. Additional probes were used to monitor air temperature and RH inside the bagged canopy. In order to avoid direct contact between probes and eventual condensed water, the probes were put in a plastic tube open at the bottom. A total of 6 probes per treatment were monitored. The *VPD* was then calculated from air temperature and RH values [27].

2.7. Data Analysis

The statistical analysis, plotting and fitting were by OriginPro 9.3 (OriginLab Corporation, Northampton, MA, USA).

3. Results

The response of OECT biosensors fabricated with organic electronic materials are increasingly tested for sensing ions or molecules within various applications and research (e.g., healthcare, environmental monitoring, healthcare products, water and food test) [6,28,29]. Nanosized transducers such as OECT based sensors are also relevant for agriculture mainly because of the role of ions and molecules in plant structure and function and food quality [30,31]. Here, an OECT based sensor installed in a living trunk has been examined to test its response to the amount of water flowing through the plant and to environmental factors (*VPD* and light) contributing to open up novel avenues for precision in plant mineral nutrition.

3.1. Sensitivity of the Bioristor to Diurnal Change of Environmental Conditions

The sensor response (R) showed a typical circadian pattern confirming previous studies employing the bioristor in annual crops [13,14]. In the present study, R trends have been analyzed against the changes of Jw and transpiration. It was observed that R quickly declined from early morning (~6:00 h) concomitantly with the beginning of plant transpiration flux (Figure 2). As soon as plant transpiration rate begun to decline (approx. 1300–1500 h), R trend reversed and continuously increased until approx. 19:00–20:00 h when the transpiration reached minimum values (Figure 2). Thereafter, while transpiration continued to sit at the minimum, R further increased and highest values were recorded around midnight or just before the beginning of the transpiration cycle of the next day (Figure 2).

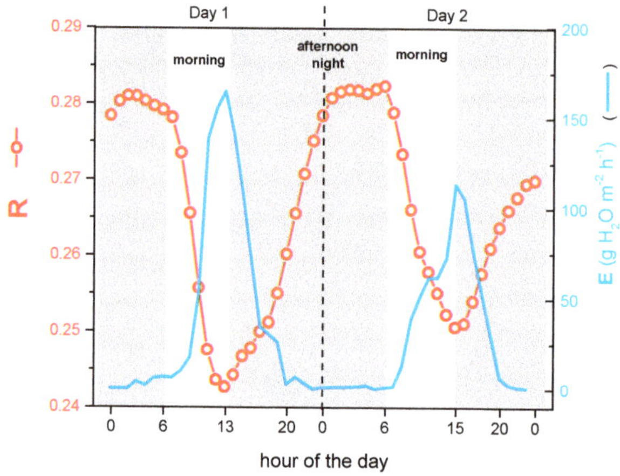

Figure 2. Diurnal variations of R (○) and plant transpiration (E) (continuous line) recorded over two consecutive days. The vertical dashed line separates the two consecutive days (8 and 9 April).

To explain increasing R values observed at null or very low transpiration (Figure 2) the main mechanisms of ions and water load into xylem should be evoked. Although the weight of active and passive loading mechanism(s) of ions (and other solutes) into the xylem has not yet clearly defined [32], the xylem loading of ions might be uncoupled from that of water because plant's membrane is differentially permeable to water and ions [20,32]. Xylem conduits are non-selective structures, hence water and solutes flow in the xylem at the same velocity through the mass flow driven by the ΔP independently by the osmotic pressure, excepting under saturating conditions when transpiration ceases (e.g., night or at very low *VPD*), or in case of low-transpiring organs (e.g., fruit,) [20–33].

Under non-limited water condition and according to the constancy of water flow through the plant [34], almost the whole volumetric water flux density taken up by root is transpired by leaf and in turn the same volume of water is loaded into xylem. Hence, variation in the Jw calculated through plant transpiration is a reliable proxy for volumetric water uptake from soil and loaded into xylem. Values of plant transpiration (E) (Figure 2) were consistent with those of sap flow (referred per unit of leaf area) measured in potted olive trees under similar *VPD* [35].

Variations of R appeared to be proportional to those of plant transpiration particularly during the first part of the day (from ~6 h to ~13 h, hereafter referred as "morning") (Figure 2). Figure 2 also shows a substantial late "afternoon" and "night" variation (increase) of R which was independent of E (and thus of *VPD*). Considering that E depends on leaf-to-air *VPD* and canopy conductance [20], R would correlate with *VPD* as suggested by previous reports [15]. Such a correlative link between *VPD* and R might be influenced by the eventual hysteretic pattern of R. That is, diurnal R might move in a hysteresis loop when plotted against *VPD* with the hysteretic magnitude differing day-by-day depending on *VPD* level (Figure 3).

Results show that in a warm day with *VPD* peaking at approx. 3 KPa, data of R collected a.m. had different slope compared to that measured p.m. (Figure 3C–D). While in a mild day (max *VPD* at ~1.5 KPa) the slopes of R data measured a.m. and p.m. were more comparable (Figure 3A–B). This might be explained considering the possible time lag effect of *VPD* on diurnal pattern of certain traits (e.g., sap flow, transpiration) [17,36,37] related to sap ionic concentration and in turn to R. It appears that disentangling the diurnal pattern of R in relation to Jw (which is triggered by *VPD*), would contribute to expand previous knowledge on the response of OECT sensors to the surrounding environment (see below). The *VPD*-R linear correlations and the evidence that *VPD* triggers the xylem stream [20] suggest that the OECT would monitor the xylem tissue although it was in contact with xylem and phloem as per installation procedure, however a specific experiment is needed to test it.

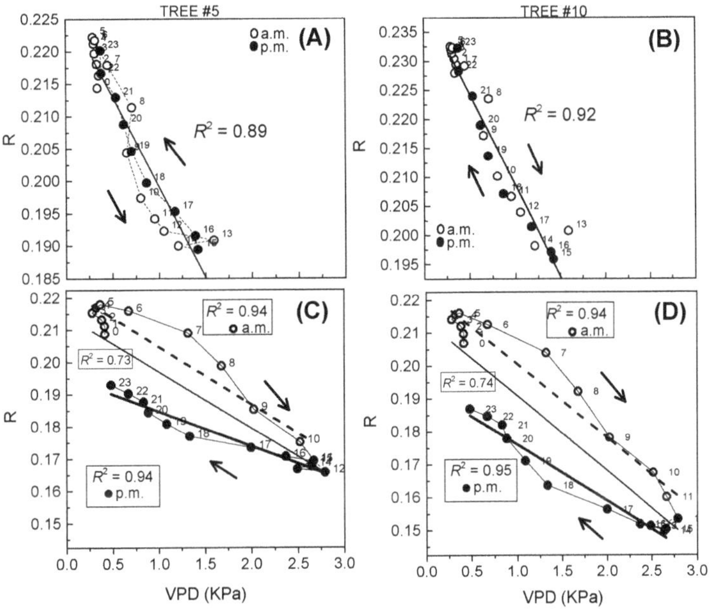

Figure 3. Example of diurnal variation of R in tree #5 (**A**,**C**) and #10 (**B**,**D**) plotted against *VPD* recorded during days with maximum high *VPD* (**C**,**D**) and low (**A**,**B**) highlighting the hysteresis of R between a.m. (○) and p.m. (•) data. Note that a.m. and p.m. values in each panel were pooled before fitting (continuous mild line) and R^2 determination. For panel (**C**,**D**) additional fittings were performed separately for a.m. (bold continuous line) and p.m. (bold dashed line) data. The numbers close to the symbol indicate the day hour when data have been recorded. Arrows indicate the time course of the day from 0 to 23 h.

Values of xylem sap concentration depend on the ratio between the solutes flux density (Js) and that of water (Jw) [38], hence factors affecting Js and Jw would be influential on output of any device measuring xylem sap concentration. This study was undertaken to test whether the bioristor response R (which is an OECT-based proxy for xylem sap concentration) changes according to Jw. Particularly, considering that in the xylem there is no membrane selectivity and that at high Jw it dominates upon other parameters influencing the xylem ions load (e.g., temperature, carriers) [20,24], it is expected that R would change (decline) with increasing Jw.

Accounting for the diurnal variation of R (Figure 2), the "morning" and the "afternoon–night" daytime parts were identified. The "morning" was characterized by a fast increase of Jw from negligible values early morning (~6 a.m.) to maximum ones recoded at around midday which induced a consistent decay of R (Figure 4). Incidentally, Figure 4 shows again the hysteretic behavior of R between a.m. and p.m. data. The Jw values detected during the "afternoon–night" stage were in general lowest than that measured "morning" and follow an exponential pattern, however the highest and often sharp increase of R was detected late in the afternoon (e.g., after 18:00 h) when Jw < 0.5 mol m^{-2} s^{-1} (Figure 4). This result is difficult to discuss due to very limited data existing on this specific issue, however it is line with the non-linear correlation between Jw and xylem sap concentration reported in the literature for soybean plant [39]. Under increasing transpiration (as during "morning") the load of solutes into the stele of xylem at root region is dominated by ions dragged in by the water flux density Jw [21,24] Hence, the increasing Jw during the "morning"

(Figure 4) induced a consistent reduction of R because of a conceivably "dilution" of the sap concentration dependent on water moving with increasing transpiration stream [24].

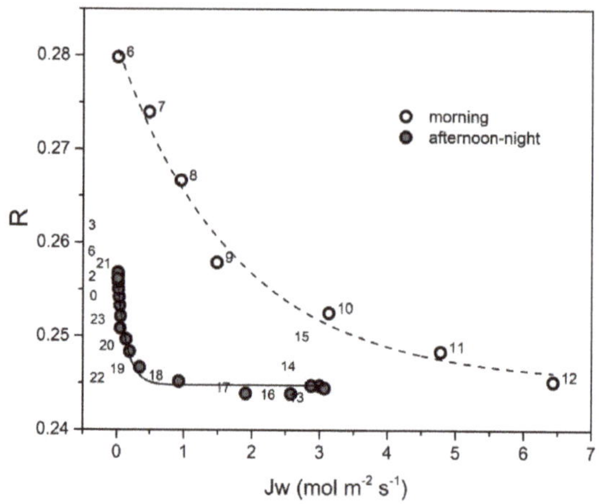

Figure 4. Example of the influence of Jw on the diurnal variation of R recorded during (○) "morning" and (•) "afternoon–night". Labels indicate the hour of the day, note that 1–5 h refer to the following day.

By contrast, when Jw is null or negligible (because of low ΔP) (e.g., during night) solutes are loaded mainly by passive diffusion or even active transport [32]. However, active transport requires availability of newly synthesized sugars/carbohydrates which are poorly available during night (White et al., 2012). Hence, the increase of R observed when transpiration is negligible (i.e., approx. from 18–19 p.m. to 4–5 a.m. of the the next day) (Figure 4) might reflect the increase of xylem sap concentration due to osmotic root pressure and unload of nutrients from saturated xylem tissues [23].

Data collected in this study allow to generalize the partitioning of R to Jw according to the daytime. Correlative information collected over an 8-day period reported in Figure 5 confirms a non-linear decay of R with increasing Jw during the "morning" (Figure 4; [38]).

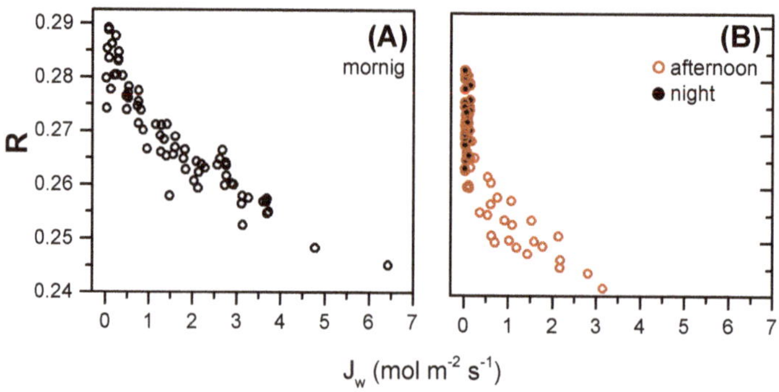

Figure 5. Correlation between R and Jw over a time series (8 consecutive days) recorded in tree #5 during the (**A**) "morning" and (**B**) "afternoon" and "night" stages. The "morning" stage begin early when transpiration start (approx. 6:00 h) and end at time n (approx. 13–15 h), when the difference between transpiration measured at time $n + 1$ and that measured at time n is <0. The "afternoon–night" span from hour $n + 1$ till the 4:00–5:00 h of the next day before the beginning of the new transpiration cycle.

The analysis of the "afternoon–night" data confirms the approx. 10% increase of R at negligible Jw (Figure 5) mainly because of the lack of the "dilution" effect which is reasonably expected to be caused by a sustained Jw. To allow comparisons across days and trees, values of R have been normalized (min–max, 0:1 range) in analogy to a study that compared daily fruit transpiration throughout the growing season [39]. Normalized R responded to Jw according to the part of the day considered (i.e., morning, afternoon–night) (Figure 6). It is confirmed also that the pattern of the "morning" dataset (increasing Jw) was substantially similar to that of the "afternoon–night" but without the vertical harm generated when Jw was stably negligible (Figures 5 and 6).

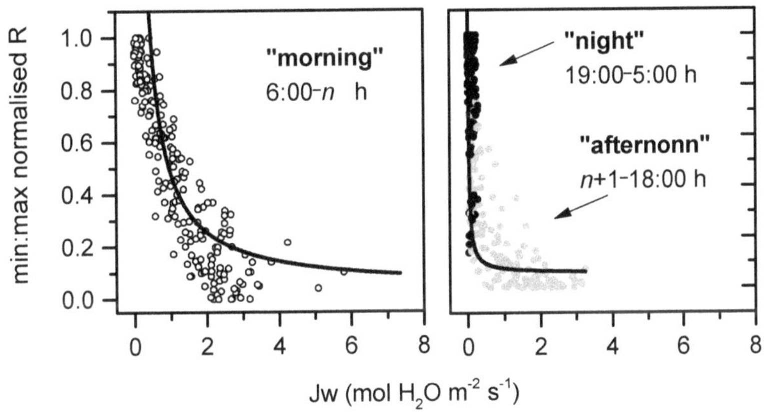

Figure 6. Correlation between R signal and the water flux density (Jw) hourly measured in olive trees during (top panel) "morning" (○) and (bottom panel) "afternoon" (•, grey filled) and "night" (•, black filled). The $n + 1$ h indicates the time of the day (between 13 and 15 h) corresponding to that when Jw become lower than that measured at time n. Note that scatters have been fitted using the model y = a + b/x consistent with that of the ions' concentration expected in the xylem, lines are illustrative only.

3.2. *Response of Bioristor to Water Flux Density*

The main driving force of the flux density of water (Jw) is the *VPD*, hence it is expected a reduction of Jw upon a reduction of *VPD*. To test the influence of Jw on R, a group of trees was bagged to bring the air surrounding the canopy close to the saturation point and in turn the *VPD* close to zero. The bagging method has been documented to effectively reduce *VPD* at plant or branch and fruit scale by reducing RH with negligible impact on air temperature [30,34,40].

The *VPD* promptly responded to the bag application (19 h, 5 DASI) and removal (19 h, 15 DASI) mainly through variation of relative humidity which approximated 90% RH (not shown). On average, the daily *VPD* of the air surrounding the bagged olive canopy was reduced by 80% compared to that of control trees (Figure 7). This induced a consistent reduction of daily water consumption in bagged trees compared to control ones (Figure 8).

Throughout the 10-day bagging experiment an 80% reduction of the *VPD* was recorded which induced a 60% reduction in Jw (Figure 7).

Under no stomatal limitation, the causal chain *VPD* → Jw → R anticipates the indirect influence of the *VPD* on R as mediated by Jw because Jw is proportional to ΔP (hydrostatic soil-plant-atmosphere pressure gradient) generated by the leaf-to-air *VPD*. Figure 9 shows diurnal oscillations of R recorded in some succeeding days with different *VPD* and consequently different Jw. It is confirmed that reducing *VPD* reduced Jw and in turn increased R. In addition, R increased by 8–12% during the time when Jw was negligible (from late afternoon until early morning of the following day). The increasing R signal at negligible Jw conceivably reflects the absence of sap dilution induced by water flux and the increasing root pressure usually occurring overnight when plants have almost null transpiration [22,23,41].

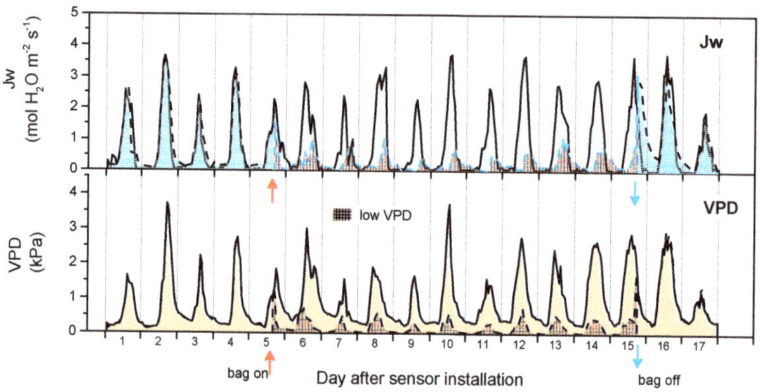

Figure 7. Values of *VPD* and Jw (at trunk level) measured throughout the experiment in olive trees under control and low *VPD* (grid gray filled). The dashed line refers to the tree enclosed in a transparent plastic bag to induce the low *VPD*. Time of bag on and off were indicated by ↑ and ↓ arrows, respectively.

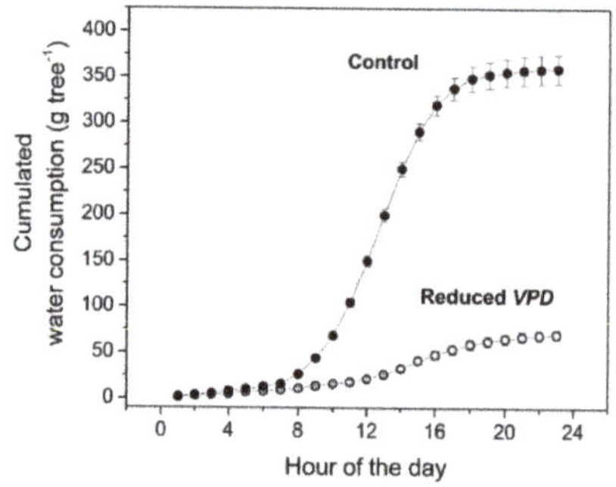

Figure 8. Cumulated diurnal plant transpiration measured in trees under control (•) and reduced *VPD* (○) by means of bag application. Each point is the average of 6 trees, bars (± SE) are visible when larger than symbol. Data relates to day 8th of experiment.

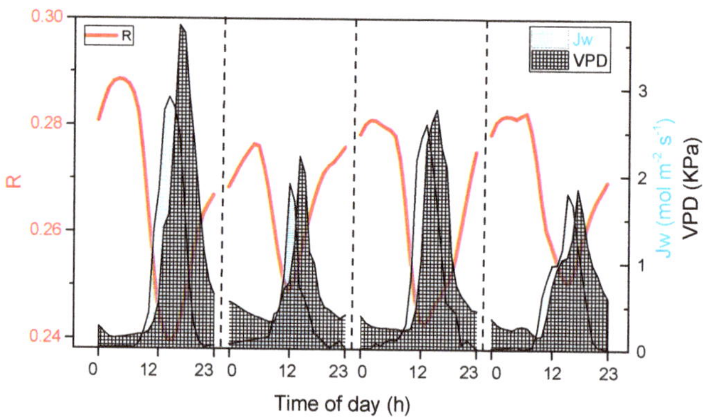

Figure 9. Oscillation of diurnal bioristor signal R (continuous line), water flux density Jw (blue filled area) and *VPD* (grid filled area) recorded over 4 consecutive days separated by vertical dashed lines.

Figure 9 shows also the general correspondence of R and Jw patterns during the morning daytime and the alignment between maximum pick of Jw and the minimum one of R. The plot highlights that the Jw lags in advance to *VPD* (Figure 9) confirming a known aspect of water relations [18,42,43] suggesting that Jw would be a robust parameter to consider for OECT-based sensors analysis.

3.3. Disentangling Bioristor Response to Environmental Conditions

Under outdoor conditions the changes of *VPD* occur simultaneously with that of other meteorological variables (e.g., light, wind) making their effect(s) on transpiration very difficult to separate. However, the present study aimed at supporting that Jw is the main driver of R, because it is a proxy of xylem sap concentration which depends on the Js/Jw ratio. Hence, trees were exposed to combinations between stable/variable *VPD* and on/off light. Figure 10 reports the values of R and Jw recorded in a tree standing in the walk-in chamber during 3 consecutive days. Results show that R responded to change of Jw induced by light-on period while the increase of *VPD* per se was not influential when it was uncoupled from light (i.e., during dark). This result is in line with the response of sap flow to variations of *VPD* and radiation observed in *Populus* spp. [43].

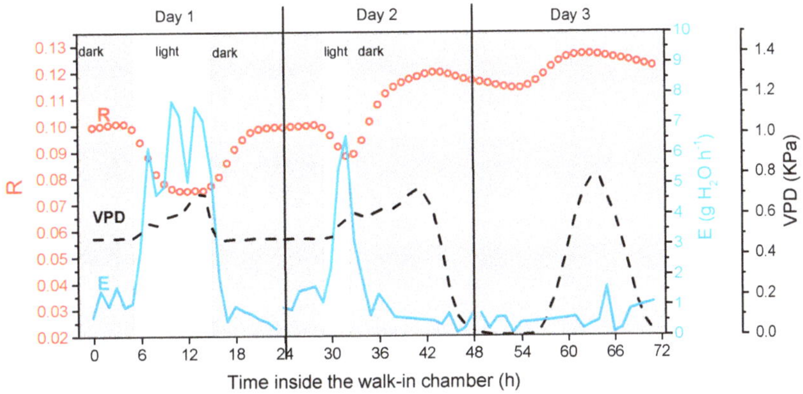

Figure 10. Diurnal pattern of bioristor signal R (circle), transpiration (E, continuous line) and *VPD* (dashed line) recorded over 3 consecutive days inside a walk-in chamber with variable light/dark and *VPD* conditions. The grey filled area indicates the dark period. Here below a detailed description of variations occurring during the 3 days analyzed.

Notably, as for the second day, also in the third day the superimposed increasing *VPD* during dark was not influential on E and in turn on R. This might be explained considering that stomata were conceivably closed during the light-off period due to the direct effect of light on stomatal opening [44].

On the third day R slightly increased (Figure 10) likely because (i) of the lack of the "dilution" effect of Jw and (ii) of the conceivable increased root pressure due to increased sap concentration determined by passive and active components of Js, moreover, increasing R might be related to reduced ions load by xylem tissues which likely saturated [21,23].

First day, light-on was set for 10 h (from 5 a.m. to 3 p.m.) and was off (dark) for the remaining time of the day. The *VPD* was set to increase concurrently to the light-on period miming its behavior under outdoor condition. In that first day, R oscillations were in line with data collected outside the chamber identifying the "morning" and "afternoon–night" stages during the light-on and light-off time, respectively.

Second day, the light-on period was restricted to just 4 h (from 5 a.m. to 9 a.m.) while the *VPD* was set to progressively increase also during the dark period until 8 p.m. As expected, R responded to change of Jw induced by light period while the increase of *VPD* was not influential when it was uncoupled with light.

Third day, light was kept off for 24 h, while the *VPD* was set to cycle from approx. 10:00 to 20:00 h. Values of transpiration sited close to negligible values during the whole 24 h period, and consequently values of R remained roughly stable (at highest value) without showing neither the "morning" nor the "afternoon–night" oscillations.

3.4. Perspective on the Use of Bioristor to Monitor Leaf Mineral Accumulation

This study also examined whether the OECT sensor might be potentially used for estimation of nutrients (i.e., Ca and K) accumulation in newly growing leaves. Considering that Jw might causes the non-diffusive xylem load of ions at root scale and their solubilization in the sap (as discussed above), the sensitivity of the bioristor to changes of Jw (Figures 9 and 10) anticipates the promising capability of the bioristor to precisely estimate ions accumulation in plant organs. Apart from concentration of nutrients in the feeding xylem sap, nutrients accumulation depends on several factors including xylem-phloem mobility of the nutrient, growth of the organ, use of nutrients along the transportation path [24]. Within the experimental period (17 days) leaf area significantly increased by approx. 30% and 15% of the initial value in trees with regular (control) and reduced Jw, respectively (Figure 11A). That increment of leaf area corresponded to approx. 21 g dry matter per tree of new leaves (control) while it was significantly 50% lower (Student's *t*-test, $p = 0.05$) in trees under reduced Jw. The Ca and K nutrient content in that dry matter was determined destructively (dry matter analysis) and through the OECT based estimates (Figure 11B) showing a good agreement between the two methods. The non-linearity between the two methods might be explained considering that the nutrient accumulation depends on several factors (see above), while OECT-based method here adopted employed only the ion sap concentration and water flux density carrying the nutrients. Coefficients used to convert R values in Ca an K molar concentrations were retrieved from [15] who measured under lab condition the values of R in response to salt solutions of known concentrations, of course these coefficients need to be validated in in vivo systems. Improved version of the bioristor accounting for in vivo condition would refine the sensor output, however, it might be preliminarily stated that the bioristor could be a promising tool for monitoring of nutrients accumulation within a precision agriculture domain.

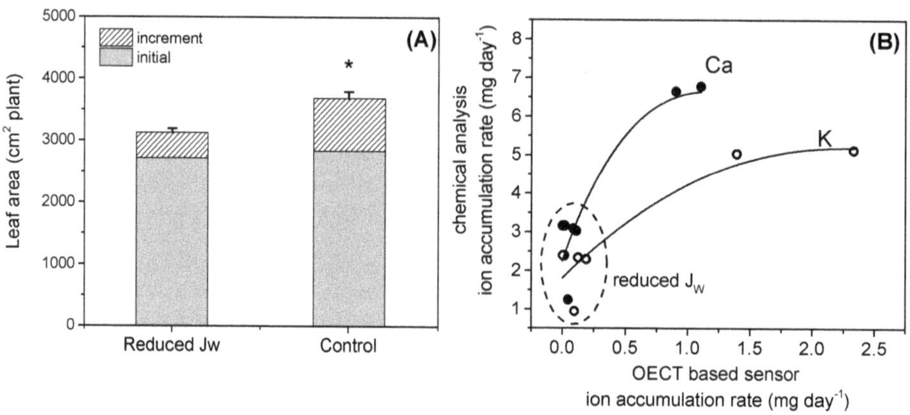

Figure 11. (**A**) Average leaf area per plant (± SE) measured at beginning and at the end of the experiment in trees under regular (control) and reduced Jw. The increment of leaf area has been determined as the difference between initial and final values. Comparing the increment of leaf area between treatments, * indicates statistically significant differences. (**B**) Accumulation rate (mg day 1) of Ca (•) and K (○) as determined through the OECT values of sap concentration and through the chemical analysis of the dry mater of new developed leaves with a 2 weeks' period. Note that dash circled values have been sampled from trees under reduced Jw.

4. Conclusions

This paper demonstrated that the signal response of an OECT sensor (a candidate proxy of sap ionic concentration) installed in the trunk of a perennial tree is inversely proportional to the water flux density (Jw) flowing-through the transpiring tree. Under negligible plant transpiration, the signal increased likely reflecting the putative increase of sap concentration due to increased osmotic pressure and lack of the "dilution" effect operated by Jw. The diurnal response of an OECT based sensor should be partitioned according to the part of the day (i.e., "morning", "afternoon", "night") also to account for possible hysteresis of its pattern due to changing interaction between Jw and *VPD*. Based on results gained in the walk-in chamber it could also be concluded that the *VPD* per se is not influential on the OECT signal unless a light source is coupled to trigger plant transpiration. Experimental results on the suitability of the sensor-based data to estimate ion accumulation in leaves open new perspectives for its application in mineral nutrition in agriculture.

Author Contributions: G.M., V.N., A.P. and M.J. conceptualized the experiment. D.A. and F.V. set up the experiment; D.A. and N.B. carried out the experiment; D.A., N.B. and F.V. carried out the data analyses; D.A., G.M. and N.B., wrote the manuscript. A.P., F.C., V.N., M.J., N.C. and A.Z. reviewed the manuscript. G.M., F.C., A.Z. obtained the funding and F.C. provided the access to the facilities. All authors have read and agreed to the published version of the manuscript.

Funding: D.A. was supported by a Smart Specialization "Industria 4.0" fellowship programme funded by Basilicata Region) of the Program "Cities and Landscapes: Architecture, Archaeology, Cultural Heritage, History and Resources" at Università degli Studi della Basilicata. This paper was partially funded by the 2014–2020 Rural Development Programme of Basilicata Region (Misura 16.2, ORGOLIO LUCANO, CUP C38I19000050006).

Institutional Review Board Statement: Not applicable.

Informed Consent Statement: Not applicable.

Data Availability Statement: The datasets generated during and/or analyzed during the current study are available from the corresponding author on reasonable request.

Acknowledgments: Authors thank A. Mossuto (Natura Informatica) for technical assistance.

Conflicts of Interest: The authors declare no conflict of interest.

References

1. Tomkiewicz, D.; Piskier, T. A plant based sensing method for nutrition stress monitoring. *Precis. Agric.* **2012**, *13*, 370–383. [CrossRef]
2. Briglia, N.; Montanaro, G.; Petrozza, A.; Summerer, S.; Cellini, F.; Nuzzo, V. Drought phenotyping in Vitis vinifera using RGB and NIR imaging. *Sci. Hortic.* **2019**, *256*, 108555. [CrossRef]
3. Lu, J.; Yang, T.; Su, X.; Qi, H.; Yao, X.; Cheng, T.; Zhu, Y.; Cao, W.; Tian, Y. Monitoring leaf potassium content using hyperspectral vegetation indices in rice leaves. *Precis. Agric.* **2020**, *21*, 324–348. [CrossRef]
4. Fallon, B.; Yang, A.; Lapadat, C.; Armour, I.; Juzwik, J.; Montgomery, R.A.; Cavender-Bares, J. Spectral differentiation of oak wilt from foliar fungal disease and drought is correlated with physiological changes. *Tree Physiol.* **2020**, *40*, 377–390. [CrossRef]
5. Zeglio, E.; Eriksson, J.; Gabrielsson, R.; Solin, N.; Inganäs, O. Highly Stable Conjugated Polyelectrolytes for Water-Based Hybrid Mode Electrochemical Transistors. *Adv. Mater.* **2017**, *29*, 1605787. [CrossRef]
6. Rivnay, J.; Inal, S.; Salleo, A.; Owens, R.M.; Berggren, M.; Malliaras, G.C. Organic electrochemical transistors. *Nat. Rev. Mater.* **2018**, *3*, 17086. [CrossRef]
7. Gentile, F.; Vurro, F.; Picelli, F.; Bettelli, M.; Zappettini, A.; Coppedè, N. A mathematical model of OECTs with variable internal geometry. *Sens. Actuators A Phys.* **2020**, *304*, 111894. [CrossRef]
8. Tarabella, G.; Villani, M.; Calestani, D.; Mosca, R.; Iannotta, S.; Zappettini, A.; Coppedè, N. A single cotton fiber organic electrochemical transistor for liquid electrolyte saline sensing. *J. Mater. Chem.* **2012**, *22*, 23830–23834. [CrossRef]
9. Gentile, F.; Delmonte, D.; Solzi, M.; Villani, M.; Iannotta, S.; Zappettini, A.; Coppedè, N. A theoretical model for the time varying current in organic electrochemical transistors in a dynamic regime. *Org. Electron.* **2016**, *35*, 59–64. [CrossRef]
10. Coppedè, N.; Villani, M.; Gentile, F. Diffusion Driven Selectivity in Organic Electrochemical Transistors. *Sci. Rep.* **2015**, *4*, 4297. [CrossRef] [PubMed]

11. Coppedè, N.; Tarabella, G.; Villani, M.; Calestani, D.; Iannotta, S.; Zappettini, A. Human stress monitoring through an organic cotton-fiber biosensor. *J. Mater. Chem. B* **2014**, *2*, 5620–5626. [CrossRef] [PubMed]
12. Alcaide Zaragoza, C.; González Perea, R.; Fernández García, I.; Camacho Poyato, E.; Rodríguez Díaz, J.A. Open source application for optimum irrigation and fertilization using reclaimed water in olive orchards. *Comp. Electron. Agric.* **2020**, *173*, 105407. [CrossRef]
13. Coppedè, N.; Tarabella, G.; Villani, M.; Calestani, D.; Iannotta, S.; Zappettini, A. n in vivo biosensing, biomimetic electro-chemical transistor with applications in plant science and precision farming. *Sci. Rep.* **2017**, *7*, 16195. [CrossRef] [PubMed]
14. Janni, M.; Coppede, N.; Bettelli, M.; Briglia, N.; Petrozza, A.; Summerer, S.; Vurro, F.; Danzi, D.; Cellini, F.; Marmiroli, N.; et al. In Vivo Phenotyping for the Early Detection of Drought Stress in Tomato. *Plant Phenomics* **2019**, *2019*, 1–10. [CrossRef] [PubMed]
15. Vurro, F.; Janni, M.; Coppedè, N.; Gentile, F.; Manfredi, R.; Bettelli, M.; Zappettini, A. Development of an In Vivo Sensor to Monitor the Effects of Vapour Pressure Deficit (VPD) Changes to Improve Water Productivity in Agriculture. *Sensors* **2019**, *19*, 4667. [CrossRef]
16. Evert, R.F.; Esau, K. *Structure and Development of the Plant Body—An Overview in Esau's Plant Anatomy: Meristems, Cells, and Tissues of the Plant Body: Their Structure, Function, and Development*; Evert, R.F., Ed.; Wiley-Interscience: Hoboken, NJ, USA, 2006; pp. 1–13.
17. Bai, Y.; Zhu, G.; Su, Y.; Zhang, K.; Han, T.; Ma, J.; Wang, W.; Ma, T.; Feng, L. Hysteresis loops between canopy conductance of grapevines and meteorological variables in an oasis ecosystem. *Agric. For. Meteorol.* **2015**, *214–215*, 319–327. [CrossRef]
18. Takagi, K.; Tsuboya, T.; Takahashi, H. Diurnal hystereses of stomatal and bulk surface conductances in relation to vapor pressure deficit in a cool-temperate wetland. *Agric. For. Meteorol.* **1998**, *91*, 177–191. [CrossRef]
19. Montanaro, G.; Dichio, B.; Xiloyannis, C. Significance of fruit transpiration on calcium nutrition in developing apricot fruit. *J. Plant Nutr. Soil Sci.* **2010**, *173*, 618–622. [CrossRef]
20. Nobel, P.S. *Solutes in Physiochemical and Environmental Plant Physiology*; Elsevier Academic Press: Cambridge, MA, USA, 2005; pp. 99–169.
21. White, P.J. *Ion Uptake Mechanisms of Individual Cells and Roots: Short-Distance Transport in Marschner's Mineral Nutrition of Higher Plants*; Marschner, P., Ed.; Elsevier: Cambridge, MA, USA, 2012; pp. 7–47.
22. White, P.J. *Ion Transport in Encyclopedia of Applied Plant Sciences*; Thomas, B., Murphy, D.J., Murray, B.G., Eds.; Academic Press: Cambridge, MA, USA, 2017; pp. 238–245.
23. Steudle, E. Water uptake by plant roots: An integration of views. *Plant Soil* **2000**, *226*, 45–56. [CrossRef]
24. White, P.J. *Long-Distance Transport in the Xylem and Phloem in Marschner's Mineral Nutrition of Higher Plants*; Marschner, P., Ed.; Elsevier: Cambridge, MA, USA, 2012; pp. 49–70.
25. Grantz, D.A.; Moore, P.H.; Zeiger, E. Stomatal responses to light and humidity in sugarcane: Prediction of daily time courses and identification of potential selection criteria. *Plant Cell Environ.* **1987**, *10*, 197–204.
26. Celano, G.; Dichio, B.; Montanaro, G.; Nuzzo, V.; Palese, A.; Xiloyannis, C. Distribution of dry matter and amount of mineral elements in irrigated and non-irrigated olive trees. *Acta Hortic.* **1999**, *474*, 381–384. [CrossRef]
27. Goudriaan, J.; van Laar, H.H. *Modelling Potential Crop Growth Processes in Current Issue in Production Ecology*; Goudriaan, J., van Laar, H.H., Eds.; Kluwer Academic Publishers: Dordrecht, The Netherlands, 1994; Volume 2, pp. 127–128. ISBN 9780792332190-238.
28. Liao, J.; Si, H.; Zhang, X.; Lin, S. Functional Sensing Interfaces of PEDOT:PSS Organic Electrochemical Transistors for Chemical and Biological Sensors: A Mini Review. *Sensors* **2019**, *19*, 218. [CrossRef]
29. Mohankumar, P.; Ajayan, J.; Mohanraj, T.; Yasodharan, R. Recent developments in biosensors for healthcare and biomedical applications: A review. *Measurement* **2021**, *167*, 108293. [CrossRef]
30. Konwarh, R.; Sharma, P.L. *Nanosensor Platforms for Surveillance of Plant Pathogens and Phytometabolites/Analytes vis-à-vis Plant Health Status in Micro and Nano Technologies, Nanomaterials for Agriculture and Forestry Applications*; Husen, A., Jawaid, M., Eds.; Elsevier: Amsterdam, The Netherlands, 2020; pp. 357–385.
31. Kundu, M.; Krishnan, P.; Kotnala, R.; Sumana, G. Recent developments in biosensors to combat agricultural challenges and their future prospects. *Trends Food Sci. Technol.* **2019**, *88*, 157–178. [CrossRef]
32. Siebrecht, S.; Herdel, K.; Schurr, U.; Tischner, R. Nutrient translocation in the xylem of poplar—diurnal variations and spatial distribution along the shoot axis. *Planta* **2003**, *217*, 783–793. [CrossRef]
33. Montanaro, G.; Dichio, B.; Lang, A.; Mininni, A.N.; Xiloyannis, C. Fruit calcium accumulation coupled and uncoupled from its transpiration in kiwifruit. *J. Plant Physiol.* **2015**, *181*, 67–74. [CrossRef]
34. Honert, T.H.V.D. Water transport in plants as a catenary process. *Discuss. Faraday Soc.* **1948**, *3*, 146–153. [CrossRef]
35. Dichio, B.; Montanaro, G.; Sofo, A.; Xiloyannis, C. Stem and whole-plant hydraulics in olive (*Olea europaea*) and kiwifruit (*Actinidia deliciosa*). *Trees* **2012**, *27*, 183–191. [CrossRef]
36. Bai, Y.; Li, X.; Liu, S.; Wang, P. Modelling diurnal and seasonal hysteresis phenomena of canopy conductance in an oasis forest ecosystem. *Agric. For. Meteorol.* **2017**, *246*, 98–110. [CrossRef]
37. Zeppel, M.J.B.; Murray, B.R.; Barton, C.; Eamus, D. Seasonal responses of xylem sap velocity to VPD and solar radiation during drought in a stand of native trees in temperate Australia. *Funct. Plant Biol.* **2004**, *31*, 461–470. [CrossRef]
38. Fiscus, E.L. Determination of Hydraulic and Osmotic Properties of Soybean Root Systems. *Plant Physiol.* **1977**, *59*, 1013–1020. [CrossRef]
39. Montanaro, G.; Dichio, B.; Lang, A.; Mininni, A.N.; Nuzzo, V.; Clearwater, M.J.; Xiloyannis, C. Internal versus external control of calcium nutrition in kiwifruit. *J. Plant Nutr. Soil Sci.* **2014**, *177*, 819–830. [CrossRef]

40. Domec, J.-C.; Noormets, A.; King, J.S.; Sun, G.; McNulty, S.G.; Gavazzi, M.J.; Boggs, J.L.; Treasure, E.A. Decoupling the influence of leaf and root hydraulic conductances on stomatal conductance and its sensitivity to vapour pressure deficit as soil dries in a drained loblolly pine plantation. *Plant Cell Environ.* **2009**, *32*, 980–991. [CrossRef]
41. Köhler, B.; Raschke, K. *Loading of Ions into the Xylem of the Root in the Apoplast of Higher Plants: Compartment of Storage, Transport and Reactions*; Sattelmacher, B., Horst, W.J., Eds.; Springer: Dordrecht, The Netherlands, 2007; pp. 181–200.
42. Roddy, A.; Dawson, T. Novel patterns of hysteresis in the response of leaf-level sap flow to vapor pressure deficit. *Acta Hortic.* **2013**, *991*, 261–267. [CrossRef]
43. Wang, H.; He, K.; Li, R.; Sheng, Z.; Tian, Y.; Wen, J.; Chang, B. Impact of time lags on diurnal estimates of canopy transpiration and canopy conductance from sap-flow measurements of *Populus cathayana* in the Qinghai–Tibetan Plateau. *J. For. Res.* **2016**, *28*, 481–490. [CrossRef]
44. Meidner, H.; Mansfield, T.A. Stomatal responses to illumination. *Biol. Rev.* **1965**, *40*, 483–508. [CrossRef]

applied sciences

MDPI

Article

Viñamecum: A Computer-Aided Method for Diagnoses of Pests and Diseases in the Vineyard

Juan Ignacio García-García [1], Daniel Marín-Aragón [1], Hanael Maciá [2] and Ana Jiménez-Cantizano [2,*]

[1] Departament of Mathematics/INDESS (Instituto Universitario para el Desarrollo Social Sostenible), University of Cadiz, 11510 Puerto Real, Spain; ignacio.garcia@uca.es (J.I.G.-G.); daniel.marin@uca.es (D.M.-A.)

[2] Department of Chemical Engineering and Food Technology, Vegetal Production Area, Faculty of Sciences, University of Cadiz, Agrifood Campus of International Excellence (ceiA3), IVAGRO, P.O. Box 40, 11510 Puerto Real, Spain; hanael_macia@hotmail.com

* Correspondence: ana.jimenezcantizano@uca.es

Abstract: Information and telecommunication technologies (ICTs) offer new opportunities to provide more timely information services to farmers. This work aims to present a progressive web app (PWA) for mobile devices, which incorporates updated technical information on the pests and diseases of grapevines. In its development, it generated a database with content related to and photographs of grapevine pests and diseases for access by users using mobile devices. In addition, using an Expert System, the application allows the diagnosis of pathologies and the identification of pests by answering questions that are asked. This PWA is mainly addressed to technicians, students, and winegrowers who want to implement more environmentally friendly crop management strategies. Viñamecum is currently freely.

Keywords: grapevine diseases; grapevine pests; diagnosis; information and telecommunication technologies; progressive web app; decision system; online systems

Citation: García-García, J.I.; Marín-Aragón, D.; Maciá, H.; Jiménez-Cantizano, A. Viñamecum: A Computer-Aided Method for Diagnoses of Pests and Diseases in the Vineyard. *Appl. Sci.* **2021**, *11*, 4704. https://doi.org/10.3390/app11104704

Academic Editors: Ginés García-Mateos, Dolores Parras-Burgos and José Miguel Molina Martínez

Received: 30 March 2021
Accepted: 18 May 2021
Published: 20 May 2021

1. Introduction

Grapevines (*Vitis vinifera* and *Vitis* spp.) are one of the most extensively grown and economically important woody perennial fruit crops in the world [1]. According to the FAO (2019) [2], the European continent, with 3.46 million ha and 26.7 million t, leads grape production worldwide and is followed by Asia (1.99 million ha and 29.51 million t), America (0.96 million ha and 14.4 millon t), Africa (0.34 million ha and 4.89 millon t), and Oceania (0.17 million ha and 1.97 millon t). Its cultivation is limited by the climatic demands imposed by optimal plant development [3] and the incidence of diseases and pests, which can compromise its productivity [4,5]. Chemical control has traditionally been used in conventional viticulture to control pests and diseases in the vineyard [6]. However, this practice is potentially leading to pollution [7] and human health issues [8]. For this reason, the use of phytosanitary products has been regulated in some countries. For example, in Europe, the Directive 2009/128/EC [9] established a framework for Community action to achieve the sustainable use of pesticides and the promotion of integrated pest management (IPM). IPM is an ecosystem-based strategy that focuses on the long-term prevention of crop pests and diseases through different techniques that include the use of resistant varieties, biological control, habitat management, modification of cultural practices and, when needed, judicious and timely use of chemical controls [10,11]. To implement this strategy in vineyard management, it is necessary to have qualified technical personnel and provide specific information to farmers so that they can act under the premises of the IPM.

The widespread growth of information and telecommunication technologies (ICTs) in rural areas offers new opportunities to provide more timely and low-cost information services to farmers, as well as assists in coordinating agricultural agents [12]. Moreover, the enormous progress of mobile devices has opened up a wide range of opportunities that were previously unfeasible. Their portability and computing performance are features that have allowed the development of innovative applications in fields such as medicine, sports, and agriculture [13]. However, in the field of viticulture, new technologies and modern management techniques are needed to ensure and facilitate a correct performance of the winegrowers in order to obtain a quality harvest and reduce the environmental impact. The goal of the present work was to develop a novel smartphone application, called Viñamecum [14], to facilitate a correct control of grapevine diseases and pests and boost IPM in vineyards. This application allows the identification of species suspected to act as pests or pathogens in the vineyard and incorporates technical information on their biological cycles, the different control strategies and images of the damages or symptoms they cause. The application was developed and implemented for mobile devices with the aim of maximizing its availability to users.

2. Solution Adopted: Viñamecum

Viñamecum is a PWA application that incorporates technical information on vineyard pests and diseases and allows the identification of pests and diseases through the use of an expert system. This PWA application is currently freely available at https://fcc.uca.es/web/vinamecum, accessed on 10 March 2021.

For the design and development of the "Viñamecum" application, the target users, the areas or places where it can be used and the applicability of the tool have been considered.

Due mainly to the fact that the native language of the users to whom this application was originally addressed is Spanish, the language used throughout the application is Spanish. Another factor to take into account in the choice of language was the lack of such applications targeted at native Spanish-speaking users. For instance, in other languages, we found applications such as "Di@gnoPlant Vigne" developed by the French National Institute for Agricultural Research (INRA) [15], "MyVigneto Care" in Italian [16], and "Pests and Diseases of Grapes" written in English, but in a non-technical language and with the downloading involving economic costs [17]. The application is planned to be translated into other languages in the future.

Furthermore, a technology was required that would allow it to run on a mobile device, be easy to install and be used with poor connections so that it could be used in places with poor mobile coverage. In addition, the application would have to be fast and resource-efficient. Other desirable features were that it should be easily upgradable, have a simple interface and offer the possibility of user notifications.

There are different tools that could be used in the development of mobile apps. In this work, PWAs were chosen [18]. A search for "PWA the future of mobile apps" in Google generates numerous outputs that illustrate the importance of this technology in the development of applications. The first result tells us: "Progressive Web Application (PWA) is truly considered the future of multi-platform development because of its application on several devices, the improved speed, and the easiness that requires no installation or updates". We could say that a PWA is an enhanced web page, although its development is not entirely straightforward since they have several features that must be implemented.

The main reasons for the choice of this technology were as follows:

1. Available offline: it allows to have the data available even if we do not have internet connection, which is very desirable due to the environment in which the application will be used.
2. Security: the idea is that a farmer can contribute to the application and introduce warning systems. PWAs have enhanced security features and for example impose the use of encrypted connections (Hypertext Transfer Protocol Secure protocol). This is essential when sending data anonymously over the internet.

Some of the features of this technology are discussed in more detail below.

2.1. Progressive Web Applications

A Progressive Web Application (PWA) is a type of application software delivered over the web [18], created using common web technologies such as HTML, CSS, and JavaScript. The origins of PWAs can be traced back to the launch of the iPhone in 2007. Steve Jobs announced that web applications, developed in HTML5 using the AJAX architecture [19], would be the standard format for iPhone applications. No software development kit (SDK) was required, and applications would be fully integrated into the device through the Safari browser engine. This model was later changed for the App Store [20], as a means to avoid jailbreakers and appease frustrated developers. In October 2007, Jobs announced that an SDK would be released the following year. As a result, although Apple continued to support web apps, the vast majority of iOS apps shifted to the App Store.

This type of application is intended to run on any platform that uses a standards-compliant browser. In addition, its includes working offline, push notifications, and device hardware access, enabling user experiences similar to native apps on mobile and desktop devices. Since a progressive web app is a web page or website known as a web app, there is no requirement for developers or users to install web apps through digital distribution systems such as the Apple App Store [20] or Google Play [21].

While web apps have been available for mobile devices since the beginning, they have generally been slower, less feature-rich, and less widely used than native apps. However, with the ability to work offline, which was previously only available to native apps, PWAs running on mobile devices can run much faster and provide more features, closing the gap with native apps, and being portable on desktop and mobile platforms (see [18] for further details).

One of the main advantages of the PWAs is that they do not require separate bundling or distribution. Publishing a progressive web app is the same process as it would be for any other web page. PWAs work in any browser, but "app-like" features, such as connectivity independent, installing on the home screen, and sending messages, depending on browser compatibility. As of April 2018, these features are supported to varying degrees in Microsoft Edge, Google Chrome, Mozilla Firefox, and Apple Safari browsers, but more browsers may support the future's necessary features.

The core of a PWA is in a program named Service Worker (SW), which is just a JavaScript program. It operates separately from the main browser to handle push notifications, synchronize background data, caches or retrieve resource requests, intercept network requests and receive centralized updates. SWs are used to give PWAs the ability to deliver the high performance and rich experience of native mobile apps, to provide real-time updates, and to improve search engine visibility of traditional web apps.

From a technical point of view, SWs go through a three-step lifecycle of registration, installation, and activation. Registration involves telling the browser the location of the service worker in preparation for installation. Installation occurs when there is no service worker installed in the browser for the web application or an update to the service worker. Activation occurs when all PWA pages are closed, so there is no conflict between the previous version and the updated version. The lifecycle also maintains consistency when switching between service worker versions, as only a single service worker can be active for a domain. For more details on these steps, the reader is referred to [18] (Chapter 4). Technically, SWs provide a programmable network proxy in the web browser to manage web/HTTP requests programmatically. Service workers sit between the network and the device to deliver content. They can use caching mechanisms efficiently and enable error-free behavior during offline periods.

Finally, we emphasize that PWAs can be installed on mobile, but this is unnecessary as the same tasks can be carried out simply through the web. In other words, they do not require installation; thanks to this feature, they offer important advantages: better performance, short loading times, interface similar to that of a native app, automatic updating, responsive design, push notifications, secure TLS protocol, and no internet connection. The technology used to develop PWAs provides a cross-platform system, independent of the browser and operating system. This means that there is no need to develop specific programming for each operating system, which greatly reduces costs.

2.2. Knowledge Base and Expert System

An expert system is a computer system emulating the decision-making ability of a human expert [22]. Expert systems are designed to solve complex problems by reasoning through knowledge bodies, represented mainly as if-then rules rather than through a conventional procedural code. The components of the expert system used in this app are the following:

- Knowledge Base: in this component, facts, and rules are represented. Knowledge in a particular domain is stored here and rules for solving a problem, procedures, and intrinsic data relevant to the domain.
- Inference engine is the brain of the expert system. Its function is to obtain the knowledge base's relevant knowledge, interpret it, and find a relevant solution to the user's problem. It contains the rules of its knowledge base and applies them to known facts to infer new facts.

The main part of this application is the Knowledge Base. There are different forms of storage for it. In this application, the form adopted has been a spreadsheet in which each row is a disease/pest, and each column is a symptom/damage. The "1" or "0" mark indicates whether, for a disease/pest, a symptom/damage appears or not. From a mathematical perspective, it is nothing more than a matrix of m rows (number of diseases/pests) and n columns (number of symptoms/damages) with zeros and ones in their entries. We denote the set of matrices of this form by $(\mathbb{Z}_2)^{m\times n}$.

2.3. User Interface

One of the most important parts of an application is the user interface. In this app, we had to make a design that made it possible for non-expert users (some of them had never worked with this kind of technology before) to interact with the program and that users could easily send all the errors found in it.

We chose to implement a very simple diagnostic system. The user only has to answer 1 (if the symptom/damage is present) or 0 (if it is not present). Once detected, the user will be redirected to a page with the characteristics of the disease/pest. The app also has many illustrations, as you can see in the example in the Section 3. The total amount of pictures used in the app is around 400. All these images make it easy for the farmer to identify the symptoms or damages caused by diseases or pests in the vineyard and they make possible to obtain a diagnosis in situ. In addition, the app includes the different strategies the farmer can use to control the diseases and pests. For instance, a link is provided to the website of the "Ministerio de Agricultura, Pesca y Alimentación" (Spain) [23] that regulates the active substances authorized for phytosanitary control.

2.4. The Efficiency of the System

We can view our knowledge base as a matrix A of m rows (diseases/pests) and n columns (symptoms/damages). Formally, we can say that we have an A matrix $(\mathbb{Z}_2)^{m\times n}$. We will see each row as the set of symptoms of a disease.

The process to determine the disease or pest consists of asking for possible symptoms/damages through the user interface. If the symptoms or damage is present, all rows (diseases/pests) that do not manifest that symptom/damage (all rows with a 0 in that column) will be deleted. If the answer is negative, it eliminates the rows that manifest the symptom/damage (all rows with a 1 in that column). The implementation in pseudo-code is showed in Algorithm 1.

The average maximum time needed to obtain an answer is $\log(m)$, where m is the number of rows (diseases and pests classified). We have to take into account that after performing the first steps of the algorithm, our matrix will be equal to the following one:

Algorithm 1: Computation of I_S.

Input : $A = (a_{ij}) \in (\mathbb{Z}_2)^{m \times n}$.
Output: the row of A verifying all the conditions.

begin
  $B \leftarrow A$;
  $(m_B, n_B) \leftarrow$ dimensions(B);
  /* we add a 0th-column to B with the row numbers */
  for $i \in \{1, \dots, m_B\}$ **do**
    $b_{i0} \leftarrow i$;
  $n_B \leftarrow n_B + 1$
  while $m_B > 2$ **do**
    $q \leftarrow$ Random$(2, n_B)$
    /* Ask question q y save the answer in r */
    $r \leftarrow$ Question(q)
    /* Remove the rows not fulfilling r */
    $B \leftarrow \{b_{0j} | j \in \{0, \dots, n_B\}\} \cup \{b_{ij} | b_{iq} == r\}$
    /* Remove the qth column */
    $B \leftarrow \{b_{ij} \mid j \neq q\}$
    $(m_B, n_B) \leftarrow$ dimensions(B);
  return b_{10};

$$B = \begin{pmatrix} 0 & \text{Diseases} & \text{Symtomp 1} & \dots & \text{Symtomp } q & \dots \\ 1 & \text{disease 1} & 1 & \dots & 0 & \dots \\ 2 & \text{disease 2} & 0 & \dots & 1 & \dots \\ \dots & \dots & \dots & \dots & \dots & \dots \\ n & \text{disease n} & 0 & \dots & 0 & \dots \end{pmatrix}.$$

In each step, the algorithm eliminates on average half of the rows. The actions performed at each step with the matrix consist of eliminating one of the rows that do not meet the condition. If the random value is q, the resulting matrix would be as follows:

$$\begin{pmatrix} 0 & \text{Diseases} & \text{Symtomp 1} & \dots & \text{Symtomp q} & \dots \\ \hline \sout{1} & \sout{\text{disease 1}} & \sout{1} & \sout{\dots} & \sout{0} & \sout{\dots} \\ 2 & \text{disease 2} & 0 & \dots & 1 & \dots \\ \dots & \dots & \dots & \dots & \dots & \dots \\ \sout{n} & \sout{\text{disease n}} & \sout{0} & \sout{\dots} & \sout{0} & \sout{\dots} \end{pmatrix}$$

The efficiency of the algorithm is based on the search for the elements of a column that are equal to 1 or 0 (depending on the user's answer).The column to be examined is obtained directly from the matrix and to determine the diseases or pests we are left with, only the rows that have not been discarded so far have to be examined. A tree design could have been used to store all the information, but making it in the form of a matrix makes it much easier to maintain. For example, adding new diseases is as simple as adding a new item to our disease database (the current list of pests and diseases is shown in Table 1).

On the other hand, this form of storage allows us to see the existing relationships between the symptoms of each of the diseases and to establish between them different chains of diseases and symptomatology. From a mathematical point of view, what we have is a graph [24] that relates the different diseases/pests and establishes a distance and a partial order between each of them. The detection of nearby diseases/pests in adjacent areas could enable us to detect diseases with very similar symptomatologies.

Table 1. List of pests and diseases detected by Viñamecum.

Pests	Diseases
Agrotis segetum	Bacterial blight of grapevine
Brevipalpus lewisi McGregor	Black-rot
Calepitrimerus vitis Nalepa	Downy mildew
Ceratitis capitata	Esca of grapevine
Colmerus vitis Pagenstecher	Eutypa dieback
Drepanotrips reuteri Uzel	Grapevine fanleaf disease
Drosophila melanogaster	Grapevine leafroll disease
Eotetranychus carpini Oud	Grey mould disease of grapevine
Epicometis hirta	Petri disease
Eulecanium cori Bouché	Phomopsis Dieback
Frankliniella occidentales	Phytoplasma diseases
Haltica ampelophaga	Pierce's disease
Jacobisca lybica y Empoasca viti	Powdery mildew
Lobesia botrana	
Melolontha melolontha	
Panonychus ulmi Koch	
Planococcus citri Risso	
Schistocerus bimaculatus	
Sparganothis pilleriana	
Tetranychus urticae Koch	
Vesperux xatarti	
Xylotrechus arvicola	

3. A Complete Example

This application can be accessed at the site Viñamecum, accessed on 10 March 2021. From there, the installation is quite simple. When selecting the option "Viñamecum" from the menu, the first three images in the Figure 1 will be seen. Then, Algorithm 1 will be executed.

When answered, the program will ask us a question that will make the algorithm perform a loop, and as many questions as necessary will be asked to determine the disease from the symptoms. Once a solution is obtained, the program will display it, and a message will be sent with a link to the description of the disease and its treatment, as shown in Table 2.

Figure 1. Screenshots of "Viñamecum" for the diagnosis of a disease.

Table 2. Conclusion and steps taken by "Viñamecum" from the data provided by the user.

Steps Performed in the App	Result of `Viñamecum`
• 1st question: Do you notice a stop in leaf growth? Yes • 2nd question: The vine shows rickets, weakening and low cropping? No • 3rd question: Do the leaves dry from the edges inward and curl slightly? Yes • 4th question: Does the plant show weakening with small shoots, short internodes, small and chlorotic leaves? No • Diagnosis: THRIPS (*Drepanotrips reuteri* Uzel)	

As can be seen, all the diseases/pests have been illustrated completely so that the specialist can appreciate the validity of the result obtained in the application. In Figure 2, we can see the images availables in the app for *Drepanotrips reuteri* Uzel.

Figure 2. Description of *Drepanotrips reuteri* Uzel.

4. Conclusions

Based on our experience with the application from its development to its implementation, we have seen that this type of app has a practical application in viticulture. The use of these applications can facilitate integrated pest management.

It is important to highlight that these applications must have the following characteristics:

- Ease of use for farmers;
- Easy installation on mobile devices. PWAs make all this work much easier;
- Automatic updates. PWAs provide simple mechanisms for this purpose;
- They have to provide clear solutions as quickly as possible. The farmers are experts, although it is sometimes difficult to handle so much information on diseases and pests. That is why this type of initiative is so well received on their part;
- They must be available in places with no or poor connections. Crop sites are sometimes far away from communication towers. This is why PWAs are the most appropriate solution;
- The use of mobile devices and the lack of connection make it so that large databases cannot be accessed. Thus, the knowledge base and inference engine used are simple, but in this context, they are a quite effective tool. In addition, the simplicity of the inference engine makes mobile devices suitable for this type of application.

Finally, we would like to emphasize that this paper is just an introductive contribution to the diffusion and the knowledge of the software Viñamecum. Our goal is to have its content continuously updated over time as it is already being used and we are receiving a lot of feedback from the users to improve it.

5. Patents

The "Viñamecum" application was registered on 27 November 2018 with reference CA-256-18 in the "Registrador Territorial de la Propiedad Intelectual de la Comunidad Autónoma de Andalucia", Spain.

Author Contributions: Conceptualization, H.M. and A.J.-C.; methodology, J.I.G.-G. and A.J.-C.; software, J.I.G.-G. and D.M.-A.; validation, J.I.G.-G., H.M., D.M.-A. and A.J.-C.; formal analysis, J.I.G.-G. and D.M.-A.; investigation, H.M. and A.J.-C.; resources, J.I.G.-G. and A.J.-C.; data curation, H.M., D.M.-A. and A.J.-C.; writing—original draft preparation, J.I.G.-G. and A.J.-C.; writing—review and editing, J.I.G.-G. and A.J.-C.; visualization, J.I.G.-G. and A.J.-C.; supervision, J.I.G.-G., D.M.-A.

and A.J.-C.; project administration, J.I.G.-G. and A.J.-C.; funding acquisition, J.I.G.-G. and A.J.-C. All authors have read and agreed to the published version of the manuscript.

Funding: The first and third authors were supported partially by Junta de Andalucía research group FQM-366 and by the project MTM2017-84890-P (MINECO/FEDER, UE). We would like to thank all those who have generously collaborated in the development of this application by allowing the use of their pictures. Among the contributors we would like to highlight the following institutions: UC Davis, University of California: Agricultura y Recursos Naturales (California), Unit State Forest Service, University of Georgia, Consejo Superior de Investigaciones Científicas, INRA (France), Ministério da Agricultura, Pecuária e Abastecimento.

Conflicts of Interest: The authors declare no conflict of interest.

Abbreviations

The following abbreviations are used in this manuscript:

API	Application Programming Interface
CSS	Cascading Style Sheets
ESYRCE	Encuesta sobre Superficies y Rendimientos Cultivos
FAO	Food and Agriculture Organization
ha	hectares
HTML	HyperText Markup Language
IPM	Integrated Pest Management
PWA	Progressive web application
SW	Service Worker
t	tons
TLS	Transport Layer Security
W3C	World Wide Web Consortium

References

1. Gramaje, D.; Urbez-Torres, J.R.; Sosnowski, M.R. Managing grapevine trunk diseases with respect to etiology and epidemiology: Current strategies and future prospects. *Plant Dis.* **2018**, *102*, 12–39. [CrossRef] [PubMed]
2. FAO. Food and Agriculture Data. Available online: http://www.fao.org/faostat/ (accessed on 1 March 2021).
3. Barnuud, N.N.; Zerihun, A.; Gibberd, M.; Bates, B. Berry composition and climate: Responses and empirical models. *Int. J. Biometeorol.* **2014**, *58*, 1207–1223. [CrossRef] [PubMed]
4. Bois, B.; Zito, S.; Calonnec, A. Climate vs grapevine pests and diseases worldwide: The first results of a global survey. *OENO One* **2017**, *51*, 133–139. [CrossRef]
5. Provost, C.; Pedneault, K. The organic vineyard as a balanced ecosystem: Improved organic grape management and impacts on wine quality. *Sci. Hortic.* **2016**, *208*, 43–56. [CrossRef]
6. Aubertot, J.N.; Barbier, J.M.; Carpentier, A.; Gril, J.J.; Guichard, L.; Lucas, P.; Savary, S.; Voltz, M.; Savini, I. *Pesticides, Agriculture et Environnement: Réduirel'Utilisation des Pesticides et en Limiter les Impacts Environnementaux*; Sab-bagh, C., de Menthière, N., Eds.; INRA/Cemagref Report; INRAE: Paris, France, 2005; 68p.
7. Komárek, M.; Čadková, E.; Chrastný, V.; Bordas, F.; Bollinger, J.-C. Contamination of vineyard soils with fungicides: A review of environmental and toxicological aspects. *Environ. Int.* **2010**, *36*, 138–151. [CrossRef] [PubMed]
8. Ntzani, A.; Chondrogiorgi, M.; Ntritsos, G.; Evangelou, E.; Tzoulaki, I. Literature review on epidemiological studies linking exposure to pesticides and health effects (External Scientific Report No. 2013:EN–497). *Eur. Food Saf. Auth.* **2013**, *10*, 497E.
9. EUR-Lex-32009L0128-EN-EUR-Lex. Available online: https://eur-lex.europa.eu/eli/dir/2009/128/oj (accessed on 1 March 2021).
10. Flint, M.L.; Van den Bosch, R. Introduction to Integrated Pest Management. In *Weed and Pest Control*; InTech Publisher: Rijeka, Croatia, 2013; pp. 105–115.
11. Wilson, H.; Daane, K.M. Review of ecologically-based pest management in California vineyards. *Insects* **2017**, *8*, 108. [CrossRef] [PubMed]
12. Aker, J.C.; Ghosh, I.; Burrell, J. The promise (and pitfalls) of ICT for agriculture initiatives. *Agric. Econ.* **2016**, *47*, 35–48. [CrossRef]
13. Aquino, A.; Millan, B.; Gaston, D.; Diago, M.P.; Tardaguila, J. vitisFlower®: Development and testing of a novel an-droid-smartphone application for assessing the number of grapevine flowers per inflorescence using artificial vision techniques. *Sensors* **2015**, *15*, 21204–21218. [CrossRef] [PubMed]
14. Viñamecum. Available online: https://fcc.uca.es/web/vinamecum (accessed on 1 March 2021).
15. Di@gnoPlant Vigne. Available online: https://play.google.com/store/apps/details?id=com.inra.DiagnoPlant.Vigne&gl=ES (accessed on 1 March 2021).

16. MyVigneto Care. Available online: https://play.google.com/store/apps/details?id=com.dooparts.ruzz.myvignetocare&hl=en (accessed on 1 March 2021).
17. Pests and Diseases of Grapes. Available online: https://play.google.com/store/apps/details?id=com.rahmakusumaningrum408 (accessed on 1 March 2021).
18. Ater, T. *Building Profressive Web Apps*; O'Reilly: Sebastopol, CA, USA, 2017; ISBN 9781491961650.
19. Ajax—Web Developer Guides | MDN. Available online https://developer.mozilla.org/en-US/docs/Web/Guide/AJAX (accessed on 15 March 2021).
20. Apple Store. Available online: https://apps.apple.com/es/app/apple-store/id375380948 (accessed on 1 March 2021).
21. Google Play. Available online: http://play.google.com/ (accessed on 1 March 2021).
22. Peter, J. *Introduction To Expert Systems*, 3rd ed.; Addison Wesley: Boston, MA, USA, 1998; p. 2, ISBN 978-0-201-87686-4.
23. [Registro de Productos Fitosanitarios]—Agricultura—magrama.gob.es. Available online: https://www.mapa.gob.es/es/agricultura/temas/sanidad-vegetal/productos-fitosanitarios/registro/menu.asp (accessed on 25 March 2021).
24. West, D.B. *Introduction to Graph Theory*, 2nd ed.; Prentice Hall: Hoboken, NJ, USA, 2001.

applied sciences

MDPI

Article

High-Density Wi-Fi Based Sensor Network for Efficient Irrigation Management in Precision Agriculture

Manuel Jiménez-Buendía [1,*], Fulgencio Soto-Valles [1], Pedro José Blaya-Ros [2], Ana Toledo-Moreo [1], Rafael Domingo-Miguel [2] and Roque Torres-Sánchez [1]

[1] Departamento de Automática, Ingeniería Eléctrica y Tecnología Electrónica (DAIETE), Technical University of Cartagena, Campus Muralla del Mar s/n, E-30202 Cartagena, Spain; pencho.soto@upct.es (F.S.-V.); ana.toledo@upct.es (A.T.-M.); roque.torres@upct.es (R.T.-S.)

[2] Departamento de Ingeniería Agronómica, Technical University of Cartagena, Campus Paseo Alfonso XIII, 48, E-30203 Cartagena, Spain; pedro.blaya@upct.es (P.J.B.-R.); rafael.domingo@upct.es (R.D.-M.)

* Correspondence: manuel.jimenez@upct.es; Tel.: +34-968-33-88-88

Featured Application: This work demonstrates the feasibility of using wireless networks with Wi-Fi technology in the deployment of sensor networks for precision agriculture.

Abstract: The application of deficit irrigation techniques is essential in arid or semi-arid areas of the southeast of Spain, where water is a scarce and very costly resource. However, to apply these techniques, it is necessary to carry out preliminary tests on the specific crop in order to develop the models that allow the optimization of water use while achieving acceptable yields. The system proposed in this article demonstrates the feasibility of using wireless technologies available in most facilities (Wireless Fidelity) to deploy a high-density network of nodes with a variety of heterogeneous sensors to collect data from the soil, plant, and atmosphere. The data are sent and stored in a cloud server for real-time visualization from any mobile device and further analysis. The nodes have been developed using low-cost processors and are equipped with batteries and solar panels, allowing their autonomy to be virtually unlimited, as shown by the consumption studies and tests carried out.

Keywords: precision agriculture; wireless sensor networks; smart agriculture; regulated deficit irrigation

Citation: Jiménez-Buendía, M.; Soto-Valles, F.; Blaya-Ros, P.J.; Toledo-Moreo, A.; Domingo-Miguel, R.; Torres-Sánchez, R. High-Density Wi-Fi Based Sensor Network for Efficient Irrigation Management in Precision Agriculture. *Appl. Sci.* **2021**, *11*, 1628. https://doi.org/10.3390/app11041628

Academic Editor: Nir Krakauer

Received: 13 January 2021
Accepted: 6 February 2021
Published: 11 February 2021

1. Introduction

In arid and semi-arid zones, such as the southeast of Spain, the agriculture yield is conditioned by limiting factors, the main one being the available irrigation water. Regulated deficit irrigation (RDI) techniques allow controlling the water applied to the crop in critical and non-critical phenological stages, determining periods where the crop can be irrigated with much less water. Therefore, these techniques have been widely used in these areas where water is a scarce resource [1–3]. However, the application of RDI strategies is conditioned by crop and climatic conditions. The development of irrigation models specifically designed for a crop would allow the application of RDI for water optimization purposes. In order to develop these models, it is necessary to carry out several comparative tests on different areas of the same crop, where the conditions of the applied RDI are modified. By studying the production and soil and plant water relations on trees subjected to the different treatments, it is possible to determine the most appropriate models for optimal irrigation management. In order to establish these RDI models, sensors that report on the state of the soil, plant and atmosphere are required. In addition, to achieve sufficient precision, it is important to have several repetitions of the measurements deploying different deficit irrigation regimes to obtain relations among the agronomic results and the water stress indicators studied [4–6].

Sensor networks are widely used when a high density of instrumentation is required on test plots. In the particular case of wireless sensor networks (WSN), their usefulness has been demonstrated in different papers [7–10], where the absence of wiring has resulted in a very important improvement in terms of installation flexibility and time savings in installation and maintenance. Several protocols have been described in the bibliography for the deployment of WSN in precision agriculture (PA) [11,12]. Zigbee is one of the most commonly used technologies in recent years with positive results [13–18], although it is claimed that ZigBee communication has the problem of short distance, complicated networking and high communication frequency [19]. From the experience of our research group in previous installations, in which we deployed WSNs with Zigbee nodes [20–23], we found significant drawbacks such as the high maintenance required by the nodes, the difficulty of programming the Zigbee stack, the high price of the software licenses and the limited availability of microcontrollers that supported this stack. Some authors such as [11], who conducted a comprehensive review of WSN technologies and their applications in PA, claim that the most suitable technology for this type of applications is narrowband internet of things (NB-IoT). A well-known example is long-range radio (LoRa), which allows very low energy consumption and long life (i.e., 10 years with battery), low cost, wide coverage and large networks (52,000 devices/channel/cell). This is why LoRa is being used extensively in IoT applications in all fields, including PA [19,24–32]. However, there are numerous WSN deployments in PA that use 802.11 wireless fidelity (Wi-Fi) [18,33–38]. It is important to highlight that the suitability of each technology in terms of range, robustness and, above all, in terms of instrumentation capacity, must be analyzed and studied to use it properly. Wi-Fi networks have also been used previously in other disciplines such as post-harvest for monitoring perishable products in transport and conservation [39].

The sensor network described in this article aims to support agronomic trials to relate criteria of deficit irrigation in almond trees with measured values of soil-plant-atmosphere. To this end, it is necessary to study different deficit irrigation treatments. Due to the size of the plot, the number of trees and parameters to measure, a large and reliable network infrastructure was needed. WSN was selected as the best option for this study, since cabling would be very expensive and unfeasible in practice (due to the need of ducts and ditches, problems with animals, soil moisture...) and a Wi-Fi network was already available at the farm. In particular, using Wi-Fi allows us to take advantage of an infrastructure that already exists in several installations, it is easy to use and the development of the nodes can be very cheap due to its widespread integration in conventional microcontrollers. In technologies such as LoRa, the chips are licensed by a unique company (Semtech Corporation, Camarillo, CA, USA) and they are subjected to Intellectual property [29,40,41]. In addition, Wi-Fi enables the integration of very heterogeneous equipment and sensors. On the other hand, there are a lot of very low-cost system on a chip (SoC) devices with Wi-Fi communication, which are very popular among IoT developers (like ESP8266 and ESP32 [38,42,43]). These devices allow easy integration of soil sensors with specific protocols in agriculture such as SDI-12. Although this technology has disadvantages including high power consumption (compared to others such as Zigbee or LoRa), the use of energy harvesting elements such as solar panels or micro wind turbines [44,45], allows the design of nodes with virtually unlimited autonomy. In fact, one of the objectives of this article is to demonstrate the feasibility of deploying WSN using Wi-Fi by properly managing the energy consumption of the nodes.

2. Materials and Methods

2.1. Experimental Plot

The deployment was carried out at the "Tomás Ferro" Experimental Agri-food Station of the Technical University of Cartagena (37°41′18.9″ N; 0°57′01.7″ W; 32 m above the sea level), located at La Palma (Murcia, Spain) during three growing seasons (2017–2019). The plant material consisted of seventeen-year-old almond trees (*Prunus dulcis* (Mill.) D. A. Webb cv. Marta, grafted on Mayor rootstock) and spaced 6 × 7 m (238 trees ha^{-1}).

The soil was a silty-clay-loam texture and the bulk density ranged 1.3–1.5 g cm^{-3}. The irrigation water, from the Tajo-Segura Water Transfer System, has low salinity (Electrical conductivity, $EC_{25\,°C}$, of 1.2 dS m^{-1}) and was applied with an automatic controller (Agronic 4000, Sistemes Electrònics PROGRÉS, S.A., Lérida, Spain) that performed control of irrigation, fertilization, pH regulation, pumping and filter cleaning with fault detection. The irrigation system was comprised of a double drip line per row with 12 pressure-compensated drippers (PC dripper 4.0 L h^{-1}, Netafim, Tel Aviv, Israel) per tree and spaced 1.0 m.

Daily agrometeorological data were recorded by a weather station at the experimental orchard owned by the Agricultural Information Network System of Murcia (SIAM—siam.imida.es (accessed on 6 April 2020)). The climate of the area was classified as Mediterranean, with dry and warm summers and mild winters. During the experimental period, most rainfall events occurred between September and May, with the average annual rainfall of 327 mm. The average annual reference evapotranspiration (ET_0) of 1320 mm with daily average temperatures ranged between 5.5 °C (January) and 29.5 °C (August) whereas air temperature rarely drops below 0.0 °C. Daily average relative humidity was ranged between 36.5% and 86.0%.

2.2. Irrigation Treatments

The irrigation scheduling was carried out in taking into account the evaporative demand (reference crop evapotranspiration, ET_0), the corresponding crop coefficients K_C [46,47] and the percentage of ground area shaded by the tree canopy (K_r). Therefore, plants irrigation requirements (ET_C) were determined as $ET_C = ET_0 \cdot K_C \cdot K_r$. In our case, the irrigation was managed remotely using a Wi-Fi infrastructure to facilitate and guarantee the reliability of the treatments.

Irrigation treatments were distributed according to a randomized complete block design with three blocks per treatment. Each experimental plot consisted of three adjacent rows of four trees per row, but only the two central trees of the central row were used for experimental measurements. A total of five irrigation treatments were applied in this experiment and the annual average of water applied was 7658, 4758, 3490, 5220 and 4770 m^3 ha^{-1} for the control treatment (this treatment is irrigated at 115% ET_C in order to obtain non-limiting soil water conditions), sustained deficit irrigation (SDI) at 80% ET_C (SDI80), and another at 50% ET_C (SDI50), regulated deficit irrigation at 45% ET_C (RDI45) and another at 30% ETc (RDI30) only in phenological stage IV and at 100% ETc during the rest of irrigation season, respectively. Irrigation timing and frequency were varied during the irrigation seasons (February–November).

2.3. WSN Infrastructure

The experimental plot already provided a basic Wi-Fi installation and data connection network, as it was included in the network infrastructure of the Technical University of Cartagena.

Figure 1 depicts the general scheme of the devices, networks and services deployed in this work for the collection and monitoring of data, as well as for the remote management of the devices. On the laboratory terrace of the farm building (spot a), a device that integrates a 19 dBi 120-degree sector antenna and router board (mANTBox 19 s, MikroTik, Riga, Latvia) was installed. This router, powered by power over ethernet standard (PoE), was connected to the university network available at the laboratory. A virtual private network (VPN) server was configured in this router to facilitate network maintenance and access to equipment such as the datalogger and the fertigation controller. In addition, the router also implements the connection via 5 GHz Wi-Fi (802.11 ac) with a second router with integrated 16 dBi drum antenna (SXT Lite5 ac, MikroTik, Riga, Latvia), located at spot b. The following devices are located at this spot (b): (i) A large weighing lysimeter (2.5 × 2.5 × 1.7 m) with forced suction at the bottom, installed in the middle of a 0.4 ha almond orchard with trees spaced at 5 × 6 m. The soil tank was laid on a platform equipped with load cells (model FX2, Sensocar, Spain) which supplied the total weight of the lysimeter. A 15-year old almond tree (*P. dulcis* (Mill) D. A. Webb cv. Marta) was used as a lysimeter tree for evapotranspiration determination (ETc). The accuracy of the whole weighing system was estimated to be ±500 g. The tree was drip irrigated by four laterals with six 4 L h^{-1} pressure compensating drippers. Irrigation management consisted of restoring water losses due to evapotranspiration and drainage from the previous day; (ii) A data-logger (CR1000, Campbell Scientific, Inc., Logan, UT, USA) used to collect data every 15 min from the sensors installed in the lysimeter: weight and drainage measurement, soil sensors for matric potential (MPS−6 Decagon Devices) and volumetric water content (VWC) (Enviroscan Sentek technologies, Stepney SA 5069 Australia), tree sensors for stress water index determination using dendrometers and thermo-radiometers. A climatic station with ET_0 and vapor pressure deficit (VPD) was also installed; (iii) an Ethernet switch to connect the access points of the wireless links; (iv) a router (BaseBox 2, MikroTik, Riga, Latvia) and a dual polarity, 120°, 2.4 GHz 15 dB, sector antenna (AM2G15-120, Ubiquiti, New York, NY, USA) which provides 2.4 GHz Wi-Fi wireless coverage (802.11n) to the sensor nodes; (v) a 60 degrees 2 × 2 MIMO 10 dbi sector antenna with wireless on-board (SXT 2, MikroTik, Riga, Latvia) to establish an 802.11 wireless link to the irrigation controller, located at spot c. The fertigation control device (Agronic 4000) was connected to a serial Wi-Fi converter (USR-WIFI232-610, Jinan USR IOT Technology Limited, Shandong, China) through its RS-232 serial port to enable its wireless connection to the access point of spot b for the remote management of the fertigation programmer from any computer connected to the VPN of our installation.

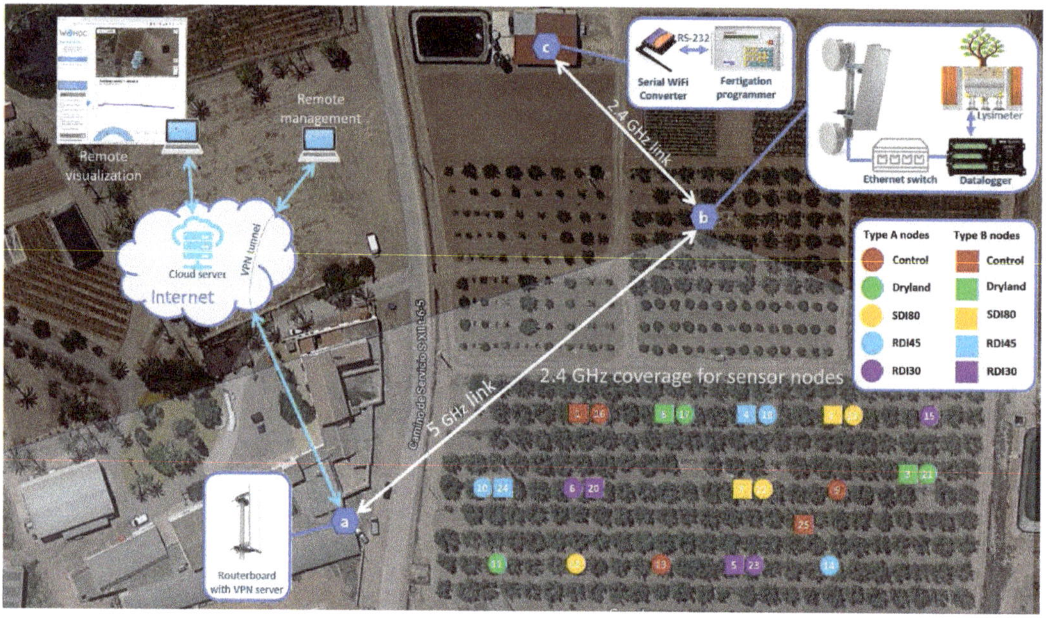

Figure 1. WSN infrastructure of the experimental orchard.

Thanks to this infrastructure, the sensor nodes of the almond plot send the data to a server hosted in the cloud, which provides all the necessary services for its visualization from any personal computer or mobile device. This application allows real-time visualization, logging and querying of historical data, and alarm reporting among other features.

2.4. Agronomic Sensors

To allow for real-time monitoring, several sensors were connected to wireless nodes (see Table 1):

- A total of four sensors per replicate (twelve in total) deployed as follows: two volumetric soil water content sensors (10HS, Decagon devices, Inc., Pullman, WA, USA) and two matric potential of soil water sensors (MPS-6, Decagon devices, Inc., Pullman, WA, USA), installed at 23 cm from each dripper and 25 and 45 cm depth located under the canopy and along the tree drip lines.
- A total of two linear variable displacement transducer sensors (LVDT, Model DF ± 2.5 mm, Solartron Metrology, Bognor Regis, UK) per replicate (six in total) were installed on the main tree branch for measuring micrometric branch diameter variations.
- A total of two infrared radiometer sensors (IR120, Campbell Scientific Inc., USA) per treatment were mounted on wooden poles and installed at 0.5–1.0 m over the trees canopy for monitoring canopy temperature.
- A total of four climatic sensors (VP4, Decagon devices, Inc., Pullman, WA USA) were installed to record air temperature, relative humidity and vapor pressure deficit.

Table 1. Overview of the sensors used in the plot.

Sensor	Measured Data	Output	Range	Resolution	Accuracy	Supply Voltage Range	Current Consumption
10HS [1]	Soil moisture	0.3–1.25 V	0–57% VWC	Not applicable	±3% VWC	3–15 VDC	15 mA
MPS-6 [1]	Soil matric potential and temperature	SDI-12	−9 to −100000 kPa −40 to +60 °C	0.1 kPa 0.1 °C	±10% (from −9 to −100 kPa) ±1 °C	6–15 VDC	10 mA
LVDT DF2.5 [2]	Displacement sensor	750 mV/mm	±2.5 mm	Not applicable (analog sensor)		10–24 VDC	10 mA
VP-4 [1]	Relative humidity, temperature, vapor pressure and barometric pressure	SDI-12	0 to 100% RH −40 to 80 °C 0 to 47 kPa 50 to 110 kPa	0.1% RH 0.1 °C 0.001 kPa 0.01 kPa	from ±2 to ±8% from ±0.3 to ±0.4 °C from ±0.03 to ±0.33% ±0.4 kPa	3.6–15 VDC	5 mA
IR120 [3]	Leaf temperature	20 mV per °C (difference from sensor body)	−25 to +60 °C	Not applicable (analog sensor)	±0.2 °C	+2 to +3.5 V (Thermopile) −2.5 V or +400 mV (Thermistor)	0.4 mA

[1] https://www.metergroup.com (accessed on 29 October 2020), [2] https://www.solartronmetrology.com (accessed on 29 October 2020), [3] https://www.campbellsci.eu (accessed on 29 October 2020).

2.5. Sensor Nodes

From the hardware/software point of view, there are two types of nodes: type A and type B, with different subtypes according to the sensors connected to them. The electronics of nodes A and B are housed inside a 120 × 120 × 90 mm IP65 box (G279C, Gainta AEK GmbH—Electrical Enclosures, Frankfurt, Germany). The power supply system consists of a 3.7 V, 3000 mAh Li-ion rechargeable battery, charged by an 80 × 80 mm solar panel (5 V, 0.8 W). Figure 2 shows pictures of both types of node installed in the field.

Table 2 shows the 25 nodes installed in the plot (see Figure 1), detailing, for each of them, the type it belongs to, the replicate, the irrigation treatment it is monitoring and the sensors connected to it.

Figure 2. Pictures of type-A (left) and type-B (right) nodes.

Table 2. Description of the deployed nodes.

Node Number	Node Type	Replicate	Irrigation Treatment	Sensors
1	B1	1	Control	2 10HS, 2 MPS-6, 1 VP-4 and 1 IR120
2	B1	1	SDI80	2 10HS, 2 MPS-6, 1 VP-4 and 1 IR120
3	B2	2	SDI50	2 10HS, 2 MPS-6 and 1 IR120
4	B2	1	RDI45	2 10HS, 2 MPS-6 and 1 IR120
5	B1	3	RDI30	2 10HS, 2 MPS-6, 1 VP-4 and 1 IR120
6	B2	2	RDI30	2 10HS, 2 MPS-6 and 1 IR120
7	B2	2	SDI80	2 10HS, 2 MPS-6 and 1 IR120
8	B2	1	SDI50	2 10HS, 2 MPS-6 and 1 IR120
9	A1	2	Control	2 10HS, 2 MPS-6 and 2 LVDT
10	A1	2	RDI45	2 10HS, 2 MPS-6 and 2 LVDT
11	A1	3	SDI50	2 10HS, 2 MPS-6 and 2 LVDT
12	A1	3	SDI80	2 10HS, 2 MPS-6 and 2 LVDT
13	A1	3	Control	2 10HS, 2 MPS-6 and 2 LVDT
14	A1	3	RDI45	2 10HS, 2 MPS-6 and 2 LVDT
15	A1	1	RDI30	2 10HS, 2 MPS-6 and 2 LVDT
16	A2	1	Control	2 LVDT
17	A2	1	SDI50	2 LVDT
18	A2	1	RDI45	2 LVDT
19	A2	1	SDI80	2 LVDT
20	A2	2	RDI30	2 LVDT
21	A2	2	SDI50	2 LVDT
22	A2	2	SDI80	2 LVDT
23	A2	3	RDI30	2 LVDT
24	B3	2	RDI45	1 IR120
25	B3	2	Control	1 IR120

The hardware of the type A node (see Figure 3) is based on a wireless datalogger designed by our research group for PA applications (MEWiN MainBoard) [20]. For the signal conditioning from the sensors, an interface board (SensorBoard A) has been designed, connected to the first one by means of expansion connectors. An ESP8266 microcontroller (Esspressif Systems Co. LTD., Shanghai, China) with WiFi communication capability is also plugged to the SensorBoard A.

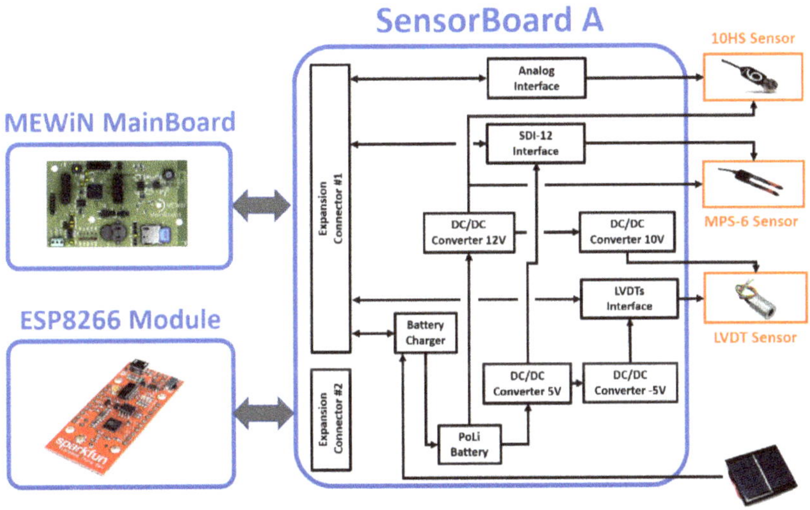

Figure 3. Block diagram of type-A node.

The SensorBoard A consists of four DC/DC voltage converters (12, 10, 5 and −5 V), required to supply the sensors and to guarantee the operation of the different interfaces on the board: one for SDI12 sensors, another for radial dendrometers and one for analogue sensors (0–3 V). Finally, the board includes a circuit for the management of the rechargeable battery that supplies the power to the node.

The type B Node is similar to type A, but it is based on the use of a Wipy 3.0 (Pycom©, London, UK) (see Figure 4).

The SensorBoard B includes four DC/DC voltage converters (12, 10, −3.3 and −2.5 V), required to supply the sensors and to guarantee the operation of the different interfaces on the board: one for SDI12 sensors, another for thermal radiometers and one for analogue sensors (0–3 V). Finally, the board includes a circuit for the management of the rechargeable battery that supplies the power to the node.

The software of the nodes was developed using the MicroPython programming language and a variety of open source libraries for Wi-Fi connectivity and sensor reading, among others. For some tasks, such as data acquisition from SDI-12 sensors, specific algorithms were developed.

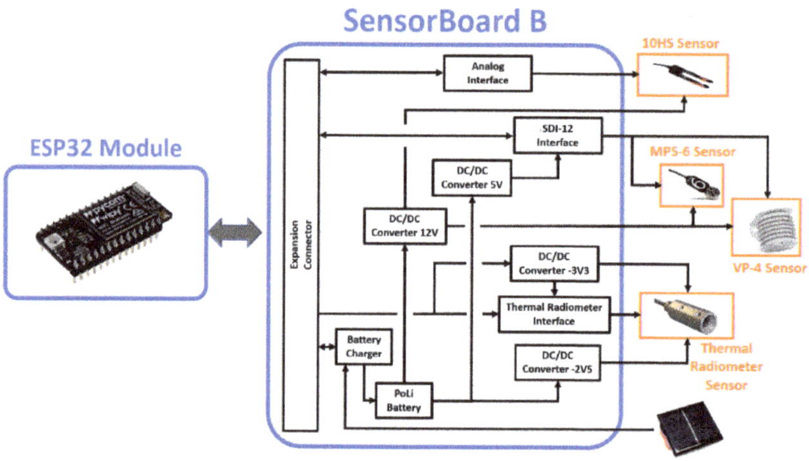

Figure 4. Block diagram of type-B node.

2.6. Time Synchronization

One of the requirements established by the agronomists for this study was that all the nodes should take the sensor measurements at the same time every 15 min so that they could be compared. Therefore, a synchronization algorithm was implemented: Once a day, the nodes connected to a time server (NTP) and update their real-time clock (RTC), which is very inaccurate (it has a daily drift of approximately 50 s). This inaccuracy, together with the drift of the internal timer in deep sleep mode (about 0.3%), generated a time drift that resulted in desynchronized measurements between nodes. To solve this problem, a time correction algorithm was developed. This algorithm determined the time the system requires to perform the measuring and data sending processes, working out the remaining time needed to be in deep sleep mode before waking up at the exact minute defined by the sampling period. In this deployment, measurements were performed every 15 min (minute 0, 15, 30 and 45 of each hour) and measurements were sent every 30 min (minute 0 and 30 of each hour).

3. Results and Discussion

After the functional validation, in view of the importance of achieving the maximum autonomy, the current consumption of each of the node types was measured in the field. The measurements were performed with a data acquisition card (USB-6008, National Instruments Corp., Austin, TX, USA) that measured the voltage drop at a 1 Ohm shunt resistor placed in series with the battery that powered the node. The analog input of the card was configured in differential mode in the range of ± 1 V and the acquisition was carried out in continuous mode at 1 kHz obtaining the averages every 0.25 s. In this mode, the card has a 12-bit resolution (equivalent to 0.48 mA) and an absolute precision at 25 °C of 1.53 mV (1.53 mA in current).

For both type A and type B nodes, consumption tests were performed for the three possible operating modes, (i) the node connects to synchronize the date and time, collects data from the sensors and sends them to the server (this is the most energy-demanding operating mode and is therefore programmed to occur only every 24 h), (ii) the node collects data and sends them to the server (occurs every 30 min) and (iii) the node collects data and stores them in the memory card waiting for the next connection to the server (occurs in the middle of the period between sending events, i.e., 15 min later).

3.1. Consumption of Type A Nodes

Node type A had a standby consumption of 1.1 mA. Figure 5 shows the current required in each of the states through which the node passes in operating mode (i). The measured current and the duration of each of these states are shown in Table 3. With these values, taking into account the standby current, the total consumption of the mode has been calculated with Equation (1).

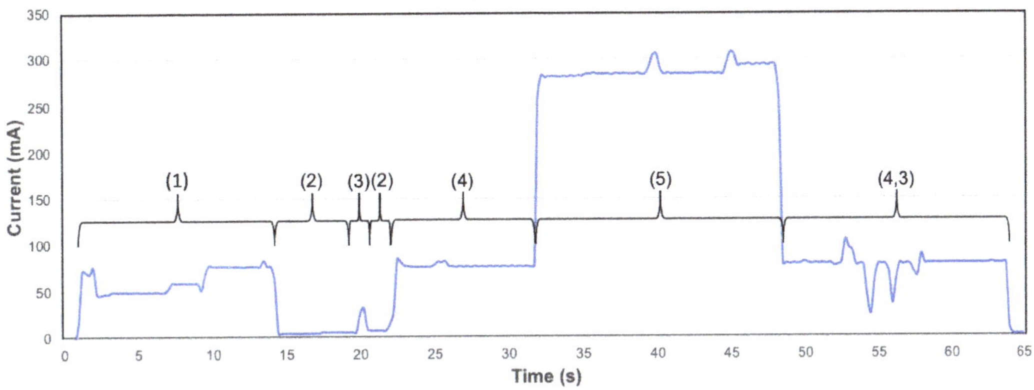

Figure 5. Node A—mode (i): NTP sync, sensor reading and data sending (once every 24 h).

The function of each state is indicated in its description (Table 3). In state 2 (task signaling), the node displays information about the task it is executing by means of a lighting sequence of four LEDs. This information helps the user to perform maintenance and commissioning tasks in the field.

Table 3. Description of functional states of node A—mode (i).

State	1	2	3	2	4	5	4, 3
Description	Time synchronization	Task signaling	Access to microSD	Task signaling	Data sending	Sensors reading	Access to microSD and data sending
$I_{average}$ (mA)	61	4.6	21	7.1	74	282	75
Duration (s)	13	6	0.5	1	10	16.5	15

$$\overline{I}_{A\_total\_i} = \frac{\sum_{A\_state=1}^{A\_state=4,3}\left(\overline{I}_{A\_state} - \overline{I}_{A\_standby}\right)\cdot t_{A\_state}}{24\cdot 60\cdot 60} = 0.084\ \text{mA} \tag{1}$$

Figures 6 and 7, Tables 4 and 5, and Equations (2) and (3) show similar calculations for the other two modes of operation of the node: modes (ii) and (iii). Note that mode (i) occurs once a day, i.e., every 86,400 s (24·60·60) and modes ii and iii every 30 min, i.e., every 1800 s (30 × 60).

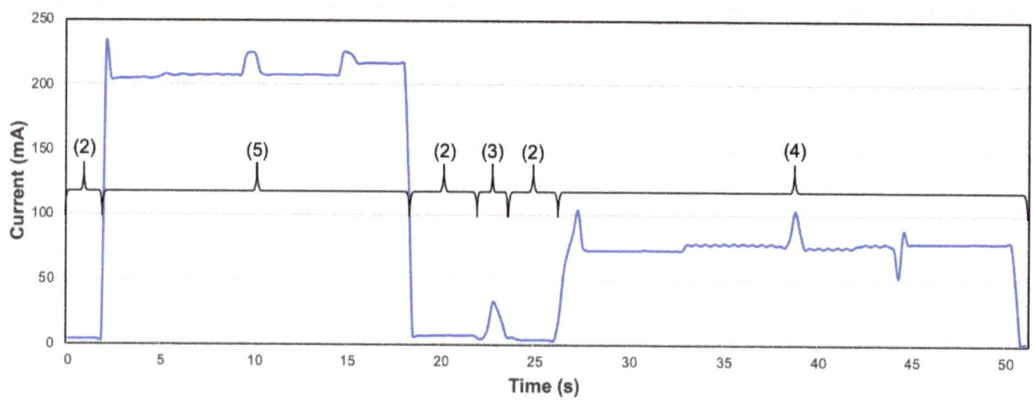

Figure 6. Node A—mode (ii): Sensor reading with data sending (every 30 min).

Table 4. Description of functional states of node A—mode (ii).

State	2	5	2	3	2	4
Description	Task signaling	Sensors reading	Task signaling	Access to microSD	Task signaling	Data sending
$I_{average}$ **(mA)**	4	206	7	23	4.5	76
Duration (s)	2	16.5	3.5	1	4	15

$$\overline{I}_{A\_total\_ii} = \frac{\sum_{A\_state=2}^{A\_state=4}\left(\overline{I}_{A\_state} - \overline{I}_{A\_standby}\right)\cdot t_{A\_state}}{30\cdot 60} = 2.54\text{ mA} \quad (2)$$

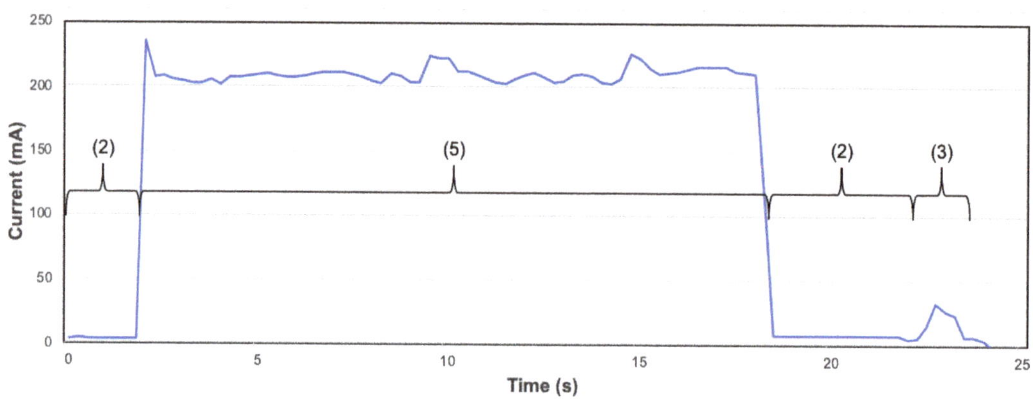

Figure 7. Node A—mode (iii): Sensor reading without data sending (every 30 min).

Table 5. Description of functional states of node A—mode (iii).

State	2	5	2	3
Description	Task signaling	Sensors reading	Task signaling	Access to microSD
$I_{average}$ (mA)	4	208	7	16
Duration (s)	2	16	3.5	1.5

$$\overline{I}_{A\_total\_iii} = \frac{\sum_{A\_state=2}^{A\_state=3}\left(\overline{I}_{A\_state} - \overline{I}_{A\_standby}\right) \cdot t_{A\_state}}{30 \cdot 60} = 1.86\ \text{mA} \tag{3}$$

Once the total consumption current of each of the operating modes has been determined, taking into account the standby current and the capacity of the battery installed in the nodes, it is possible to calculate (Equation (4)) the autonomy (in days) provided that the battery is not charged by any means.

$$Number\ of\ days = \frac{3000\ \text{mA}\cdot\text{h}}{\left(\overline{I}_{A\_total\_i} + \overline{I}_{A\_total\_ii} + \overline{I}_{A\_total\_iii} + \overline{I}_{A\_standby}\right)\cdot 24\ \text{h}} = 22.38\ \text{days} \tag{4}$$

3.2. Consumption of Type B Nodes

The same measurements and calculations made for node A were carried out for type B node. These calculations are shown in Figures 8–10, Tables 6–8, and Equations (5)–(7). In this case, node B had a standby current of 1.5 mA.

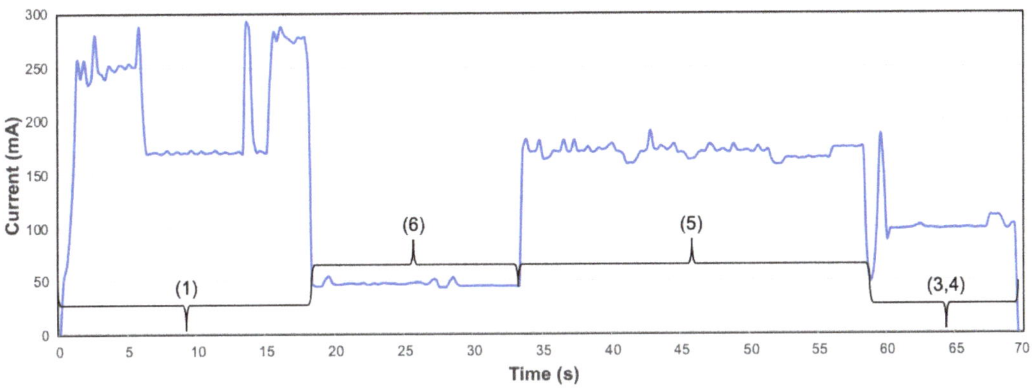

Figure 8. Node B—mode (i): NTP sync, sensor reading and data sending (once every 24 h).

Table 6. Description of functional states of node B—mode (i).

State	1	6	5	3, 4
Description	Time synchronization	Sensors preheating	Access to microSD	Access to microSD and data sending
$I_{average}$ (mA)	61	4.6	21	74
Duration (s)	13	6	0.5	10

$$\overline{I}_{B\_total\_i} = \frac{\sum_{B\_state=1}^{B\_state=3,4}\left(\overline{I}_{B\_state} - \overline{I}_{B\_standby}\right) \cdot t_{B\_state}}{24 \cdot 60 \cdot 60} = 0.111 \text{ mA} \quad (5)$$

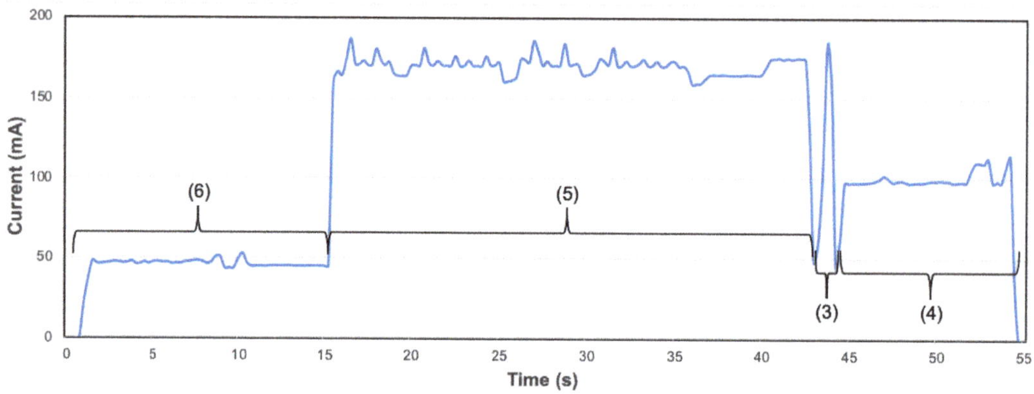

Figure 9. Node B—mode (ii): Sensor reading and data sending (every 30 min).

Table 7. Description of functional states of node B—mode (ii).

State	6	5	3	4
Description	Sensors preheating	Sensors reading	Access to microSD	Data sending
$I_{average}$ (mA)	46	170	150	100
Duration (s)	14	17	1	10

$$\overline{I}_{B\_total\_ii} = \frac{\sum_{B\_state=2}^{B\_state=4}\left(\overline{I}_{B\_state} - \overline{I}_{B\_standby}\right) \cdot t_{B\_state}}{30 \cdot 60} = 2.57 \text{ mA} \quad (6)$$

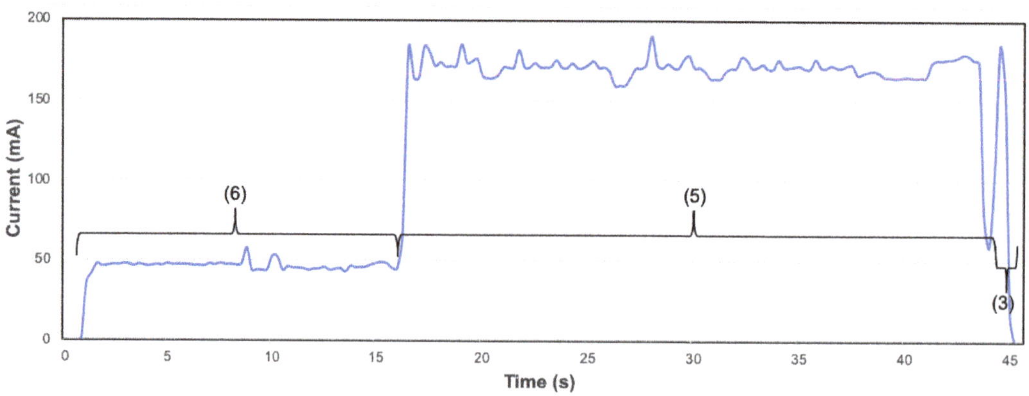

Figure 10. Node B—mode (iii): Sensor reading without data sending mode (every 30 min).

Table 8. Description of functional states of node B—mode (iii).

State	6	5	3
Description	Sensors preheating	Sensors reading	Access to microSD
$I_{average}$ **(mA)**	46	170	150
Duration (s)	14	17	1

$$\overline{I}_{B\_total\_iii} = \frac{\sum_{B\_state=2}^{B\_state=3}\left(\overline{I}_{B\_state} - \overline{I}_{B\_standby}\right)\cdot t_{B\_state}}{30\cdot 60} = 2.02\ \text{mA} \quad (7)$$

The autonomy of node B, provided that the battery is not charged by any means, is depicted in Equation (8).

$$Number\ of\ days = \frac{3000\ \text{mA}\cdot\text{h}}{\left(\overline{I}_{B\_total\_i} + \overline{I}_{B\_total\_ii} + \overline{I}_{B\_total\_iii} + \overline{I}_{B\_standby}\right)\cdot 24\ \text{h}} = 20.15\ \text{days} \quad (8)$$

3.3. Sensor Data Validation

In order to demonstrate the usefulness and viability of the deployed network of nodes, data registered over a significant period of time from an agronomic and sensor network reliability point of view are shown below.

Figure 11 shows the evolution of the VWC at both 25 and 45 cm depth, measured by 10HS sensors (node 04). Irrigation was scheduled every two or three days at 11 a.m. and it took 2 h. VWC values ranged between 0.413 and 0.336 $m^3\ m^{-3}$ during the test period. The maximum values were reached during the irrigation events. After the irrigation events, VWC values fell due to infiltration and they stabilized at values close to 0.358 $m^3\ m^{-3}$. The information registered by MPS-6 sensors (Figure 12) was complementary to that obtained by 10HS sensors (Figure 11). In each irrigation event and depth, the value increased until it was close to -7 kPa, being considered this as field capacity [48]. The falls of MPS-6 sensors were steeper than those registered by 10HS sensors. Both types of sensors detected the irrigation events between 1 and 2 h after irrigation events started in both depths. During the test period, evaporative demand increased strongly, going from a reference crop evapotranspiration of 2.5 to 4.0 mm d^{-1} at the end of the study period, which accelerated the falls of the VWC and soil matric potential values, as well as it showed a greater drop in values. The values showed both types of sensors at saturation level were similar to those reported by [49] on similar texture soil.

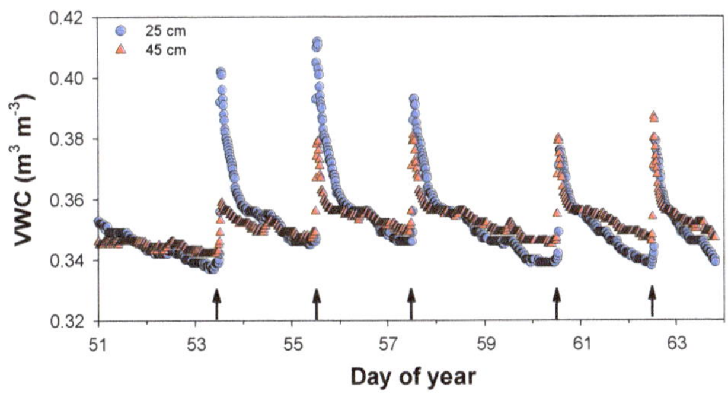

Figure 11. Evolution of VWC sensors at 25 and 45 cm depth measured by node 04 with 10HS sensors (year 2020). The arrows indicate the irrigation events.

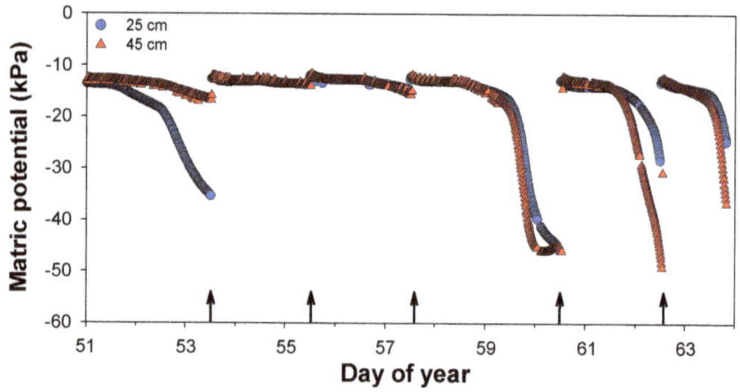

Figure 12. Evolution of matric potential at 25 and 45 cm depth measured by node 04 with MPS-6 sensors (year 2020). The arrows indicate the irrigation events.

Additionally, the nodes automatically and continuously monitored plant water status with LVDT sensors. Figure 13 shows trunk diameter fluctuations (TDF) where the daily cycle of shrinkage for RDI30 for three consecutive days can be observed. The maximum trunk diameter (MXTD) was reached at the beginning of the day and the minimum trunk diameter (MNTD) was recorded during the afternoon. From data obtained from TDF, two indices were calculated to establish plant water status. The daily trunk shrinkage values, calculated as the difference between MXDT and MNTD on the same day, were 271, 411 and 292 μm for three consecutive days, respectively. The trunk growth rate values were calculated as the difference between MXTD of two consecutive days. In general, they were close to 0 μm throughout the period considered. The trend and values obtained were in agreement with those reported by [50,51].

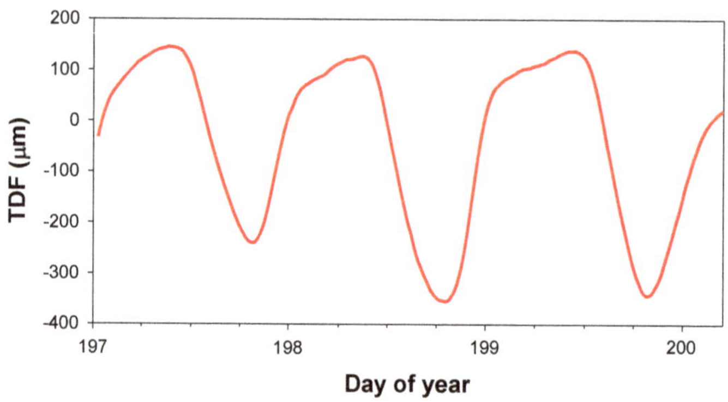

Figure 13. Evolution of trunk diameter fluctuations (TDF) of the RDI30 treatment trees during deficit irrigation period (stage IV). The data was stored by node 23 equipped with linear variable displacement transducer sensors (year 2019).

3.4. Reliability of the Nodes

In general terms, the performance of the nodes was satisfactory. Obviously, the installation required periodic maintenance tasks, such as cleaning the solar panels, and eventual interventions due to occasional failures in the hardware of the nodes.

The evolution of the battery voltage of the nodes is a useful indicator of the "health" of the node. As an example, Figure 14 shows the evolution of the battery voltage of one of the nodes (node 25) in typical conditions. As can be seen, the battery management algorithm starts the charge when the voltage falls below the lower threshold (3.95 V) and stops when the upper threshold is reached (4.1 V). In this way, it works within the optimum voltage range for the battery and reduces the charge cycles, extending its useful life. During the charge there are different voltage peaks corresponding to the daylight hours (the solar panel supplies charge) and the hours of darkness during the night (there is no photovoltaic generation and the battery discharges due to the node consumption). The "discrepancy" that appears between days 82 and 87 (22 and 27 March 2020) corresponds to days when solar radiation was lower due to the existence of cloudiness (see Figure 15).

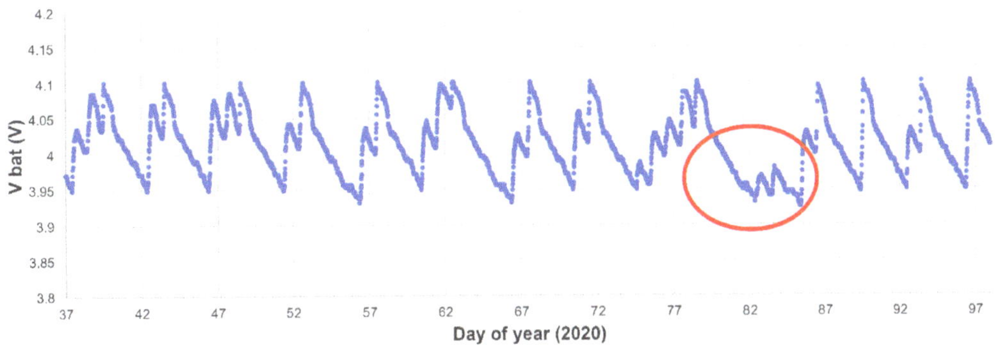

Figure 14. Typical behavior of the battery voltage in a node without operational problems.

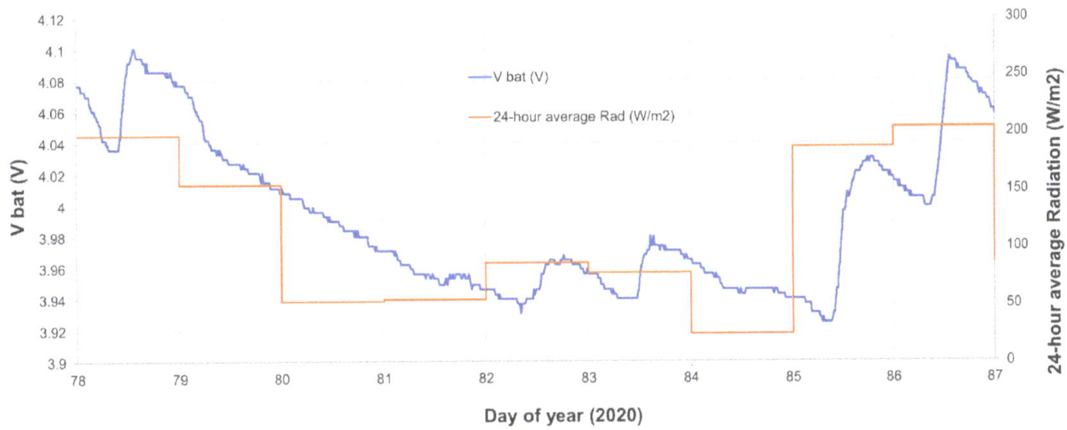

Figure 15. Battery voltage (node 25) vs. daily average radiation (source: http://siam.imida.es/ Agro-Meteorological Weather Station CA12-La Palma, accessed on 6 April 2020). Cloudy days.

3.5. Typical Node Failures

Among the common problems detected during the operation of the deployment, the most frequent were those that led to the exhaustion or destruction of the battery. The usual reasons were (i) overheating due to the high temperatures reached inside the box in summer, (ii) dirt or ageing of the solar panels and (iii) natural ageing of the batteries as they reached the end of their lifespan.

Figure 16 presents the evolution of the battery voltage of node 23. As can be seen, between 7 and 27 March there was a prolonged discharge due to the ageing of the panel, which stopped charging (this was later verified in a field intervention). Once the solar panel was replaced on the 27th, the node started charging again. It is worth noting that from this graph it is possible to extract very interesting information: due to the failure of the panel, the node showed an autonomy of approximately 20 days, which validates the theoretical calculations indicated in the section of consumption and autonomy estimation of the nodes.

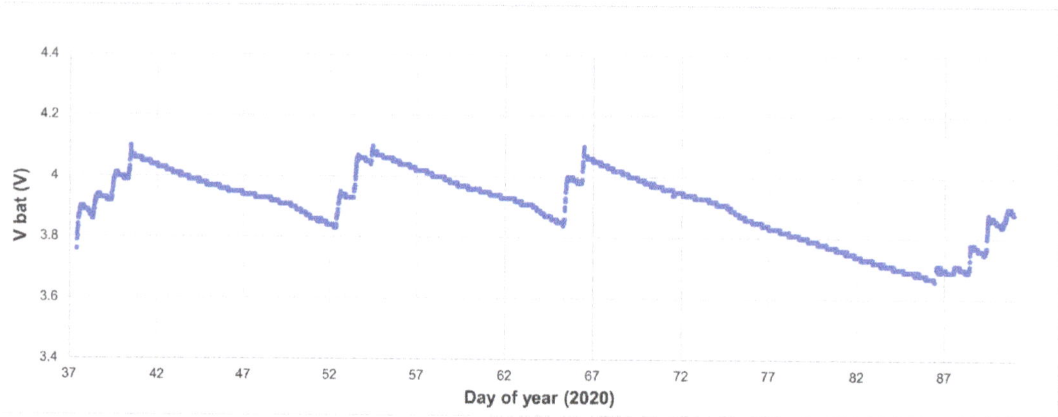

Figure 16. Evolution of the battery voltage of node 23.

Figures 17 and 18 depict failures in two other nodes due to dirt on the solar panels (mainly dust and bird droppings). In the case of node 5 (Figure 17), it can be seen that since 15 March, there were successive loads that did not reach the upper threshold. From 21 March, the panel did not supply the required current to increase the battery charge, whose voltage started a downward trend until 25 March, the day on which the panel was cleaned, and the node recovered. In the case of node 4 (Figure 18), the node never managed to reach the upper threshold because the panel current was not sufficient, so on 24 March, the voltage fell below the minimum operating voltage and the node stopped transmitting.

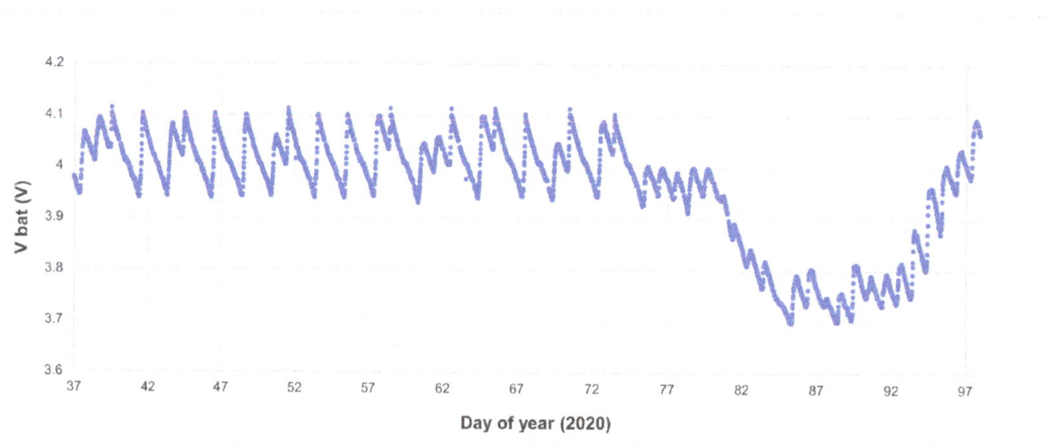

Figure 17. Evolution of the battery voltage of a node with a dirty panel (node 05).

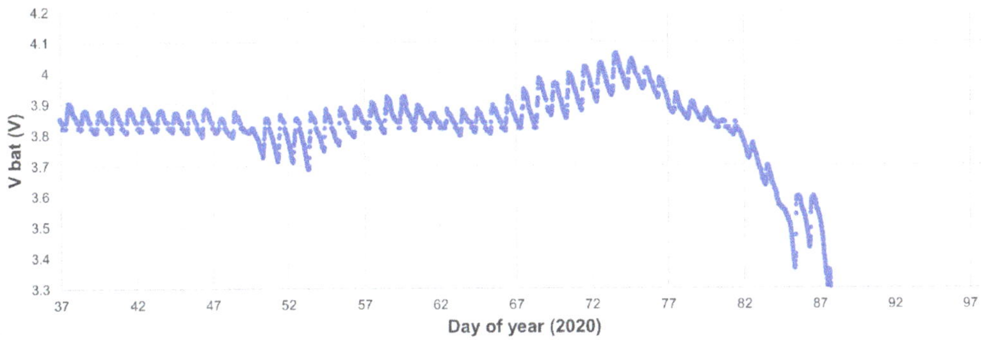

Figure 18. Evolution of the battery voltage of a node with a dirty panel (node 04).

4. Conclusions

The results described in this work demonstrate the viability of using new or existing Wi-Fi infrastructures for the collection of agronomic data with low-cost nodes in deployments for taking measurements over long periods of time. Although some authors argue that Wi-Fi is not the most suitable technology for WSN in PA, the application of techniques to reduce consumption and the use of energy harvesting methods (solar panels) allows for virtually unlimited node operating autonomy. Consequently, this work demonstrates that, properly managing the energy demand, WSN deployment using Wi-Fi is feasible and even advantageous in installations where Wi-Fi coverage is already available or a major investment in the communications infrastructure is not desired. In addition, the use of Wi-Fi is particularly convenient when using multi-node deployments on the same plot (high density), e.g., in research trials where it is necessary to replicate measurements in several locations to minimize variability, and also to perform them with different parameters to evaluate the performance of irrigation strategies on a given crop.

In addition, time synchronization techniques have been implemented to ensure time alignment of the data measured by the sensors. This facilitates its subsequent analysis for the realization of agronomic studies that relate the different irrigation treatments with the state of the plant and the crop yield.

Likewise, the experimental installation described in this work has demonstrated the reliability of the nodes, sensors and communications infrastructure used. The experimental results have shown that, although wireless networks require less installation and maintenance tasks than wired ones, they do need programmed servicing interventions and occasional troubleshooting.

Next steps would involve the integration of nodes with other communication technologies such as LoRa to carry out studies to compare the advantages and disadvantages of each technology.

Author Contributions: Conceptualization, R.T.-S. and M.J.-B.; methodology, R.D.-M. and P.J.B.-R.; software, A.T.-M., F.S.-V. and M.J.-B.; validation, P.J.B.-R., F.S.-V. and A.T.-M.; formal analysis, F.S.-V. and P.J.B.-R.; investigation, R.D.-M. and P.J.B.-R.; resources, R.T.-S. and R.D.-M.; data curation, M.J.-B. and A.T.-M.; writing—original draft preparation, M.J.-B.; writing—review and editing, M.J.-B., P.J.B.-R., F.S.-V. and R.T.-S.; visualization, M.J.-B.; supervision, R.T.-S. and R.D.-M.; project administration, R.T.-S. and R.D.-M.; funding acquisition, R.T.-S. and R.D.-M. All authors have read and agreed to the published version of the manuscript.

Funding: This research was funded by the Spanish Science and Innovation Ministry (MICIIN) and the European Agricultural Funds for Rural Development, grant numbers: AGL2016-77282-C3-3-R and PID2019-106226RB-C22/AEI/10.13039/501100011033.

Conflicts of Interest: The authors declare no conflict of interest.

References

1. Food and Agriculture Organization of the United Nations. *Deficit Irrigation Practices*; Food and Agriculture Organization of the United Nations: Rome, Italy, 2002; Volume 22.
2. Yu, L.; Zhao, X.; Gao, X.; Siddique, K.H.M. Improving/maintaining water-use efficiency and yield of wheat by deficit irrigation: A global meta-analysis. *Agric. Water Manag.* **2020**, *228*, 105906. [CrossRef]
3. Pardo, J.J.; Martínez-Romero, A.; Léllis, B.C.; Tarjuelo, J.M.; Domínguez, A. Effect of the optimized regulated deficit irrigation methodology on water use in barley under semiarid conditions. *Agric. Water Manag.* **2020**, *228*, 105925. [CrossRef]
4. Kang, S.; Zhang, L.; Liang, Y.; Hu, X.; Cai, H.; Gu, B. Effects of limited irrigation on yield and water use efficiency of winter wheat in the Loess Plateau of China. *Agric. Water Manag.* **2002**, *55*, 203–216. [CrossRef]
5. Du, T.; Kang, S.; Zhang, J.; Davies, W.J. Deficit irrigation and sustainable water-resource strategies in agriculture for China's food security. *J. Exp. Bot.* **2015**, *66*, 2253–2269. [CrossRef]
6. Blanco, V.; Torres-Sánchez, R.; Blaya-Ros, P.J.; Pérez-Pastor, A.; Domingo, R. Vegetative and reproductive response of 'Prime Giant' sweet cherry trees to regulated deficit irrigation. *Sci. Hortic.* **2019**, *249*, 478–489. [CrossRef]
7. Haule, J.; Michael, K. Deployment of wireless sensor networks (WSN) in automated irrigation management and scheduling systems: A review. In Proceedings of the 2nd Pan African International Conference on Science, Computing and Telecommunications, PACT 2014, Arusha, Tanzania, 14–18 July 2014; pp. 86–91.
8. Wang, N.; Li, Z. Wireless sensor networks (WSNs) in the agricultural and food industries. In *Robotics and Automation in the Food Industry: Current and Future Technologies*; Elsevier Ltd.: Amsterdam, The Netherlands, 2012; pp. 171–199, ISBN 9781845698010.
9. Balendonck, J.; Hemming, J.; van Tuijl, B.A.J.; Incrocci, L.; Pardossi, A.; Marzialetti, P. Sensors and wireless sensor networks for irrigation management under deficit conditions (FLOW-AID). In Proceedings of the International Conference on Agricultural Engineering (AgEng 2008), Hersonissos, Crete, 23–25 June 2008; pp. 3–452.
10. Elijah, O.; Rahman, T.A.; Orikumhi, I.; Leow, C.Y.; Hindia, M.N. An Overview of Internet of Things (IoT) and Data Analytics in Agriculture: Benefits and Challenges. *IEEE Internet Things J.* **2018**, *5*, 3758–3773. [CrossRef]
11. Jawad, H.M.; Nordin, R.; Gharghan, S.K.; Jawad, A.M.; Ismail, M. Energy-efficient wireless sensor networks for precision agriculture: A review. *Sensors* **2017**, *17*, 1781. [CrossRef]
12. Aqeel-ur-Rehman; Abbasi, A.Z.; Islam, N.; Shaikh, Z.A. A review of wireless sensors and networks' applications in agriculture. *Comput. Stand. Interfaces* **2014**, *36*, 263–270. [CrossRef]
13. De Oliveira, K.V.; Castelli, H.M.E.; Montebeller, S.J.; Avancini, T.G.P. Wireless Sensor Network for Smart Agriculture Using ZigBee Protocol. In Proceedings of the 2017 IEEE 1st Summer School on Smart Cities, S3C 2017—Proceedings, Natal, Brazil, 6–11 August 2017; pp. 61–66.
14. Prakash, S. Zigbee based wireless sensor network architecture for agriculture applications. In Proceedings of the 3rd International Conference on Smart Systems and Inventive Technology, ICSSIT 2020, Tirunelveli, India, 20–22 August 2020; pp. 709–712.
15. Rodríguez, S.; Gualotuña, T.; Grilo, C. A System for the Monitoring and Predicting of Data in Precision Agriculture in a Rose Greenhouse Based on Wireless Sensor Networks. In *Proceedings of the Procedia Computer Science*; Elsevier B.V.: Amsterdam, The Netherlands, 2017; Volume 121, pp. 306–313.
16. Sahitya, G.; Balaji, N.; Naidu, C.D. Wireless sensor network for smart agriculture. In Proceedings of the 2016 2nd International Conference on Applied and Theoretical Computing and Communication Technology, iCATccT 2016, Bengaluru, Karnataka, India, 21–23 July 2016; pp. 488–493.
17. Samijayani, O.N.; Darwis, R.; Rahmatia, S.; Mujadin, A.; Astharini, D. Hybrid ZigBee and WiFi Wireless Sensor Networks for Hydroponic Monitoring. In Proceedings of the 2nd International Conference on Electrical, Communication and Computer Engineering, ICECCE 2020, Istanbul, Turkey, 12–13 June 2020.
18. Xiao, J.; Li, J.T. Design and implementation of intelligent temperature and humidity monitoring system based on ZigBee and WiFi. In *Proceedings of the Procedia Computer Science*; Elsevier B.V.: Amsterdam, The Netherlands, 2020; Volume 166, pp. 419–422.
19. Wang, Z.; Jiang, Z.; Hu, J.; Song, T.; Cao, Z. Research on Agricultural Environment Information Collection System Based on LoRa. In Proceedings of the 2018 IEEE 4th International Conference on Computer and Communications (ICCC), Chengdu, China, 7–10 December 2018; pp. 2441–2445.
20. Albaladejo, C.; Soto, F.; Torres, R.; Sánchez, P.; López, J.A. A Low-Cost Sensor Buoy System for Monitoring Shallow Marine Environments. *Sensors* **2012**, *12*, 9613–9634. [CrossRef]
21. López, J.; Soto, F.; Sánchez, P.; Iborra, A.; Suardiaz, J.; Vera, J. Development of a Sensor Node for Precision Horticulture. *Sensors* **2009**, *9*, 3240–3255. [CrossRef]
22. Navarro-Hellín, H.; Torres-Sánchez, R.; Soto-Valles, F.; Albaladejo-Pérez, C.; López-Riquelme, J.A.; Domingo-Miguel, R. A wireless sensors architecture for efficient irrigation water management. *Agric. Water Manag.* **2015**, *151*, 64–74. [CrossRef]
23. Riquelme, J.A.L.; Soto, F.; Suardíaz, J.; Sánchez, P.; Iborra, A.; Vera, J.A. Wireless Sensor Networks for precision horticulture in Southern Spain. *Comput. Electron. Agric.* **2009**, *68*, 25–35. [CrossRef]

24. Alahi, M.E.E.; Pereira-Ishak, N.; Mukhopadhyay, S.C.; Burkitt, L. An Internet-of-Things Enabled Smart Sensing System for Nitrate Monitoring. *IEEE Internet Things J.* **2018**, *5*, 4409–4417. [CrossRef]
25. Bhattacherjee, S.S.; Shreeshan, S.; Priyanka, G.; Jadhav, A.R.; Rajalakshmi, P.; Kholova, J. Cloud based Low-Power Long-Range IoT Network for Soil Moisture monitoring in Agriculture. In Proceedings of the 2020 IEEE Sensors Applications Symposium (SAS), Kuala Lumpur, Malaysia, 9–11 March 2020; pp. 1–5.
26. Jing, L.; Wei, Y. Intelligent Agriculture System Based on LoRa and Qt Technology. In Proceedings of the 31st Chinese Control and Decision Conference, CCDC 2019, Nanchang, China, 3–5 June 2019; pp. 4755–4760.
27. Kokten, E.; Caliskan, B.C.; Karamzadeh, S.; Soyak, E.G. Low-Powered Agriculture IoT Systems with LoRa. In Proceedings of the 2020 IEEE Microwave Theory and Techniques in Wireless Communications (MTTW), Riga, Latvia, 1–2 October 2020; pp. 178–183.
28. Ma, Y.W.; Chen, J.L. Toward intelligent agriculture service platform with lora-based wireless sensor network. In Proceedings of the 4th IEEE International Conference on Applied System Innovation 2018, ICASI 2018, Chiba, Japan, 13–17 April 2018; pp. 204–207.
29. Mohammed, T.S.; Khan, O.F.; Al-Bazi, A. A Novel Algorithm Based on LoRa Technology for Open-Field and Protected Agriculture Smart Irrigation System. In Proceedings of the 2019 2nd IEEE Middle East and North Africa COMMunications Conference, MENACOMM 2019, Manama, Bahrain, 19–21 November 2019.
30. Nursyahid, A.; Aprilian, T.; Setyawan, T.A.; Helmy; Nugroho, A.S.; Susilo, D. Automatic sprinkler system for water efficiency based on lora network. In Proceedings of the 2019 6th International Conference on Information Technology, Computer and Electrical Engineering, ICITACEE 2019, Semarang, Indonesia, 26–27 September 2019.
31. Sarker, V.K.; Queralta, J.P.; Gia, T.N.; Tenhunen, H.; Westerlund, T. A survey on LoRa for IoT: Integrating edge computing. In Proceedings of the 2019 4th International Conference on Fog and Mobile Edge Computing, FMEC 2019, Rome, Italy, 10–13 June 2019; pp. 295–300.
32. Usmonov, M.; Gregoretti, F. Design and implementation of a LoRa based wireless control for drip irrigation systems. In Proceedings of the 2017 2nd International Conference on Robotics and Automation Engineering, ICRAE 2017, Shanghai, China, 29–31 December 2017; Volume 2017, pp. 248–253.
33. Ahmed, N.; De, D.; Hussain, I. Internet of Things (IoT) for Smart Precision Agriculture and Farming in Rural Areas. *IEEE Internet Things J.* **2018**, *5*, 4890–4899. [CrossRef]
34. Liang, M.H.; He, Y.F.; Chen, L.J.; Du, S.F. Greenhouse Environment dynamic Monitoring system based on WIFI. *IFAC-PapersOnLine* **2018**, *51*, 736–740. [CrossRef]
35. Ndzi, D.L.; Harun, A.; Ramli, F.M.; Kamarudin, M.L.; Zakaria, A.; Shakaff, A.Y.M.; Jaafar, M.N.; Zhou, S.; Farook, R.S. Wireless sensor network coverage measurement and planning in mixed crop farming. *Comput. Electron. Agric.* **2014**, *105*, 83–94. [CrossRef]
36. Sadowski, S.; Spachos, P. Wireless technologies for smart agricultural monitoring using internet of things devices with energy harvesting capabilities. *Comput. Electron. Agric.* **2020**, *172*, 105338. [CrossRef]
37. Tzounis, A.; Katsoulas, N.; Bartzanas, T.; Kittas, C. Internet of Things in agriculture, recent advances and future challenges. *Biosyst. Eng.* **2017**, *164*, 31–48. [CrossRef]
38. Bartolini, A.; Corti, F.; Reatti, A.; Ciani, L.; Grasso, F.; Kazimierczuk, M.K. Analysis and Design of Stand-Alone Photovoltaic System for precision agriculture network of sensors. In Proceedings of the 2020 IEEE International Conference on Environment and Electrical Engineering and 2020 IEEE Industrial and Commercial Power Systems Europe, EEEIC/I and CPS Europe 2020, Madrid, Spain, 9–12 June 2020.
39. Torres-Sánchez, R.; Martínez-Zafra, M.T.; Castillejo, N.; Guillamón-Frutos, A.; Artés-Hernández, F. Real-Time Monitoring System for Shelf Life Estimation of Fruit and Vegetables. *Sensors* **2020**, *20*, 1860. [CrossRef]
40. Montori, F.; Bedogni, L.; Di Felice, M.; Bononi, L. Machine-to-machine wireless communication technologies for the Internet of Things: Taxonomy, comparison and open issues. *Pervasive Mob. Comput.* **2018**, *50*, 56–81. [CrossRef]
41. SigFox, V.s. LoRa: A Comparison between Technologies & Business Models. Available online: https://www.link-labs.com/blog/sigfox-vs-lora (accessed on 16 November 2020).
42. Doshi, J.; Patel, T.; Bharti, S.K. Smart Fanning using IoT, a solution for optimally monitoring fanning conditions. In *Proceedings of the Procedia Computer Science*; Elsevier B.V.: Amsterdam, The Netherlands, 2019; Volume 160, pp. 746–751.
43. Frei, M.; Deb, C.; Stadler, R.; Nagy, Z.; Schlueter, A. Wireless sensor network for estimating building performance. *Autom. Constr.* **2020**, *111*, 103043. [CrossRef]
44. Sharma, H.; Haque, A.; Jaffery, Z.A. Maximization of wireless sensor network lifetime using solar energy harvesting for smart agriculture monitoring. *Ad. Hoc. Netw.* **2019**, *94*, 101966. [CrossRef]
45. Sardini, E.; Serpelloni, M. Self-powered wireless sensor for air temperature and velocity measurements with energy harvesting capability. *IEEE Trans. Instrum. Meas.* **2011**, *60*, 1838–1844. [CrossRef]
46. Allen, R.G.; Pereira, L.S.; Raes, D.; Smith, M. *Crop Evapotranspiration: Guidelines for Computing Crop Water Requirements*; FAO Irrigation and Drainage Papers; Food and Agriculture Organization of the United Nations (FAO): Rome, Italy, 1998; ISBN 978-92-5-104219-9.
47. Todorovic, M. Crop Evapotranspiration. In *Water Encyclopedia*; John Wiley & Sons, Inc.: Hoboken, NJ, USA, 2005.
48. Jones, H.G. Monitoring plant and soil water status: Established and novel methods revisited and their relevance to studies of drought tolerance. *J. Exp. Bot.* **2007**, *58*, 12. [CrossRef]
49. González-Teruel, J.; Torres-Sánchez, R.; Blaya-Ros, P.; Toledo-Moreo, A.; Jiménez-Buendía, M.; Soto-Valles, F. Design and Calibration of a Low-Cost SDI-12 Soil Moisture Sensor. *Sensors* **2019**, *19*, 491. [CrossRef]

50. Ortuño, M.F.; Conejero, W.; Moreno, F.; Moriana, A.; Intrigliolo, D.S.; Biel, C.; Mellisho, C.D.; Pérez-Pastor, A.; Domingo, I Ruiz-Sánchez, M.C.; et al. Could trunk diameter sensors be used in woody crops for irrigation scheduling? A review of currer knowledge and future perspectives. *Agric. Water Manag.* **2010**, *97*, 1–11. [CrossRef]
51. Blanco, V.; Domingo, R.; Pérez-Pastor, A.; Blaya-Ros, P.J.; Torres-Sánchez, R. Soil and plant water indicators for deficit irrigatio management of field-grown sweet cherry trees. *Agric. Water Manag.* **2018**, *208*, 83–94. [CrossRef]

MDPI
St. Alban-Anlage 66
4052 Basel
Switzerland
www.mdpi.com

Applied Sciences Editorial Office
E-mail: applsci@mdpi.com
www.mdpi.com/journal/applsci

www.ingramcontent.com/pod-product-compliance
Lightning Source LLC
Chambersburg PA
CBHW061808210926
51850CB00006B/645
9783036597584